T0348610

Cork:

Biology, Production and Uses

Cork:

Biology, Production and Uses

Helena Pereira
Centro de Estudos Florestais,
Instituto Superior de Agronomia,
Universidade Técnica de Lisboa,
Lisboa, Portugal

ELSEVIER

Amsterdam • Boston • Heidelberg • London • New York • Oxford
Paris • San Diego • San Francisco • Singapore • Sydney • Tokyo

ELSEVIER
Radarweg 29, PO Box 211, 1000 AE Amsterdam, The Netherlands
The Boulevard, Langford Lane, Kidlington, Oxford OX5 1GB, UK

First edition 2007

Library of Congress Cataloging-in-Publication Data
A catalog record for this book is available from the Library of Congress

British Library Cataloging in Publication Data
A catalogue record for this book is available from the British Library

ISBN-13: 978-0-444-52967-1

For information on all Elsevier publications
visit our website at books.elsevier.com

Transferred to Digital Print 2008
Printed and bound by CPI Antony Rowe, Eastbourne

Working together to grow
libraries in developing countries

www.elsevier.com | www.bookaid.org | www.sabre.org

ELSEVIER   BOOK AID
             International   Sabre Foundation

Contents

Preface

Cork is a biological material with unique properties. It has attracted the attention of man since antiquity and has been used since then in various applications. Today, cork is widely known as the closure of wine bottles and, as such, many certainly wonder about the origin of the cork that pops out of the wine bottle. The ongoing discussion among wine makers and enologists on the differences between cork and alternative closures has also triggered attention to the wine–closure interaction and to the role of cork in the bottle aging of quality wines.

Cork is obtained from the cork oak tree as a renewable and sustainable product during the tree's life. It supports a biologically rich and important environmental system of the western Mediterranean. Hence, cork and its role in conservation of nature are certainly drawing the attention of environmental organizations. In spite of cork's historical record and its current fame, there is no comprehensive work on the cork that summarises its biology, properties, industrial processing and applications. Research has been active since the 1990s and numerous international publications have enlarged the body of knowledge on cork and on the cork oak. However, these publications mainly focus on specific areas and the general reader may encounter difficulty in finding information on the whole cork chain, from fundamentals to production and uses.

The aim of this book is to fill this gap and to present an up-to-date synthesis on the universe of cork, crossing over different disciplines, i.e. biology, forestry, chemistry, materials science, industrial engineering. The book describes the basics of cork formation, structure and chemistry, the characteristics of the cork oak and of cork production, as well as the different cork properties including its macroscopic aspects, physical, mechanical and thermal behaviour. Industrial processing as well as applications are described, from bottle closures to spacecraft insulation and designer objects. The relation of cork and wine is detailed in the last chapter.

As a compilation of available scientific and technical knowledge on cork, the book is directed to students, researchers and professionals involved in any way with cork. *Cork: Biology, Production and Uses* may serve as a reference work to them.

Cork is a fascinating material. I have the privilege of witnessing the creation of new knowledge, the innovation in the industry and the renewal of scientific interest in this material. Many aspects are still not known and fundamental as well as applied research are required for better understanding of its formation, properties and uses.

Cork is part of our history and culture. In this regard, it was also my aim to write a book that might be interesting to the non-specialist and so, while keeping the technical jargon to a minimum, it is illustrated with many photographs and schematic drawings.

I hope that next time the reader uses a cork screw, he/she will experience a new intimacy with the cork.

H. Pereira

Preface

Acknowledgements

This book would not have been possible without research done by many persons in institutions in Portugal and other countries in Europe. Financial support of underlying research projects was mostly given, in Portugal, by the Fundação para a Ciência e Tecnologia (formerly Junta Nacional de Investigação Científica e Tecnológica) and Ministério da Agricultura, Desenvolvimento Rural e das Pescas. Support was also received from the European Commission.

I thank all those who have been involved in my research life and who stimulated my interest on the cork and the cork oak. This book is also a tribute to all my co-authors in publications, my partners in research projects and especially my past and present post graduate and graduate students. I learned a lot on cork as a material from my colleagues and friends, Manuel Amaral Fortes and Maria Emília Rosa. My colleagues at the Forestry Department of the university helped me discover the world of trees and forests.

I also thank the cork industry, forest owners and their associations for allowing numerous visits as well as for providing industrial or field facilities for several of our studies.

The making of the book had precious help from my co-workers in the preparation of figures and tables: Vicelina Sousa and Augusta Costa were unforgettable, as also Sofia Leal, Sofia Knapic, Jorge Gominho and Isabel Miranda. Ana Rita Alves made the hand drawings of Chapter 1, and my son João Campos also helped. I am indebted to all who gave me their loyal and friendly support.

I thank Pieter Baas for the photography of van Leeuwenhook's drawing (Fig. 2.2), Marisa Molinas for Figures 1.7 and 3.21, Mark Bernards for Figure 3.6, Hachemi Merouani for Figures 4.5b,c and 4.15, Ofélia Anjos for Figure 9.10, the designer Daniel Michalik for Figure 11.4, and the artist Dieter Coellen for Figure 11.5. The original photographs used for Figures 1.5, 1.6 and 1.14 were taken by José Graça and the spectra in Figures 3.24 and 7.5 were recorded by José C. Rodrigues. The reproduction of Hooke's drawing (Fig. 2.1) was provided by the Herzog August Bibliothek Wolfenbuttel (38.2 Phys.2°-1 plate).

Finally, a word of appreciation for Nancy Maragioglio and Mara Vos-Sarmiento, of Elsevier, for their understanding and interest in this book.

Lisboa, August 2006

Acknowledgements

This book would not have been possible without so much time, energy, and help we obtained in England and other countries in Europe.

Introduction

Cork has attracted the curiosity of man since ancient times when some of its main properties were reckoned and put to use. Cork is light and does not absorb water, so it was an adequate material for floats. It is compressible and impermeable to liquids and therefore used to plug liquid-containing reservoirs. The very low thermal conductivity made it a good insulator for shelter against the cold temperatures and its energy-absorbing capacity was also put to practical applications.

Some of these uses have stayed practically unchanged through times until the present. It was only with the boom of the chemical industry that synthetic polymers have substituted cork in some applications, either totally such as in fishing devices and buoyancy equipment, or to a large extent such as in cold and heat insulation.

But cork as a sealant for liquid containers has remained in its essence practically unchanged albeit the automation and technological innovation introduced in the industrial processing. Cork is the material that one thinks of when in need to plug an opening and the cork stopper is the symbol of a wine bottle. Therefore it left its mark in the English language when it gave birth to the verb "to cork" and the noun "cork". Although the use of plastic stoppers and aluminium screw caps was started by some wine cellars, the natural cork stopper remains unquestionably "the" closure for good quality red wines.

Innovation also occurred in the development of new cork materials, i.e. in composites and in high-performance insulators. Space vehicles or complex structures under vibration and dynamic loads are examples of their high-tech applications.

Cork is a natural product obtained from the outer bark of an oak species, the cork oak. It is a Mediterranean-born species with a natural distribution that has been restricted to the western part of the Mediterranean basin and the adjoining Atlantic coasts. Some of the cork oak's distinct properties have been known to man since antiquity. The cork layers that are produced in its bark form a continuous envelope with an appreciable thickness around stem and branches. This cork may be stripped off from the stem without endangering the tree vitality and the tree subsequently rebuilds a new cork layer. This is the basis for the sustainable production of cork during the cork oak's long lifetime.

Albeit its restricted area of production, cork soon travelled the world. In Europe first, i.e. in England and the Low Lands, later extending to Russia and crossing the ocean to North America, cork was traded in planks and as manufactured stoppers. Cork also attracted the attention of early researchers: in the 17th century the observations of its cellular structure, first by Hooke and then by van Leeuwenhook, were important steps in plant anatomy development, while in the late 18th century it was also a study material for chemists who started to elucidate plant chemical composition.

Interest in cork oaks was triggered, not only because of the cork but also due to their dense wood which shows high resistance to impact and friction. Before cork, it was the cork oak wood that discovered the world as bows and keels in the hulls of Portuguese sailing ships.

The animal nutritional quality of the acorns also contributed to the tree's multifunctionality. Roosevelt tried to introduce cork oaks in the United States, and attempts to plant the species in several countries occurred from Australia to Bulgaria and South America. However, the difficult seed conservation and field establishment of the young plants, coupled with the slow growth and the long life cycle, never allowed more than a few stands and scattered trees. Because of its cork bark, the tree is considered to have a high ornamental value and as such it is found in many parks and botanical gardens all over the world.

Most of the cork oaks under exploitation for cork can be found in the Southern European countries of Portugal and Spain, where they constitute the tree component within a multifunctional managed system combining forest, agriculture and cattle, named "montado" in Portugal and "dehesa" in Spain. The cork oak landscapes are highly distinctive. The newly debarked tree stems shining in bright yellow-reddish colour under Portugal's Alentejo and Spain's Extremadura and Andalucia sun and blue skies are a unique and intriguing sight. Cork oaks are also exploited in Sardinia (and to a very limited extent in other parts of Italy and in France), and in northern African Morocco and Tunisia (and in Algeria, presently only to a limited extent). The cork that is produced feeds an important industrial sector that exports its products all over the world.

Despite their distinctive characteristics and diverse applications, cork and the cork oak have not been systematically researched until the late 1990s. The recognition of the important role of cork oaks in the ecologically fragile regions of southern Europe and northern Africa as a buffer to soil erosion and desertification drew the attention of present environmentalists and researchers. The fact that the cork oak supports a socio-economic chain in regions where other crops and activities are scarce also enhanced the recent scrutiny that already allowed recognizing the complexity of the present cork oak agro-forestry systems. The overall sustainability of the cork oak lands as well as the economic soundness of what is the most important non-wood forest product of Europe are key issues.

The investigation of properties of cork from a materials science point of view started in the 1980s and studies on the chemical composition of cork enhanced. Important findings on the chemical elucidation of its structural components were obtained since the late 1990s. However, many uncertainties and gaps of knowledge still remain both on the functioning of the cork oak in relation to cork formation and on the understanding of the fundamentals of cork properties. The structural and chemical features of cork are not fully exploited and the present applications do not cover the many possibilities offered by the special properties of this natural cellular material.

This book synthesizes the present status of knowledge in a comprehensive way. It intends to contribute to the further development of studies on cork and the cork oak. The organization of its contents is as follows. First, cork and its formation are described (Chapter 1), structural features (Chapter 2) and chemistry (Chapter 3), which constitute the basis for understanding the material's properties and applications. Cork cannot be dissociated from the tree and its exploitation: the species' characteristics are presented in Chapter 4, the extraction of cork in Chapter 5 and the management and sustainability issues of the cork oak forest systems in Chapter 6. The properties of cork are presented in the following chapters: the macroscopic appearance and quality in Chapter 7, density and moisture relations in Chapter 8, the mechanical properties in Chapter 9 and the surface, thermal and other properties in Chapter 10. The uses of cork start with a retrospective of historic references to cork and go until prospective applications (Chapter 11). The processing chain, from the tree to the industrial production of natural cork stoppers and discs, is detailed in Chapter 12 and cork agglomerates and composites in Chapter 13. The role of cork as a sealant in wine bottles, as well as references to the recent challenges regarding wine bottling and alternative closures are detailed in Chapter 14.

The aim was to make a scientifically based book but accessible to readers of different disciplines and interests. Therefore, each chapter starts with a small introduction and ends with concise conclusions that summarize the main points of the chapter. It was also the aim to make each chapter as self-contained as possible, allowing the reader to use only some parts as required, i.e. the non-chemist may overlook the molecular details in Chapter 3.

The cork oak and the cork material are complex and intriguing subjects that have occupied the major part of my research life. I hope that some of the fascination they entrain has passed into the book.

Part I
Cork biology

Chapter 1

The formation and growth of cork

The cork that we know from wine bottles is extracted from the bark of the cork oak tree (*Quercus suber* L.). In plant anatomy, cork is a tissue named phellem and is part of the periderm in the bark system that surrounds the stem, branches and roots of dicotyledonous plants with secondary growth. Cork is a protective tissue that separates the living cells of the plant from the outside environment.

The formation of cork in the periderm is the result of the activity of a secondary meristem, the cork cambium or phellogen. The cellular division of the phellogen is linked to the physiological cycle of the tree and to the factors that influence it, namely the environmental conditions.

In relation to other trees, the periderm of the cork oak has special characteristics of development, regularity, growth intensity and longevity that have singularised this species. Upon the death of the phellogen, even if in large areas of the stem, as it occurs during the man-made extraction of cork, there is a rapid formation of a traumatic phellogen that resumes its functions as a producer of the protective cork layer. This response of the tree is repeated whenever necessary. These features open up the possibility of using the cork oak tree as a sustainable producer of cork throughout its lifetime and they are the basis for the use of cork as an industrial raw material.

A general introduction to tree barks and to the formation of cork tissues in barks is made in the beginning of this chapter following the general description of plant anatomy (Esau, 1977; Fahn, 1990). The formation of the first periderm in the cork oak is detailed by showing the formation of the phellogen, the initiation of its meristematic activity, the differentiation and maturing of cork cells as well as the differentiation and activity of the traumatic phellogens that originate the successive traumatic periderms that are formed during the tree exploitation. The formation and development of the lenticular channels in the cork tissue are also presented. The analysis of the successive annual growth of cork is made in conjunction with a discussion on the sustained cork production.

1.1. Bark and periderm in trees

When looking at a cross-section of a tree stem, the distinction between two main parts can be made: the inner part is the wood (anatomically named xylem), located to the inside of

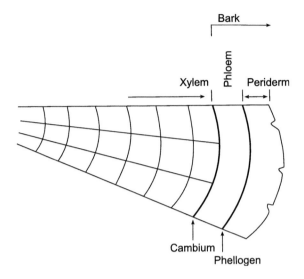

Figure 1.1. Schematic drawing of a cross-section of a tree stem showing the wood, the phloem and the periderm and the location of the lateral meristems (cambium and phellogen).

the cambium, and the outer part to the outside of the cambium constitutes what is named bark (Fig. 1.1).

The bark is not homogeneous and is constituted by two types of tissues, from inside to outside: the phloem, produced by the cambium, and the periderm that is the result of the activity of another meristem, the cork cambium or phellogen. The phloem is the principal food-conducting tissue in the vascular plants, and it may be divided into an inner functional phloem and an outer non-functional phloem (also called non-collapsed and collapsed phloem, respectively). The periderm is a system with a three-part layered structure: (a) the phellogen, or cork cambium, is the meristematic tissue whose dividing activity forms the periderm; (b) the phellem, or cork, is formed by the phellogen to the outside; and (c) the phelloderm is divided by the phellogen to the interior.

In most tree stems, one periderm is only functional during a limited period, and it is substituted by a new functioning periderm located to the inner side. Therefore the bark accumulates to the outside of the functioning periderm a succession of dead tissues containing the previous periderms that have become non-functional. This region is called the rhytidome and is defined as the outermost tissues that are situated to the outside of the functioning periderm.

The periderm is a protective tissue that is formed during secondary growth in the stems, branches and roots of most dycotyledons and gymnosperms. It substitutes the epidermis in its functions of protection and confinement, when this tissue no longer can accompany the radial growth of the axis and fractures. A periderm is also formed during wound healing in the process of protection from exposure and infection. Figure 1.2 schematically represents the location of epidermis and periderm in cross-section.

The epidermis is a one-cell layer that constitutes the outermost layer of cells of the primary plant body. It is made up of compactly arranged epidermal cells without intercellular voids where stomata and their guard cells are dispersed. The epidermal cells are externally

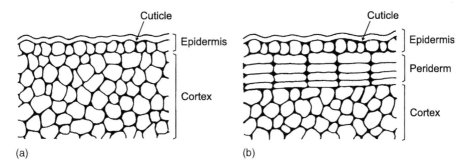

Figure 1.2. Schematic diagram showing the epidermis (a) and the periderm (b).

lined by a cuticle, a layer with variable thickness in different species that contains a special chemical component, cutin, which is the responsible for the mass-exchange restrictions of the epidermal cells. Cutin is an aliphatic polymer of glycerol esterified to long-chain fatty acids and hydroxyacids, similar in composition with the suberin contained in the cork cell walls (Graça et al., 2002; see Chapter 3). In plant organs such as leaves, the epidermis is maintained during their life, but in others such as stems and roots it is destroyed by the tangential growth stress and is functionally substituted by the periderm.

The periderm starts with the formation of the phellogen. The phellogen forms underneath the epidermis in the living cells of the primary tissues that become meristematic. The phellogen is structurally simple and is composed of only one type of cells. These cells are rectangular in transverse section, flattened radially and polygonal in tangential section. There are no intercellular spaces in the phellogen except where lenticels develop. The division of the phellogen cells is mostly periclinal, parallel to the tangential section, by which the number of cells increases radially. The division also occurs anticlinal, in a radial plane, allowing for the extension in perimeter of the phellogen.

Most often, the divisions of a phellogen mother cell are periclinal and may be described as follows (Fahn, 1990). With the first periclinal division two cells are formed that are similar in appearance, but while the inner cell is capable of further division it does not do so, and is regarded as a phelloderm cell. The outer cell undergoes a second periclinal division resulting in the formation of two cells. The outer of these cells differentiates into a cork cell and the inner cell constitutes the phellogen initial and continues to divide. In most species the phelloderm consists of only one to three layers of cells. Therefore the number of divisions from one phellogen cell that originate phellem cells is much higher than of those that originate phelloderm cells. In a single growth season the number of layers of phellem cells that are produced varies in the different species, and may be very large, as in the case of *Quercus suber*.

The phelloderm cells are living cells with non-suberised walls. They are similar to parenchyma cells of the cortex but they can be singularised due to their arrangement in radial rows under the phellogen initials. The phellem cells are also arranged in radial rows to the outside of each phellogen initial, forming a compact tissue without empty spaces. The phellem cells are dead cells that are characterised by the formation of a relatively thick layer containing suberin with a lamellated structure that is deposited internally to the cell primary wall. A cellulosic tertiary wall lines the cell to the interior. The protoplasm of the phellem cells is lost after the various wall layers have been formed and the cell lumen becomes filled with air.

Suberin is the characteristic component of phellem cells. It is an aliphatic polymer of glycerol esterified to long-chain fatty acids and alcohols, as described in detail in Chapter 3. The suberin is deposited onto the primary wall through the protoplasmatic activity but its biosynthetical pathway and deposition process are still a matter of research (Bernards, 2002). Instead of what occurs in the formation process of the secondary cell wall of wood cells where lignin synthesis is the last step, in the phellem cells the deposition of suberin and lignin and other aromatics may be simultaneous (Bernards and Lewis, 1998). The suberised wall is very little permeable to water and gases and it is resistant to the action of acids. The protective and insulating properties of the cork layers in barks directly result from this cellular structure and the chemical composition of the cell wall. The waxes, also present as non-structural cell wall components, contribute to decrease the water permeability (Groh et al., 2002).

The activity of the phellogen is seasonal with periods of activity and dormancy in response to environmental conditions of light and temperature, much in the way of the cambium, although not necessarily with the same rhythm and number of periods (Fahn, 1990).

The overall duration of the activity of the phellogen is variable between species from less than 1 year to a few years. At the end of its lifespan, a new phellogen develops leading to the formation of subsequent periderms after the first one. The first periderm remains functional during the tree's life only in a few species. The first phellogen is formed usually below the epidermis but in some species it may develop in the epidermis or in the phloem. The new phellogens are formed each time deeper inside the living tissues of the phloem and therefore the subsequent periderms are layered one inside the other either as continuous cylinders or as short segments making up a scale-type arrangement (Fig. 1.3). These assembly of tissues to the outside of the innermost functional periderm is the rhytidome. The term outerbark is also commonly used for this purpose, while the term innerbark is used to designate the tissues between cambium and active phellogen.

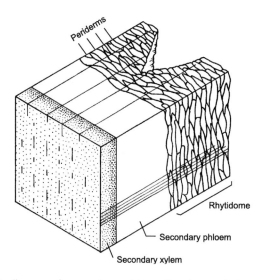

Figure 1.3. Schematic diagram of successive periderms forming a sale-type rhytidome.

1.2. Periderm and cork formation in the cork oak

1.2.1. The epidermis

In the cork oak, as in other species, the leaves and the shoots in their initial phase of growth are externally protected by an epidermis. In the leaves, the epidermis is a thin layer of epidermal cells with a width of 20–21 μm covered by a 4–5 μm thick cuticle. Trichomas (epidermic hairs) are very common in xerophytes and they are also numerous on the abaxial surface of cork oak leaves as stellate multicellular hairs (Fig. 1.4).

The epidermis in the young shoots of cork oak has similar characteristics with those of leaf epidermis, as shown in Figure 1.5. The cuticle may vary in thickness from a thin layer to a substantial external lining of the epidermal cells. The stomata are present and the multicellular hairs are abundant.

The epidermis accompanies the radial growth of the shoot by stretching the cells tangentially but this only occurs for a short period of time. After that the epidermis fractures under the stress of perimeter increase, especially when the periderm is formed underneath

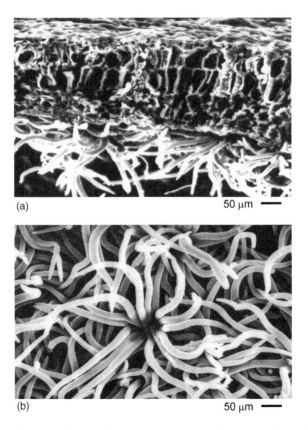

(a)   50 μm ▬

(b)   50 μm ▬

Figure 1.4. Cork oak leaves observed in scanning electron microscopy: (a) section showing the general leaf structure and the epidermis with a thick cuticle and stomata, and trichoma in the abaxial surface; (b) abaxial surface showing the numerous stellate multicellular hairs.

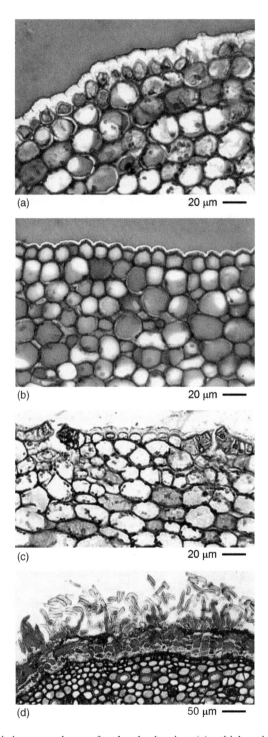

Figure 1.5. Epidermis in young shoots of cork oak, showing: (a) a thick cuticle; (b) a thin cuticle; (c) stomata; and (d) numerous multiple hairs.

and the phellogen starts to divide into numerous phellem cells. This occurs in the first year of the shoot.

1.2.2. The first periderm

The phellogen forms in the cell layer immediately below the epidermis during the first year of the shoot (Graça and Pereira, 2004). Figure 1.6 shows the different phases of the formation of the first periderm, from the initiation of the phellogen to the resulting cork layer after some years of growth.

The meristematic activity is initiated by a periclinal division of a few cells, first in small fractions around the perimeter, but very rapidly all the cell layers exhibit this meristematic activity and in cross-section a continuous ring of dividing cells can be found. In the first division, the inner cell constitutes a phelloderm cell and the outer cell divides and forms the outside phellem cells. The newly formed phellem cells maintain the tangential form of the phellogen initial and are flattened radially. With the continuing formation of new cells, the phellem cells are pushed outwards, compressed against the epidermis, and appear somewhat distorted and compacted, especially in the outermost few layers. Occasionally anticlinal divisions occur that increase the number of phellogen initials and radial rows of phellem cells. Overall the phellogen and the phellem cells make up a regular cylindrical sheath around the axis, concentric with the cambium.

The phellogen cells are polygonal shaped in the tangential section and very similar to each other, only varying dimensionally to a certain extent. This can be clearly seen in Figure 1.7, showing a scanning electron microscopic image of the phellogen surface after separation of the cork tissue.

During the first year only a few layers of phellem cells are formed, and in the second year they still keep their radial flattened appearance. It is only in the subsequent years that the cork cells enlarge in the radial direction and attain the typical appearance of cork cells. During this process, the epidermis stretches and soon fractures. The rate of perimeter increase is very high in this early development of the shoots and the corresponding tangential stress causes the fracture of not only the epidermis but also the first layers of cork cells. By the third or fourth year of the stem, longitudinal running fractures already show the underlying cork layers.

In the cork oak the first phellogen maintains its activity year after year, producing successive layers of cork. The cork in the first periderm is called virgin cork. With age and the radial enlargement of the tree, the virgin cork develops deep fractures and cracks that extend irregularly but mostly longitudinally oriented and give to the cork oak stems and branches their typical grooved appearance (Fig. 1.8).

The phellogen in the cork oak may be functional for many years, probably during the tree's life, although the intensity of its activity decreases with age. Natividade (1950) observed one sample with 140 years of cork growth attaining 27 cm of width: in the initial years of the phellogen activity, the annual width of the cork layer was about 3–4 mm, after 80–100 years it was reduced to 0.3–0.5 mm and the last years before the age of 140 years only included 12–15 cells in one radial row. Our observations of a 190-year old virgin cork allows to estimate an average radial growth of 0.92 mm per year. Figure 1.9 shows a photograph of a sample of virgin cork with 55 years of age, where the average annual width of cork was 1.9 mm.

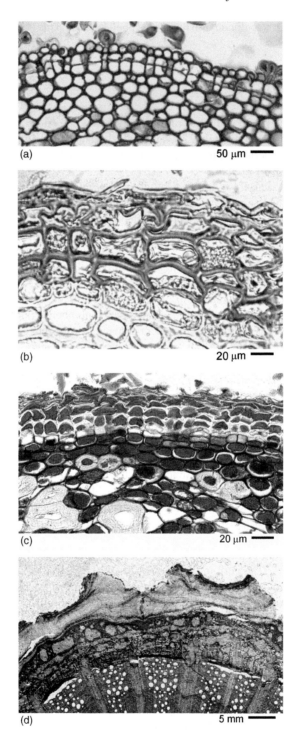

(a)   50 μm

(b)   20 μm

(c)   20 μm

(d)   5 mm

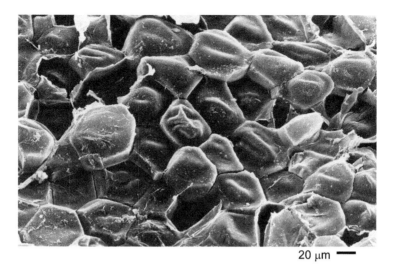

20 μm ——

Figure 1.7. Image of the tangential surface of the phellogen after separation of the cork layers.

1.2.3. Traumatic periderms

When the first phellogen ceases to be functional, a new phellogen is formed in the inner tissues of the phloem. In the cork oak, the death of the phellogen may be the result of natural or accidental aggression: a very dry and hot period, the occurrence of fire, a biological attack or wounding. It may also be the result of a deliberate removal of the cork layer. The virgin cork may be separated from the underlying bark tissues during the period of activity of the phellogen when the young cells are turgid and fragile. The removal of the cork layer exposes the phellogen to the atmosphere, and it dies and dries out as well as the underneath cells. When this happens, a new periderm is formed (traumatic periderm) as a wound response to the death of the initial phellogen and the unprotected exposure to the environment of the living tissues of the phloem. The process of traumatic periderm formation is synthesised in Figure 1.10 in schematic form.

The new phellogen is formed after about 25–35 days by a process of meristematic activation within the non-functional phloem near the limit to the functional phloem, and it starts its activity much in the same way as it occurred with the formation of the first periderm (Machado, 1935). The process is initiated first in some cells and then extends tangentially to form a continuous layer. Some obstacles are encountered along the path of the circular development of this layer: for instance, the sclerified cells that are spread in the

←————————————————————————————————————

Figure 1.6. Formation of the first periderm in the cork oak: (a) the first phellogen cell divides in the cell layer below the epidermis and forms a continuous layer of dividing cells; (b) initial phase of the formation of phellem cells showing their compression against the epidermis and the tangentially stretched epidermal cells; (c) a young periderm with some phellem layers; the result of anticlinal division of the phellogen initial can be observed (see Colour Plate Section); (d) the cork layer fractures due to the tangential growth stress and longitudinal fractures appear in the cork oak stem after some years of growth showing the underlying cork layers and the patches of remaining epidermis.

Figure 1.8. Photograph of a young cork oak stem showing the areas still covered with the epidermis (smooth appearance) and longitudinally running fractures that expose the cork tissue underneath (see Colour Plate Section).

outer phloem make up a barrier to the process of meristematic activation. In this case, the phellogen formation has to pass underneath and the meristematic line acquires a more or less wavy development along the circumference.

The activity of the traumatic phellogen proceeds as was described for the formation of the first periderm and after about 50 days a new layer of cork cells can already be observed (Fig. 1.11). As it occurred in the first periderm, the traumatic phellogen and the cork layers are disposed as a regular cylindrical envelop around the stem.

The new periderm isolates to the exterior the tissues that were located to the outside of the place of phellogen formation: they include the phelloderm remaining from the previous periderm and the non-functional phloem. These unprotected tissues dry out and fracture easily upon the radial growth of the new periderm. They form the external part of the cork layer of the traumatic periderm and consequently are named the cork back.

In the case of the cork oak, the cork produced by this second periderm is called second cork. When the radial growth of the stem is important, as is the case in young ages, the external regions of the cork layer are subject to a large tangential stress that may result

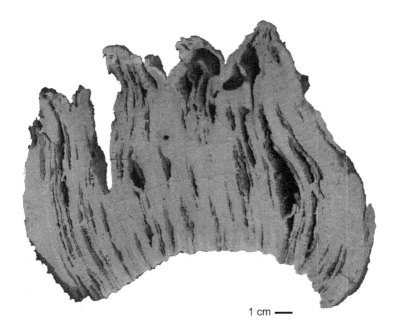

Figure 1.9. Photograph of a sample of virgin cork with 80 years of age.

into deep fractures (Fortes and Rosa, 1992). If this second cork is removed, the process is repeated in a similar way with the formation of another new traumatic periderm in the non-functional phloem and the production of a new layer of cork, now named reproduction cork. The external surface of reproduction cork shows few fractures since the cork tissue can already resist the tangential growth stress that decreases with tree age and stem diameter. Figure 1.12 shows the appearance of tree stems covered with second cork and virgin cork.

There are also some differences in the circumferential development of the phellogen between the first traumatic phellogen (leading to the second cork) and the following ones (leading to reproduction cork) in what regards its regularity. The formation of the first traumatic phellogen finds many obstacles to its development caused by the numerous sclerified nodules that are present in the non-functional phloem, and therefore its surface is irregular; in the subsequent formation, the phloem has less sclerids and therefore the phellogen-formation path is less accidented and develops in a smoother way. This may be seen in Figure 1.13 showing the imprint of the phellogen on the tree stem after the removal of a second cork (upper part) and a reproduction cork (lower part); a small collar in the upper part of the stem refers to the removal of virgin cork. It is clear that the part corresponding to the removal of the second cork is much more irregular.

The procedure may be repeated with successive removals of reproduction cork, and each time there is the formation of the new traumatic periderm as described. This is the basis for the exploitation of the cork oak as a producer of cork in a sustainable way during the tree life, with successive removals of the cork layer and formation of traumatic periderms (see Chapter 5).

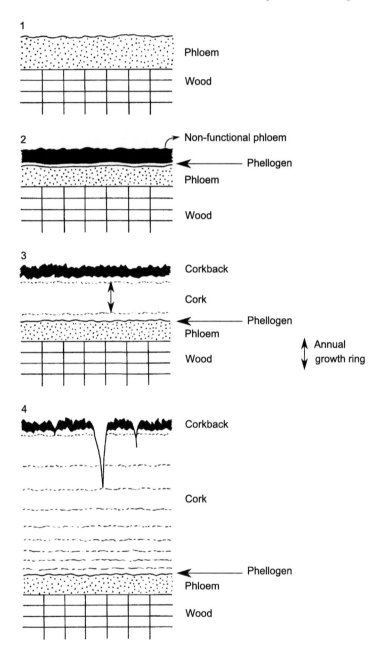

Figure 1.10. Schematic representation of the formation of the traumatic periderm in the cork oak: (1) after removal of the cork layer, the phloem is the outermost layer; (2) the traumatic phellogen forms after approximately 30 days in the phloem at the boundary between functional and non-functional phloem; (3) the phellogen produces cork layers that develop radially with an annual physiological rhythm; one complete annual growth ring is represented here; the phloemic tissue that remains to the outside make up the cork back; (4) after 10 years from the year of the removal in (1), the cork layer is thick and some fractures may occur at the cork back and outer cork regions due to the tangential tensile stress.

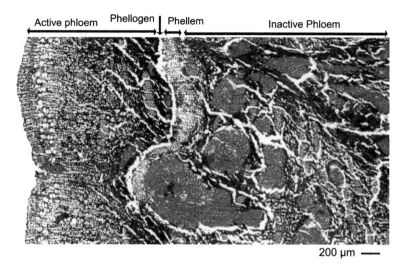

Figure 1.11. Formation of a traumatic periderm in the phloem of the cork oak showing the circumferential path of the phellogen and the first layers of cork cells (see Colour Plate Section).

1.3. The formation of lenticular channels

The periderm of most plants includes small regions of a different looking tissue made up of relatively loosely arranged cells, mostly non-suberised, and usually more numerous than in the surrounding periderm. These areas are called lenticels, and they are often conspicuous on the stems and branches because they protrude above the periderm. The cells in the lenticels constitute the lenticular filling or complementary tissue. In contrast to the surrounding periderm, the lenticels have many intercellular open spaces and it is assumed that their function is connected with gas exchange with a role similar to that of the stomata in the epidermis. The number and form of lenticels differ in various species.

The lenticels are formed by the activity of specific zones of the phellogen, called the lenticular phellogen. They appear below a stoma or group of stomata, where the cells start to divide in different directions progressing inwards into the cortex and causing the bulging of the epidermis. The divisions become more periclinal until the lenticular phellogen is formed underneath. The formation of the lenticular phellogen is made in deeper zones in the cortex and its course is concave. The activity of the lenticular phellogen is intense with a high rate of periclinal divisions leading to the formation of numerous cells. These soon cause the rupture of the epidermis so that they are pushed out and rise above the neighbouring surface level. The lenticels may remain active for many years and in this case they extend transversely in the periderm by the anticlinal division of the lenticular phellogen initials in conjunction with the radial growth.

The cork oak periderm has lenticels that are formed as described. The formation of the lenticular phellogen appears to be the initiation step in the process of the first periderm formation. Figure 1.14 shows a sequence of the formation of lenticels: under a stoma occurs the division of cells that extend to the interior; the lenticular phellogen is formed underneath and makes the contour of this cell mass, thereby acquiring a concave aspect; the

Figure 1.12. Cork oak stem with virgin (upper part) and second cork (lower part).

lenticular phellogen joins to the phellogen formed below the epidermis; the division of the lenticular phellogen initials is higher than that of the adjoining phellogen and the epidermis and superficial cell layers fracture and expose the complementary tissue. The borders of the lenticular phellogen where it joins to the normal phellogen are somewhat pushed upwards to the surface.

The activity of the lenticular phellogen is maintained year after year and it has the same longevity as the normal phellogen. Therefore the lenticels extend radially from the phellogen to the external surface of the periderm forming approximate cylinders of complementary tissue that are usually referred to as lenticular channels. In the cork oak the lenticels do not increase tangentially and the number of anticlinal divisions of the phellogen initials is small.

Lenticels also form in the traumatic periderms and remain active during their lifetime. The radial lenticular channels crossing the cork layer from the phellogen to the external

Figure 1.13. Cork oak after extraction of cork showing the differences in the regularity of phellogen development around the stem. The finger points out to the limit of extraction of reproduction cork (lower part) and second cork (upper part). In the upper part a small collar corresponds to the extraction of virgin cork.

surface are one of the characteristic features of cork (Fig. 1.15), their number and dimensions being variable between different trees. The controlling factors for the formation of the lenticular phellogen and of the lenticels are not known but evidence points out that they must be controlled genetically to a large extent. The lenticular channels are of very high practical and economic importance since they relate directly to the quality and value of the cork material, as discussed in detail in Chapter 7 (Pereira et al., 1996). They constitute what is called the porosity of cork.

Figure 1.16 shows scanning electron photographs of lenticular channels in tangential sections and in transverse section of cork. The cross-sectional form of the lenticular channel is approximately circular, usually elongated in the tree axial direction. The appearance of the lenticels in the inner side of a cork plank when it is separated from the tree stem is very characteristic, with slightly protuberant borders.

The filling tissue in the cork oak lenticels has a dark brown colour that is conspicuous in the light cream-brown colour of the cork tissue and it has a powdered appearance. The cells

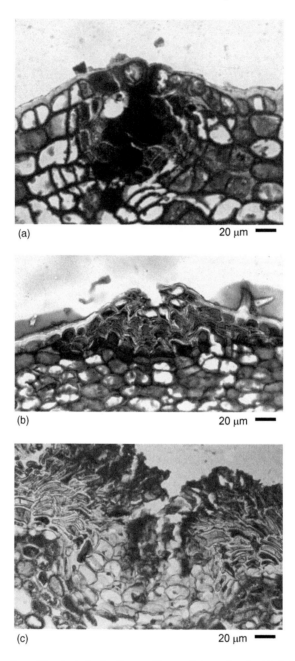

(a)   20 μm

(b)   20 μm

(c)   20 μm

Figure 1.14. Formation of lenticels in the first periderm of the cork oak: (a) division of cells under a stomata; (b) formation of the lenticular phellogen underneath the stomata; (c) initial stage of the development of the lenticel and fracture of the epidermis and outer layers.

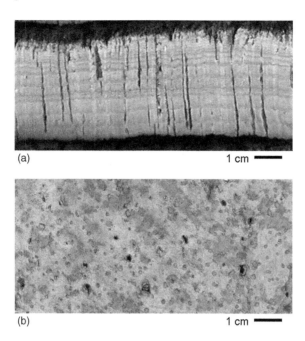

Figure 1.15. Lenticular channels crossing the cork layer: (a) in cross-section; and (b) in the tangential section of the belly.

show a loose arrangement with many intercellular voids although the radial alignment of the cells is usually partially recognised. The cells are rounded, almost spherical, with small dimensions in the range of about 10–20 µm of diameter (Fig. 1.17). Often the lenticular channels are bordered by thick-walled lignified cells that rigidify the system.

1.4. The growth of cork

1.4.1. Annual rings

After division from the phellogen initial, the cork cells are constituted by a primary wall and cytoplasm. Cell growth is made in the radial direction, and the tangential dimensions of the phellogen mother cell are maintained in the daughter cells. Cell wall thickening occurs with deposition of a highly suberised secondary wall after which the cells become practically impermeable to water. During this process the water and residues of cytoplasm are incorporated into the cell wall and the mature cell has an empty air-filled lumen.

With the climatic conditions prevailing in the cork oak area, the phellogen is functional between early April and October/November, usually with maximal activity around June and a slow growth in high summer, and it has no activity in winter. This within-the-year seasonal rhythm of physiological activity results in the formation of annual growth rings, much in the same way as it occurs in wood. The cells that are formed in spring and summer are elongated in the radial direction, and have thin cell walls, while the cells formed in the end stage

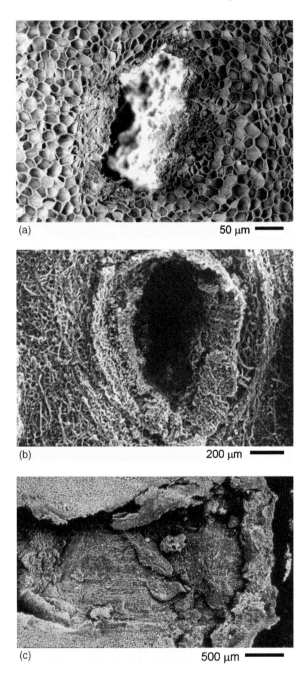

(a)   50 µm ━━━

(b)   200 µm ━━━

(c)   500 µm ━━━

Figure 1.16. Scanning electron photographs of lenticular channels in reproduction cork: (a) tangential section; (b) inner side of the cork plank after separation from the phellogen; and (c) transverse section.

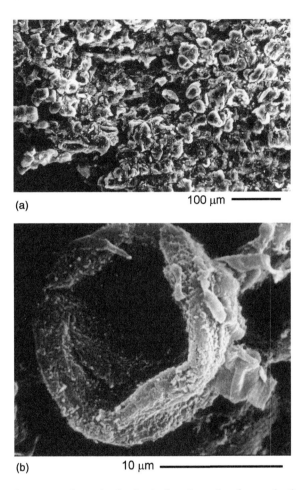

(a)   100 µm ▬▬▬

(b)   10 µm ▬▬▬▬▬▬▬

Figure 1.17. Complementary tissue in the lenticular channels of reproduction cork: (a) loose arrangement of the tissue; and (b) cell form and cell wall.

of the growth period are flattened radially and have thicker walls. Therefore the cork produced in the late growth season has a higher solid-volume fraction in comparison with the cork produced in the early growth season and it appears macroscopically differentiated as a darker strip when observed in a cross-section. This allows counting and measuring the annual rings of cork.

The distinction of annual rings is not always clear and quite often the occurrence of darker looking areas in cork does not unequivocally correspond to annual growth, especially in milder climatic areas. Variation in cell dimensions or in cell wall corrugation intensity (see Chapter 2) as well as differences in the physiological activity in the spring–summer period may lead to differences in colour shadings with ambiguous interpretation possibilities. In a study involving the sampling of 680 cork oaks in different locations in Portugal to evaluate cork characteristics, it was found that in 42% of the samples it was not possible to confidently mark the annual rings.

There are large differences in the width of annual rings in reproduction cork of trees in different geographical locations, going from thin growth rings of about 1 mm or less to large growth rings of more than 6 mm. An average value of 3.5 mm was obtained in a large sampling across Portugal. The differences in width of the annual growth rings in cork correspond to different number of cork cells in one radial row, therefore resulting from differences in the intensity of the meristematic activity of the phellogen initial, that is to say, from different number of divisions (Pereira et al., 1992).

1.4.2. Cork growth variation with age

The annual growth rings of cork produced by the first phellogen are in general small (less than 1–2 mm of width) and decrease with age. There are no studies on cork ring width variation in the first periderm, but the visual observation of stem discs with virgin cork show a decrease of ring width especially after around 20 years of age. Calculations of mean ring width for old virgin cork samples showed 0.95 mm for a sample with 93 years of age, and 0.85 mm for another sample with 183 years of age (Natividade, 1938; Machado, 1944).

The traumatic phellogen that is formed when the cork layer is removed, as a wound-healing response, has an enhanced meristematic activity and the rings of cork produced in the years that follow the extraction are wider. The intensity of cork growth subsequently decreases and after about 10–15 years the ring width is reduced to values similar to those found in virgin cork and remains subsequently rather constant. Figure 1.18 shows growth curves for a few samples of reproduction corks from different country origins with ages

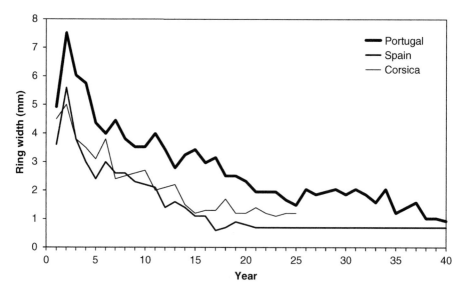

Figure 1.18. Variation of ring width with age of phellogen in a traumatic periderm (reproduction cork) for three samples: from Portugal (40 years old, Natividade, 1950), from Spain (40 years old, Gonzalez-Adrados and Gourlay, 1998) and from Corsica (25 years old).

higher than usual. This decrease of phellogen activity with age in the traumatic periderms is well established and has been repeatedly found in all corks regardless of provenance.

The absolute values of ring width vary both between different trees as well as between different sites, leading to differences in the average thickness of the cork layer for the same period. The aspects of cork productivity are discussed in more detail in Chapter 6.

In the usual exploitation practice, the cork is extracted from the tree stem around June or July. In the year when the phellogen is formed and initiates its activity, the growth period is shorter due to the time taken by the regeneration of the phellogen (about 1 month) and the time elapsed before the removal of the previous cork layer. Therefore the cork growth in the first year of the traumatic phellogen is smaller and the width of the cork layer is below that of the usual annual ring. It is common to call this year a "half year" (corresponding to only a summer/autumn growth) and similarly the last growth period in the year of extraction is also a "half year" (corresponding to the spring growth) (see Chapter 5). Therefore the growth curves of cork are analysed only in relation to the years with complete growth, and the first year of complete cork growth corresponds therefore to a traumatic phellogen in its second year of age.

Figure 1.19 shows typical curves for the variation of cork ring width with age of traumatic phellogen in some cork oaks from one stand and Table 1.1 summarises the mean values of cork growth for the 8 complete years of growth in five sites in one region in Portugal (40 trees in each site). Ring width is highest in the first year after phellogen formation and decreases steadily in the following years with a higher rate until about the fifth year and at a lower rate afterwards.

The characteristics of the traumatic phellogen activity that were referred, i.e. enhanced activity in the initial years after formation and gradual decrease of activity with age, apply

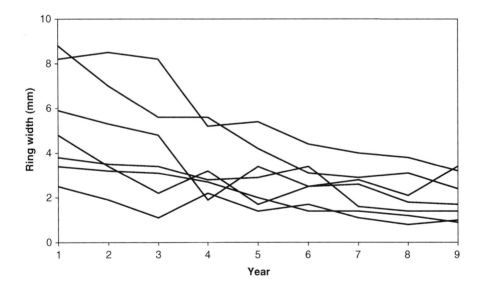

Figure 1.19. Variation of ring width with age of phellogen in a traumatic periderm (reproduction cork) in five cork oaks for the same site and period. Only the rings corresponding to complete annual growth are represented.

Table 1.1. Variation of ring width in reproduction cork during the 8 years of complete growth of one cork extraction cycle in five cork oak stands (A–E) in Portugal.

Year	Ring width (mm)				
	A	B	C	D	E
1	4.87 (1.28)	5.84 (1.57)	5.63 (1.77)	6.61 (2.01)	5.51 (1.73)
2	4.47 (0.86)	5.53 (1.46)	4.78 (1.38)	5.29 (1.20)	3.81 (1.39)
3	4.45 (1.02)	4.97 (1.39)	4.13 (1.10)	4.66 (1.32)	3.61 (1.53)
4	4.01 (1.16)	3.97 (1.00)	4.24 (1.22)	4.52 (128)	4.34 (1.83)
5	3.50 (0.83)	3.66 (1.02)	3.43 (1.17)	3.73 (1.25)	2.99 (1.09)
6	3.48 (1.21)	3.31 (1.03)	3.37 (1.19)	3.47 (0.98)	2.81 (0.75)
7	2.67 (0.86)	2.58 (0.64)	3.18 (1.10)	3.13 (0.87)	2.29 (0.65)
8	2.41 (0.69)	2.19 (0.61)	2.59 (0.97)	2.74 (1.04)	2.10 (0.78)
Average	3.78 (0.83)	4.01 (0.89)	3.92 (1.04)	4.27 (1.07)	3.43 (0.96)

Note: Mean of 40 trees in each site and standard deviation in parenthesis (Ferreira et al., 2000).

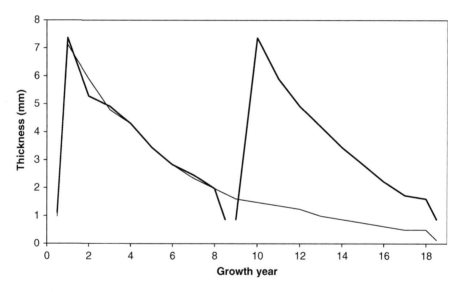

Figure 1.20. Variation of ring width in reproduction cork from one tree with a 18-year period (one periderm) and from another tree with a cork extraction in year 9 (two successive periderms) (adapted from Natividade, 1950).

not only to the first traumatic phellogen but also to the following ones. The process of periodic cork removal therefore induces the successive formation of new phellogens of which result a much higher overall production of cork. There are no research studies comparing for the same period the production of cork in one periderm and in successive periderms, and only a few singular cases are available for measurement. One such example is shown in Figure 1.20 where the reproduction cork production by one periderm during 18 years was compared with the production of two successive periderms during 9 years (Natividade, 1950). For a cork oak exploitation cycle corresponding to a tree of age

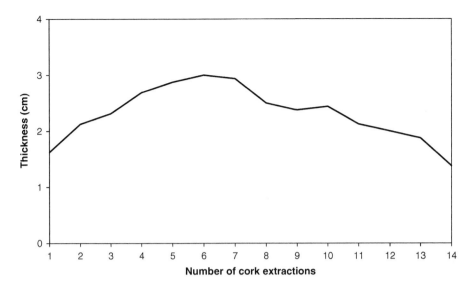

Figure 1.21. Variation of the thickness of reproduction cork from one tree in successive 10-year periderms (cork extraction was made every 10 years) (Natividade, 1950).

150–200 years, it is estimated that the accumulated production of cork in trees with periodic removals of cork and periderm renewal is 3–5 times higher than the cork produced by the first periderm of a never debarked tree of the same age.

Another aspect is related to the influence of tree age on the cork growth, i.e. how cork ring width varies in the successive periderms during the tree life. Again no systematic studies are available and the existing data refers to singular cases, as that represented in Figure 1.21 for one tree where the thickness of the cork layer is plotted for successive 10-year periderms. The meristematic activity of the traumatic phellogens increases with tree age until a maximum is attained and decreases afterwards. In the case shown the maximum was obtained in the 6th periderm, corresponding to a tree age of approximately 90 years.

1.4.3. Environmental effects on cork growth

The growth of cork is influenced by climatic conditions, as some studies have proved. Methodological approaches used in dendroecology to correlate climatic variables with ring width in wood (Fritts, 1976; Schweingruber, 1988) have also been applied to the cork oak and to ring width in cork, and the effect of precipitation and temperature in different periods within the year were tested (Caritat et al., 1996, 2000; Ferreira et al., 1998; Costa et al., 2002).

In relation to precipitation, the variable that shows the best correlation with cork growth once the age trend is eliminated is the accumulated rainfall from the previous autumn. The cork annual growth is higher in years with more rainfall and significantly reduced in drought years (annual precipitation below 500 mm).

As regards temperature, generally a negative effect is observed on cork ring width except in the period of onset of phellogen activity in April and May when cork growth is stimulated by an increase in temperature (Caritat et al., 2000). A positive effect of mean and minimum temperature in the months before the initiation of phellogen activity (January–February) was also reported (Ferreira et al., 1998).

Other factors have impact on cork growth such as biological attacks, severe edaphic conditions, fire or silvicultural practices. The occurrence of accidental wounding during the operation of cork extraction (see Chapter 6) decreases tree vitality and radial growth, and reduces the cork ring width in the initial years of the phellogen activity (Costa et al., 2004).

1.5. Conclusions

Cork is part of the periderm in the outer bark that covers the cork oak stem and branches. In relation to other species, the cork oak periderm has some characteristics that together singularise this species and are the basis for a feasible and sustainable exploitation of the tree as a cork producer. These features may be summarised as follows:

- the first periderm is formed very early, in the first year of the shoot;
- the phellogen makes up a regular cylinder around the axis, concentric with the wood cambium;
- the meristematic capacity of the phellogen is maintained during the tree's lifetime;
- each phellogen mother cell generates a high number of cork cells with a regular arrangement and without intercellular voids;
- the cork tissue is homogeneous in terms of cell types except for the presence of lenticels;
- traumatic periderms substitute the initial periderm if this becomes inactive;
- the traumatic periderms have the same characteristics of the first periderm with an increased cell-dividing activity in the initial years after formation; and
- the process of formation of traumatic periderms may be repeated numerous times during the tree's life.

The lenticular channels that cross the cork tissue in the radial direction from phellogen to the outside are biological natural features of the periderm, functionally related to gas exchange necessities.

The activity of the phellogen is periodic in the year and annual rings are observed in the cork. These can vary in width as a result of the intensity of the meristematic dividing cell capacity of the phellogen. The factors that impact on cork growth may be summarised as follows:

- cork growth decreases with age of the phellogen, more rapidly in the first initial years, and remains approximately constant for older phellogens;
- in a traumatic phellogen, cork growth is enhanced in the initial years after formation;
- cork growth probably increases with the age of tree until a maximum corresponding to tree maturity and decreases somewhat with tree decline in older ages;
- cork growth is related to climatic conditions, and drought is a limiting factor;
- silvicultural and edaphic factors, as well as accidental occurrences, that negatively impact on the tree vitality and growth, also decrease cork growth.

References

Bernards, M.A., 2002. Demystifing suberin. Canadian Journal of Botany 80, 227–240.

Bernards, M.A., Lewis, N.G., 1998. The macromolecular aromatic domain in suberized tissue: a changing paradigm. Phytochemistry 47, 915–933.

Caritat, A., Molinas, M., Gutierrez, E., 1996. Annual cork ring width variability of *Quercus suber* L. in relation to temperature and precipitation. Forest Ecology and Management 86, 113–120.

Caritat, A., Gutierrez, E., Molinas, M., 2000. Influence of weather on cork ring width. Tree Physiology 20, 893–900.

Costa, A., Pereira, H., Oliveira, A., 2002. Influence of climate on the seasonality of radial growth of cork oak during a cork production cycle. Annals of Forest Science 59, 429–437.

Costa, A., Pereira, H., Oliveira, A., 2004. The effect of cork stripping on diameter growth of *Quercus suber* L. Forestry 77, 1–8.

Esau, K., 1977. Anatomy of Seed Plants, 2nd Edition. Wiley, New York.

Fahn, A., 1990. Plant Anatomy, 4th Edition. Pergamon Press, Oxford.

Ferreira, A., Mendes, C., Lopes, F., Pereira, H., 1998. Relação entre o crescimento da cortiça e condições climáticas na região da bacia do Sado. In: Pereira, H. (Ed), Cork Oak and Cork. Centro de Estudos Florestais, Lisboa, pp. 156–161.

Ferreira, A., Lopes, F., Pereira, H., 2000. Caractérisation de la croissance et de la qualité du liège dans une région de production. Annals of Forest Science 57, 187–193.

Fortes, M.A., Rosa, M.E., 1992. Growth stresses and strains in cork. Wood Science and Technology 26, 241–258.

Fritts, H., 1976. Tree Rings and Climate. Academic Press, New York.

Gonzaléz-Adrados, J.R., Gourlay, I., 1998. Applications of dendrochronology to *Quercus suber* L. In: Pereira, H. (Ed), Cork Oak and Cork. Centro de Estudos Florestais, Lisboa, pp. 162–172.

Graça, J., Pereira, H., 2004. The periderm development in *Quercus suber* L. IAWA Journal 25, 325–335.

Graça, J., Schreiber, L., Rodrigues, J., Pereira, H., 2002. Glycerol and glyceryl esters of ω-hydroxyacids in cutins. Phytochemistry 61, 205–215.

Groh, B., Hubner, C., Lendzian, K.J., 2002. Water and oxygen permeance of phellems isolated from trees: the role of waxes and lenticels. Planta 215, 794–801.

Machado, D.P., 1935. A intensidade do crescimento da cortiça e o melhoramento da sua qualidade nas sucessivas despelas. Boletim da Junta Nacional de Cortiça 74, 57–58.

Machado, D.P., 1944. Contribuição para o estudo da formação da cortiça no sobreiro. Revista Agronómica 23, 75–104.

Natividade, J.V., 1938. O que é a cortiça. Boletim da Junta Nacional de Cortiça 1, 13–21.

Natividade, J.V., 1950. Subericultura. Ministério da Economia, Direcção Geral dos Serviços Florestais e Aquícolas, Lisboa.

Pereira, H., Graça, J., Baptista, C., 1992. The effect of growth rate on the structure and compressive properties of cork. IAWA Bulletin 13, 389–396.

Pereira, H., Lopes, F., Graça, J., 1996. The evaluation of the quality of cork planks by image analysis. Holzforschung 50, 111–115.

Schweingruber, H., 1988. Tree Rings. Kluwer Academic, Dordrecht.

Chapter 2

The structure of cork

Cork is a cellular material with closed cells. In materials science, a cellular material is defined as a material made up of empty cellular elements, either open or closed, with a solid fraction under 30% of the total volume. In materials with closed cells, these are poly-hedral volumes with solid faces that are in contact with the adjacent cells.

The properties of cellular solids depend on the way the solid is distributed in the cell faces and edges. In such materials, the geometry and dimensions of the cells, as well as their vari-ability, have an important role as well as the three-dimensional arrangement of the individual cells. So the first step in understanding the behaviour of such a material is to quantify and describe its structure, or as Gibson and Ashby (1997) advise: "First characterise your cells".

Cork has a regular structure of closed cells that derive from the one-cell layer of phel-logen and grow uni-directionally outwards in the tree's radial direction (Chapter 1). There are periodic variations in cell size and density resulting from the physiological rhythm of the tree that lead to the formation of growth rings. The regularity of the cellular arrange-ment is also disturbed by the occurrence of discontinuities, either of biological origin (i.e. the lenticular channels and woody inclusions) or accidental (i.e. cracks). In cork, the solid that builds up the polyhedral faces is a natural composite of several biosynthesised poly-mers (suberin, lignin, cellulose and hemicelluloses, see Chapter 3).

In this chapter the structure of the cork tissue will be characterised in relation to the geometry and size of the individual cells, their two- and three-dimensional arrangement, the occurrence of specific features, as well as the fine structure of the cell wall. But first an historical approach is made highlighting the part played by cork in the scientific deve-lopment of biology.

2.1. The early observations of cork

It was in the 16th century, a period of curious questioning and rich experimental research, that the microscope was perfected. Cork was one of the first materials to be observed and it was described by Robert Hooke (1635–1703), who made rigorous drawings of thin slices of cork as he saw them in the microscope. The findings were published in 1665 (Hooke, 1665), in the research monograph *Micrographia*, and for the first time it was shown that this natural plant material was made up of tiny and hollow structures that he named "cells" from the Latin *cella* meaning small room, or "pores". Hooke considered that cork cells were passages for the fluid involved in plant growth. But the name of cells was born and took its place in history, since it was later adopted by cytologists as the unifying concept for the structure of biological materials.

A reproduction of Hooke's drawings is shown in Figure 2.1. Many of the topological characteristics of what we now know for cork cells were recorded at that time. Two sections were shown, one where the cells are approximately rounded hexagons and the other with rectangular cells aligned in parallel rows (respectively, A and B in the drawing).

Similar research and experimentation was carried out by others, namely by the Dutch Antoni van Leeuwenhoek (1632–1723), from Delft, who also made observations of cork thin sections and registered his findings, usually in letters to the Royal Society of London. One is a compilation on the anatomy of several barks (van Leeuwenhoek, 1705) where he used a wine bottle cork stopper and made drawings from two sections showing the cells hexagonal shape and the row alignment of the rectangular shaped cells, as reproduced in Figure 2.2. Recently, in 1988, the original thin (<20 µm) slices of cork made by Leeuwenhoek were found in the Royal Society of London and were observed using an early microscope as well as modern optical and scanning electron microscopy, proving the

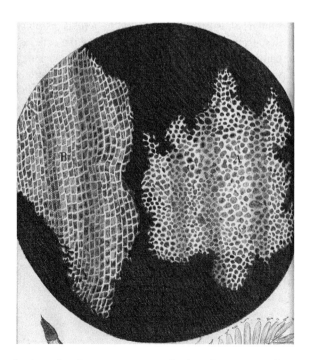

Figure 2.1. Reproduction of cork structure as seen in the microscope and represented by Robert Hooke in *Micrographia* (1665).

quality of the early observations (Ford, 1982). Leeuwenhoek also reasoned on the three-dimensional arrangement of cork and identified the growth rings as such (Baas, 2001).

Much more recently, cork was observed using the optical microscope in thin slides by the Portuguese researcher Joaquim Vieira Natividade (1899–1968), who compiled the knowledge gathered on cork and the cork oak in his book *Subericultura* (Natividade, 1950). Thin slices of cork are difficult to make with the microtome and his numerous pictures of anatomical characteristics of the cork oak certainly demonstrate an outstanding technical expertise. Figure 2.3 reproduces the sections of cork showing the two-dimensional cellular structure already depicted by the early microscopists: the hexagonal-shaped cells in a honeycomb-type arrangement and the rectangular cells arranged in rows in a brick-layered-type arrangement. This researcher was the first to describe the cork structure in two and three dimensions and to relate it to its physiological formation and the resulting properties and uses. It was only in 1981 (Gibson et al., 1981) and later in 1987 (Pereira et al., 1987) that cork structure was further analysed and attention was called to some

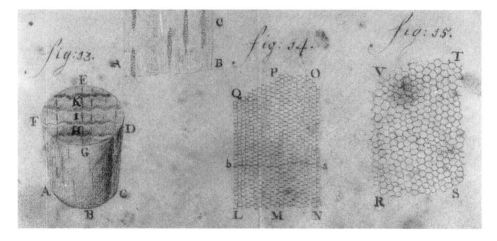

Figure 2.2. Reproduction of a cork specimen and of cork structure as seen in the microscope and represented by van Leeuwenhoek (1705).

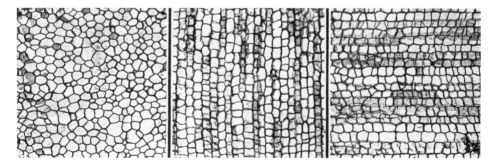

Figure 2.3. Micrographs obtained by Natividade (1950) for microtomed sections of cork observed with an optical microscope in the tangential, transverse and radial sections (from left to right).

characteristics of cells and their arrangement that play an important role in the material's unusual properties.

2.2. Cell shape and topology

The description of cork structure requires its location in space and in relation to its original position in the tree. The nomenclature used in plant anatomy, namely by wood anatomists, to refer to the different directions and sections will be used here. It is a system that is associated to the tree (or plant) growth, as represented in Figure 2.4. The axial direction (the z-axis) is the direction of apical growth and in a tree it is represented by the line of pith development; in the usual cases of straight standing stems it corresponds to the vertical direction. The directions in a plane perpendicular to the axial direction crossing the pith are radial directions (the y-axis) and correspond to the direction of stem thickening (radial growth) of a tree. The direction perpendicular to a radial and an axial direction is called tangential (the x-axis). The sections are named as follows:

• the transverse section is perpendicular to the axial direction (and contains the radial and tangential directions, defined by the $x–y$ lines);
• the tangential section is perpendicular to a radial direction (and contains the tangential and axial directions, defined by the $x–z$ lines);
• the radial section contains the axial direction and is perpendicular to the tangential direction (contains the axial and radial directions, defined by the $z–y$ lines).

The structure of cork observed by scanning electron microscopy in the three principal sections is shown in Figure 2.5. In the tangential section (Fig. 2.5a) the cork cells are

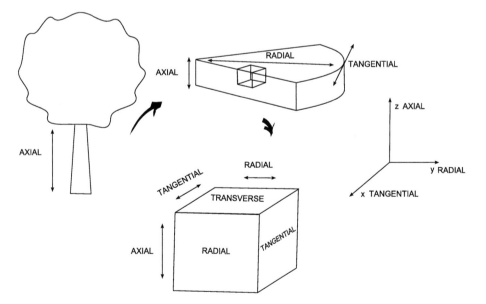

Figure 2.4. Diagram for the spatial description of cork structure showing the axis system and sections nomenclature as used in plant anatomy.

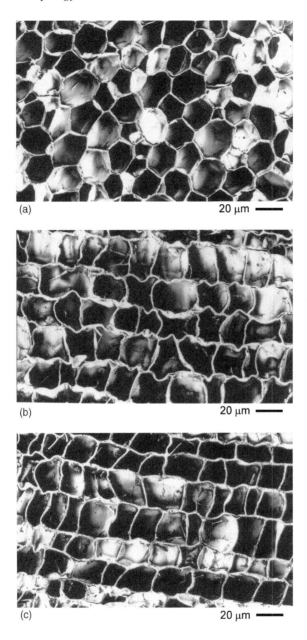

Figure 2.5. Scanning electron micrographs of sections of reproduction cork: (a) tangential; (b) radial; and (c) transverse sections.

polygons arranged in a honeycomb-type structure, making up a bi-dimensional network made up by the edges of the cells. Three cells meat at each vertex of the network and about half of the cells have six sides, with five- and seven-sided polygons making up most of the other cases (Table 2.1). The average number of sides in the polygons is six

Table 2.1. Distribution function of the number of edges of cells in the tangential, radial and transverse sections of reproduction cork.

Number of edges	Tangential	Transverse	Radial
4	0.021	0.026	0.024
5	0.249	0.226	0.203
6	0.478	0.526	0.562
7	0.216	0.178	0.172
8	0.034	0.041	0.038
9	0.002	0.004	0

Source: Pereira et al. (1987).

and the dispersion around the average of the number of polygonal edges is low, reflecting a large homogeneity of cell shape. Large cells have more sides than small cells.

The radial and transverse sections of cork look similar and different from the tangential section (Fig. 2.5b and c). Their appearance resembles a brick wall with the cells aligned in parallel rows. The individual cells look rectangular but their topological description shows that three sides meet at each vertex and the number of sides of each cell is on average six with a distribution that is not substantially different from the tangential section (Table 2.1).

In the radial and transverse sections of cork it can be observed that the sides that are roughly oriented along the radial direction in most cases are not straight but show a cell wall undulation. The number and amplitude of the cell wall corrugations is very variable within and between samples, as discussed later on. The corrugation pattern is also easily changed during cork processing (see Chapters 8 and 10).

In the tangential section the sides of the cells do not usually show corrugations, although some buckling may occur.

2.3. 3D structure

The sections observed in cork allow visualising its three-dimensional structure. In general cork is described as being formed by cells that are hexagonal prisms that are stacked base-to-base forming rows; within one row the prisms have bases with the same dimension but the prism height varies. The cell rows are assembled parallel in a compact space-filling arrangement. In adjacent rows the prism bases are not coincident and lay in staggered positions. The rows are aligned in the radial direction and therefore the individual cells have the prism height oriented in the radial direction and the prism base in a tangential plane.

Figure 2.6 shows a schematic diagram of cork's three-dimensional structure. Topologically the cork cells are polyhedra that contact in faces, edges and vertices. Two cells contact each other in a face, three cells contact in an edge and four cells in a vertex. In an ideal structure with equal prisms, each cell would contact with 14 cells (each lateral face would be in contact with two cells) as in Figure 2.7a. The cells in cork are therefore polyhedral with 14 faces (tetrakaidecahedron), with 8 faces limited by 6 edges (hexagonal) and 6 faces by 4 edges (square), topologically equivalent to the Kelvin's polyhedron (Fig. 2.7b).

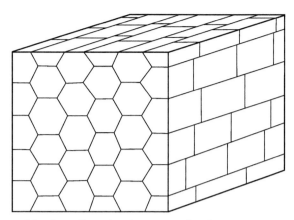

Figure 2.6. Diagram of the three-dimensional structure of cork.

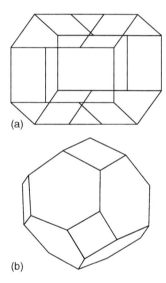

Figure 2.7. (a) Individual cell in cork showing edges and vertices, (b) the regular tetrakaidecahedra (Kelvin's polyhedron).

2.4. Cell dimensions

The cork cells are very small, much more than in normal foamed plastics, and the cell walls are thin. In an ideal average cell, the dimensions would be as follows: prism height 40 μm, base edge 20 μm, cell wall thickness 1 μm with two to three corrugations per prism lateral face with a wavelength of 15 μm and amplitude of 5–6 μm (Gibson et al., 1981). However, the cork cell dimensions are far from uniform and there are large variations within the sample and between samples.

One factor of variation is related to the seasonal growth rhythm of the tree and it results into a dimensional cell variation within one annual growth ring. One can distinguish the cells formed in the first period of growth from the cells formed at the end of the growing season, named, respectively, earlycork and latecork cells, by adapting the nomenclature used for the similar occurrence in wood.

Earlycork cells are larger and have thinner walls while latecork cells have thicker walls and a much smaller prism height (Fig. 2.8). Average cell dimensions as reported by Pereira et al. (1987) are summarised in Table 2.2. The aspect ratio of the cells (h/l) is about 2–3 in earlycork cells and close to 1 in latecork cells.

Growth rate variations between different trees also translate into different cell sizes in the radial direction. In larger growth rings, the average prism height was found to be about 22% above the prism height of cells in thinner growth rings (Pereira et al., 1992).

The solid distribution in faces and edges is very uniform and the wall thickness is constant in the different directions. The cell edges have substantially the same cell wall

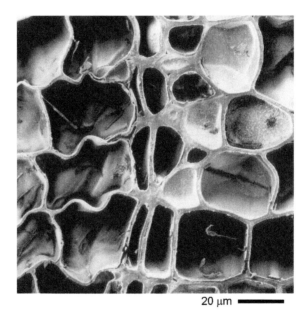

20 μm ▬▬▬

Figure 2.8. Micrograph of a transverse section of cork showing the first earlycork cells formed at the onset of the physiological activity (left) and the last latecork cells (right) formed in the previous season before the winter rest period.

Table 2.2. Dimensional characteristics of cork cells.

	Earlycork	Latecork
Prism height (h)	30–40 μm	10–15 μm
Prism base edge (l)	13–15 μm	13–15 μm
Average base area	4×10^{-6} to 6×10^{-6} cm^2	4×10^{-6} to 6×10^{-6} cm^2
Cell face thickness	1–1.5 μm	2–3 μm
Number of cells per cm^3	4×10^7 to 7×10^7	10×10^7 to 20×10^7

Source: Pereira et al. (1987).

thickness as faces, apart from the smooth rounding that is observed at the inner side of cells at face junction.

Considering the average dimensions of the cellular units, it can be calculated how much of the cork volume is occupied by the solid. The individual cell is taken as a hexagonal prism, and the solid volume (V_s) as the difference between total volume (V) and the empty volume (V_0), as given by

$$V = 3\sqrt{\frac{3}{2}}\, l^2 h$$

$$V_0 = 3\sqrt{\frac{3}{2}}\left(l - \frac{e}{\sqrt{3}}\right)^2 (h - e)$$

$$V_s = 3\sqrt{\frac{3}{2}}l^2 h - 3\sqrt{\frac{3}{2}}\left(l - \frac{e}{\sqrt{3}}\right)^2 (h - e)$$

with l as the base edge, h the prism height and e the wall thickness.

The solid fraction in the cork calculated in percent volume is approximately 8–9% in the earlycork and 15–22% in the latecork region.

2.5. Cork structure in relation to cell formation and growth

The cork cells originate from the meristematic one-cell layer named phellogen, as seen in Chapter 1. In a tangential section, the phellogen cells have the space-filling honey-comb-type arrangement with on average hexagonal cells as in Figure 1.7. The cellular division that originates the daughter cell that will develop as a cork cell is done through division in a tangential plane (periclinal division). The newly formed cell will enlarge in the radial direction and outwards, which is the only available direction for expansion.

The form of the cork cell in the tangential section is the same as its mother cell, and therefore all cells in one radial row will have the prism bases with the same dimensions and equal to the phellogenic cell. During cell radial elongation the cell wall thickens by biosynthesis and deposition of the structural components.

In periods of high physiological activity, the process of cell division and expansion proceeds at a fast rate and the resulting cells will be numerous, long and having thin walls (the earlycork cells). This corresponds to the period of high radial growth in the cork oak in April–July (Costa et al., 2002). By the end of the growing season, the rate of cell division and expansion decreases, resulting in the formation of a few thicker-walled cells (the latecork cells). The number of latecork cells remains approximately constant in a row along the successive years, only amounting to about 4–8 cells in one annual growth, and independent of the cork ring width. The number of earlycork cells in one annual row varies from about 40 to 200 cells depending on the cork growth rate (Pereira et al., 1992).

The cork may be schematically represented as a layered material composed by the rep-etition of an earlycork layer and a latecork layer, as in Figure 2.9. The earlycork layer is

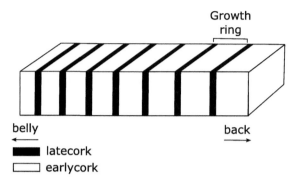

Figure 2.9. Schematic representation of cork as a multi-layered material composed of alternating earlycork and latecork layers.

less dense, wider and shows a within- and between-sample variation in width while the latecork is denser and of thinner and more uniform width.

The cellular division and expansion occurs in a similar way in the different cells of the phellogen, therefore, giving way to the space filling parallel-lying rows of their daughter cells. Because the process does not occur simultaneously in time in adjacent phellogen cells, the tangential cellular faces (the prism bases) are not in the same plane but are located at different radial distances, resulting in the staggered arrangement of prism bases in adjacent rows (Fig. 2.5b and c).

The cell walls in cork are flexible to a large extent and under compression may undulate or corrugate with variable intensity. The short and thick-walled latecork cells are much more rigid and stronger in comparison to the earlycork cells, and do not show any undulations.

When the phellogen starts its meristematic activity at the onset of a new growing season, the first cork cells that are formed are pushed outwards against the overlying cork layers produced previously and are compressed directly against the previous year's latecork cells. The cell walls are flexible to a large extent and deform under stress. The result of this radial compression is an undulation of the cell walls much like a concertina that can be exaggerated into a heavy corrugation or even the collapse of the mechanically more fragile cells (Fig. 2.10).

Enhanced corrugation of cells may also occur in a region of the cork sample, appearing for instance as a corrugation band in the transverse section (Fig. 2.11).

The occurrence of stresses in directions that deviate from the radial direction will distort the cells and the radial alignment of their rows. The tangential stresses due to the radial thickening of the tree stem are an example especially in young trees when the diameter growth rate is high. Figure 2.12 represents a transverse section of virgin cork from a young tree, where the distortion of the cell rows in relation to the radial direction is notorious due to the tangential stress as well as the corrugation of individual cells or group of cells. The stress may be higher than the strength of the cork tissue and in this case rupture occurs. Fracture of cork is a distinctive feature of the first cork produced by the growing stem or branches, the virgin cork (see Fig. 1.8). It is also very frequent in the second cork produced after removal of the virgin cork because the cork growth is high and the stem diameter increase is still important (see Fig. 1.12).

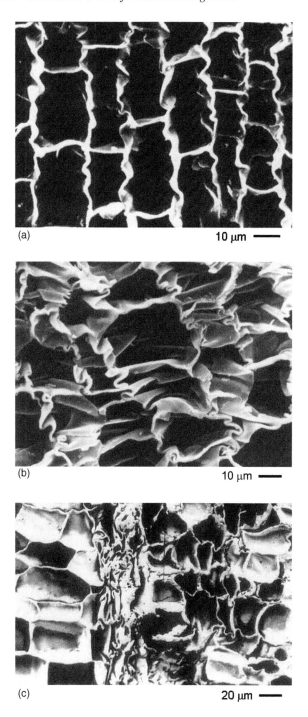

Figure 2.10. Different types of cellular undulations: (a) regular undulation of cell walls; (b) heavy corrugation; and (c) cell collapse.

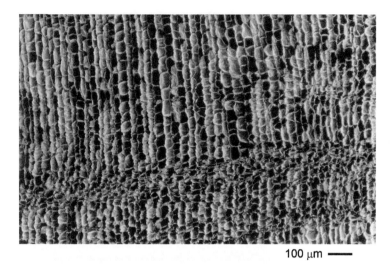

100 μm ⎯⎯

Figure 2.11. Enhanced corrugation of cork cells as a corrugation band, shown in a transverse section of reproduction cork.

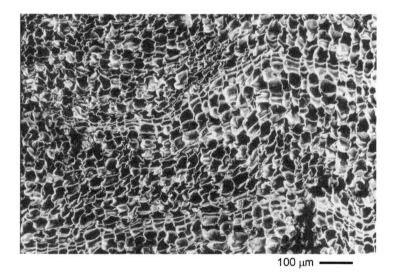

100 μm ⎯⎯

Figure 2.12. The structure of virgin cork in transverse section showing the distortion of the radial alignment of the cell rows and the corrugations of the cells.

2.6. The cell wall

The solid material in cork is contained in the faces of the prismatic units. Due to its biological origin, each of these units is one cell: it is surrounded by a cell wall and the hollow interior space corresponds to the cell's lumen. The structure is built by the junction of such units that are glued together by the biosynthetical deposition of a structural polymer (lignin, see Chapter 3), in the space between faces, and at cell corners and edges making up a continuum of solid material that spans three-dimensionally in the cork tissue.

The solid material of cells is however not uniform. It shows an ultrastructure at the below micrometer level derived from its composite nature in terms of chemical components and their spatial arrangement and resulting from its biological development process.

Similarly to the growth of wood cells, it is thought that the newly formed cells after division of the phellogen are surrounded by a cellular membrane, the primary wall containing cellulose and pectins. During cell enlargement, the cell wall material is biosynthesised and deposited onto the cell primary wall, making up the secondary wall that will be lined by a thin tertiary wall to the interior. The incrustation with lignin will fill in the open spaces within the wall structure, namely in the between-cell region named middle lamella.

Figure 2.13 shows a section cut over a few cork cells observed in a scanning electron micrograph where the main structural features of the solid material may be visualised. In the case of cork cells, the secondary wall is made up by suberin layers that are deposited from the protoplasts directly on the inner side of the cell wall. No polysaccharides are included in this secondary wall. Figure 2.14a is a schematic representation of the cell wall construction in cork and it is similar to the proposal made for the first time by Von Hohnel (1877) for suberised cells (Fig. 2.14b and c) and later by Wisselingh (1925). The solid walls of the structure are therefore made up of two cell walls separated by the common middle lamella, building up what is sometimes called the double cell wall.

A finer structure of the secondary wall of the cork cells was found using the higher resolution given by transmission electron microscopy: concentric layers of thin lamellae that alternately appear in the microscope as dark and light shaded lamellae, electron-opaque and electron-translucent, respectively, as shown in Figure 2.15. These correspond to a cell wall topochemical arrangement of alternating layers containing the suberin aliphatic polymer (the light shaded lamellae) and the associated phenolic components (the dark shaded lamellae), as described in more detail in Chapter 3.

The lamellar ultrastructure of the suberised cell wall was already described by Von Hohnel (1877) who made microscopic observations in cells after swelling and chemical treatment with potassium hydroxide. Figure 2.14b shows one of his representations. The layered suberin deposition has been found repeatedly in many other species (Wattendorf, 1980).

For *Quercus suber* cork cells the secondary wall was described by Sitte (1962) working with a sample of virgin cork from a young cork oak. Using this multi-lamellate description for the suberin deposits, he reported about 30–40 lamellae, with the light

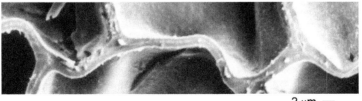

2 μm ▬

Figure 2.13. Section cut through a few cork cells observed in a scanning electron micrograph to show the cell wall cross section.

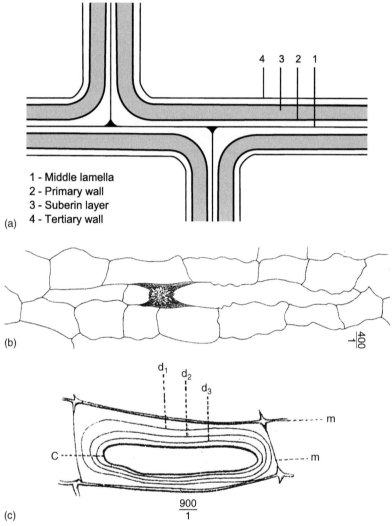

1 - Middle lamella
2 - Primary wall
3 - Suberin layer
4 - Tertiary wall

(a)

(b)

(c)

Figure 2.14. Structure of cork cells as proposed for the first time by Von Hohnel in 1878: (a) Schematic representation of his description of cell wall construction; (b) the cork cells in *Quercus suber*; and (c) the lamellar structure of the cell wall in cork cells of *Cytisus laburnum* after treatment with potassium hydroxide.

shaded lamellae maintaining a rather constant thickness of 3 nm and the dark lamellae varying between 2.5 and 20 nm. However his proposal that waxes make up the alternating layer with suberin and his assignment of the lucent layers to waxes and the dark layers to suberin have proved inaccurate. His often reproduced schematic representation of the cell wall therefore does not take into account the chemical knowledge on the cork cell wall components. There are only a few additional measurements of the lamellae in the cork cell wall. Rosa et al. (1991) refer 22 pairs of light and dark lamellae, with a

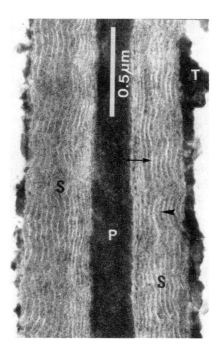

Figure 2.15. Micrographs made in the electron transmission microscope of cell walls of *Quercus suber* cork, showing the alternating dark and light lamellae in the secondary wall (Rosa et al., 1991).

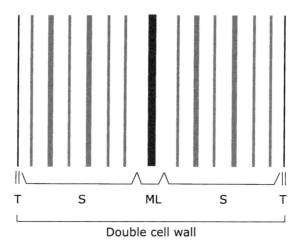

Figure 2.16. Model for the ultrastructure of the double cell wall from cork of *Quercus suber*.

thickness less than 10 nm for the translucent lamellae and values varying from 5 to 20 nm for the dark electron-dense lamellae.

Figure 2.16 is a schematic representation of the ultrastructure of the cell wall from cork of *Quercus suber* incorporating the available knowledge on the cork chemical composition (see Chapter 3). Plasmosdesmata, when present, are stuffed with an electron opaque material, probably callose, but the extent of their occurrence in the cork cells is still a matter that requires further investigation.

2.7. Structural discontinuities

The cork tissue is not completely homogeneous and the cellular structure that has been described contains discontinuities to some extent. These can be either from natural biological origin, and therefore present in all corks, with the only variation being their relative amount, or accidental features mostly of an external origin. These discontinuities in the cork cell structure are influential in several properties of the material and on the performance in use of the cork products. They are closely related with the so-called macroscopic quality of cork that plays a determining role in the commercial value of cork and cork products.

The most important feature that characterises cork heterogeneity is of biological origin and it is associated with the formation of cork and the activity of the phellogen: the lenticular channels. Also related to the physiological activity of the phellogen is the occasional inclusion of woody cells within the cork tissue. Both will be presented here. The accidental occurrence of other discontinuities may be grouped in the general concept of defects and it will be analysed when discussing the quality of cork (Chapter 7).

2.7.1. Lenticular channels

The lenticular channels run radially across the cork planks from the phellogen side to the outside. Macroscopically the lenticular channels are conspicuous due to their dark brown colour in contrast with the light brown cork tissue. In the three sections of cork they appear differently shaped: (a) in the transverse and radial sections they are elongated rectangular channels with a variable width from less than 0.1 mm to a few millimetres; and (b) in the tangential section they have an approximately circular to elliptical form.

The lenticular channels are usually referred to as the "pores" or the "porosity" of cork. The number and dimensions of these pores vary between samples and it is their visual appreciation that forms the basis for the quality classification of the cork stoppers (see Chapter 14). The aspect of the lenticular pores as seen in the transverse and in the tangential sections of cork has already been presented in Figures 1.15 and 1.16. Most of the recent results on the characterisation of the cork porosity were obtained using image analysis, as detailed in Chapter 7 (Pereira et al., 1996; Gonzalez-Adrados and Pereira, 1996).

The number and the dimensions of the pores have a large variation between samples as also in different parts of the same cork plank. Table 2.3 summarises the range found in several variables characterising the lenticular channels in the tangential and the transverse/radial section of numerous cork planks (Pereira et al., 1996).

In the tangential section, the pores range from minute spots at the detection limit of 0.1 mm^2 to values over 100 mm^2, with average areas in a sample between 0.4 and 2.6 mm^2. The number of pores is large but their distribution by area shows a large dominance of small pores (<0.8 mm^2) that correspond to 85% of all pores on average.

Table 2.3. Range of values (minimum–maximum) for some variables that characterise the porosity of cork tangential and transverse/radial surfaces.

	Tangential	Transverse/radial
Number of pores per 100 cm^2	285–1297	142–508
Number of pores >0.8 mm^2 per 100 cm^2	13–292	49–170
Average pore area (mm^2)	0.4–2.6	1.7–3.2
Maximum pore area (mm^2)	3.9–132.2	18.5–114.6
Porosity coefficient (%)	1.1–18.9	2.1–16.4

Source: Pereira et al. (1996).

The pores are elongated with a length-to-width ratio on average of 2 and oriented nearly in the axial direction.

The appearance and the characteristics of pores in the transverse and radial sections are similar. They look as thin rectangles oriented radially with average areas ranging 1.7 to 3.2 mm^2. Their length is the result of how the lenticular channels are sectioned by the observation plane: they will cross the full width of the cork plank in case the section completely contains the channel axis, but are usually much shorter. Mean widths of 0.4–0.6 mm and lengths of 1.7–2.2 mm were reported in a study characterising cork production (Ferreira et al., 2000).

The lenticular channels are also of cellular nature, filled up by an agglomerate of cells with many intercellular spaces and lacking a regularly ordered structural arrangement. These cells make up the so-called lenticular filling tissue, or complementary tissue (see Chapter 1). The cells are approximately spherical with overall dimensions in the range of or below those of the cork cells, and the cell wall is thicker (Fig. 1.17). Very little is known on the characteristics of this filling tissue apart from the fact that it chemically differs from the cork cells by being not suberised (see Chapter 7). At the border of the lenticular channels, the cork tissue occasionally includes some cells with thick lignified walls (sclerified cells) and may show some alteration in the structure such as cell deformation and cell anomalous corrugation, as exemplified in Figure 2.17.

The lenticular channels originate from some cells of the phellogen that have differentiated as lenticular phellogen. As seen in Chapter 1, this lenticular phellogen develops initially under the epidermal stomata and divides to the outside, filling the cells at a rate similar to the surrounding phellogen, thereby building up radially aligned cylinders of this loosened cellular tissue. The dimension of the regions of lenticular phellogen may vary from a few hundred cells (corresponding to the smallest lenticular channels with a transverse section below 1 mm^2) to a few thousand cells for the larger channels.

2.7.2. Woody inclusions

The cork tissue includes occasionally lignified woody type cells. These cells may occur isolated in the phellem mass in aggregates of a few cells, designated as sclereids or sclerified cells. They have very thick cell walls, and almost no lumen. Lignified cells also frequently appear in the border of the lenticular channels, as shown in Figure 2.17b.

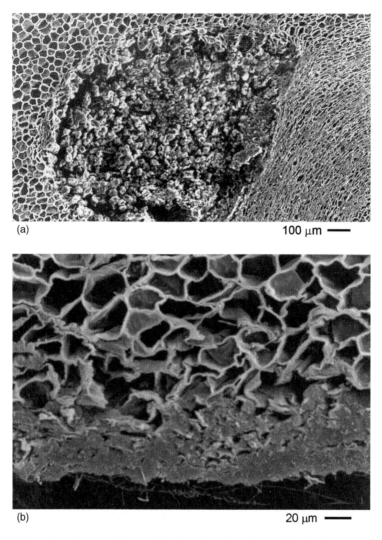

(a)   100 µm ━━

(b)   20 µm ━━

Figure 2.17. Scanning electron micrographs of cork sections through lenticular channels showing: (a) disturbed cellular structure at the border of the pore; and (b) lignified thick-walled cells.

Another reason for the presence of woody inclusions in the cork tissue is related with the occurrence of the death of the phellogen in a small portion of its area. In this case a fragment of a new phellogen is formed underneath in the phloem and it rapidly joins to the neighbouring phellogen. The overall periderm continuity is not disturbed but the phloem portion that was isolated by this new phellogen fragment is pushed to the outside into the phellem mass, and appears as a woody inclusion. It is a process similar to the traumatic regeneration of the phellogen in the phloem after the removal of cork and to the occurrence of the ligneous cork back on the external surface of the cork planks (see Chapter 1). This is shown in Figure 2.18 where the arrow points to the area of the death of the phellogen under which the phloemic inclusion is located.

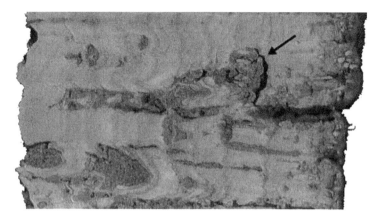

Figure 2.18. Lignified cells included in the cork tissue as phloemic inclusions (the arrow marks the area of phellogen death).

These woody inclusions are denser and harder than the cork tissue and are conspicuous because of their darker colour. They are called "nail" by the cork industry and if extensive they constitute a negative factor for cork quality.

2.8. Structural anisotropy

Anisotropy in cellular solids can arise from two different reasons: the anisotropy in their structure, resulting from cell shape, and the anisotropy in the cell wall itself. The structural anisotropy and the material anisotropy are additive and, in principal, independent. Most natural cellular solids are anisotropic because of the way they are formed and how they grow: the cells are usually elongated and the chemical construction of the cell walls also differs in its distribution, meeting natural mechanical needs in an efficient way. One obvious example is wood, which is strongly anisotropic at both levels.

The anisotropy in cell shape may be measured by the ratio of the cell largest dimensions to the smallest: the shape–anisotropy ratio R (for instance, $R_{12} = L_1/L_2$, with L_1 the largest principal direction). The properties of a material depend strongly on R, and the anisotropy cannot be ignored in predicting properties and in engineering design.

In the case of cork, it has been shown that the cells are not equidimensional, and therefore a structural anisotropy is present. The shape–anisotropy ratio calculated from the transverse and radial sections has values of about 1.5–1.7, showing the average elongation of the cells. In the tangential section, the calculated anisotropy ratio is about 1.0–1.1, meaning a practical dimensional isotropy. As a curiosity, even if cork is structurally anisotropic, its shape–anisotropy ratios are one order of magnitude less than those of wood where R_{12} vales are about 15–20 or above.

Like almost all cellular solids, cork has orthotropic symmetry, meaning that the structure has three perpendicular mirror planes. However cork may be approximated to a material with circular symmetry in the tangential plane; that is, cork cellular structure is roughly isotropic in the tangential plane, and it is axisymmetric in relation to the radial direction. These are in fact the characteristics of a regular solid with closed cells that are hexagonal prisms.

The material anisotropy of cork is largely unknown due to the uncertainties that still exist regarding the topochemistry of the cell wall. However as a first approximation it can be considered as roughly isotropic, taking in account the small relative amount of cellulose as a structural framework in the cell wall (see also Chapter 3).

Two other features in the structure of the cork material also have some role in the anisotropy. One is the regular occurrence of growth rings, with a thin layer of latecork cells with different cell shape: in the case of latecork cells R_{13} is about 0.8. However the axisymmetry in the radial direction is not disturbed. The other is the occurrence of lenticular channels. Although clearly directed in the radial direction, the lenticular channels do not have exactly the same dimensions in the different directions in the tangential plane, and on average the largest dimension is oriented axially (R_{12} about 2). This may introduce a factor of directional variation in the tangential plane of cork.

2.9. Conclusions

Cork is a natural closed-cell foam composed by tiny hollow cells of hexagonal prismatic shape arranged in a space-filling structure without intercellular voids. Each cell is formed by one mother cell in the phellogen and the direction of growth (cell elongation) is radial, building up parallel aligned rows. The cell walls are composed of a lamellate suberinic secondary wall and are flexible enough to undulate or corrugate under the growth compressive stress.

A summary of the main structural characteristics of cork is made in Table 2.4. In comparison to other cellular materials, the cells of cork are smaller and with thinner cell walls, with an overall uniformity in cell wall thickness.

The structure is approximately axisymmetric in the radial direction. Specific features of cork are two fold: the regular formation of growth rings by deposition of a thin layer of denser latecork cells, and the occurrence of lenticular channels crossing radially

Table 2.4. Main characteristics of cork structure.

Material	Natural suberised lignocellulosic composite
Density	120–170 kg/m^3
Type of cells	Closed
Mean edges/face	$n=6$
Mean faces/cell	$f=14$
Individual cell shape	Hexagonal prism
Symmetry of structure	Axisymmetric
Cell thickness	1–1.5 μm
Fraction of solid material	10%
Largest principal cell dimension	40 μm
Smallest principal cell dimension	20 μm
Intermediate principal cell dimension	30 μm
Shape anisotropy ratios	$R_{13}=1.5–1.7$, $R_{12}=1–1.1$
Other specific features	Growth rings, lenticular channels

the cork tissue. The lenticular channels or the porosity of cork introduce a factor of variation and randomness in the structure and may increase the anisotropy in the tangential plane.

References

Baas, P., 2001. Leeuwenhoek's observations on the anatomy of bark. Holzforschung 55, 123–127.

Costa, A., Pereira, H., Oliveira, A., 2002. Influence of climate on the seasonality of radial growth of cork oak during a cork production cycle. Annals of Forest Science 59, 429–437.

Ferreira, A., Lopes, F., Pereira, H., 2000. Caractérisation de la croissance et de la qualité du liège dans une région de production. Annales des Sciences Forestières 57, 187–193.

Ford, B.J., 1982. The origins of plant anatomy – Leeuwenhoek's cork sections examined. IAWA Bulletin n.s. 3, 7–10.

Gibson, L.J., Ashby, M.F., 1997. Cellular Solids. Structure and Properties, 2nd Edition. Cambridge University Press, Cambridge.

Gibson, L.J., Easterling, K.E., Ashby, M.F., 1981. The structure and mechanics of cork. Proceedings of the Royal Sociey of London, A377, 99–117.

Gonzalez-Adrados, J.R., Pereira, H., 1996. Classification of defects in cork planks using image analysis. Wood Science and Technology 30, 207–215.

Hooke, R., 1665. Micrographia, or Some Physiological Descriptions of Minute Bodies Made by Magnifying Glasses. With Observations and Inquiries Thereon. Martyn and Allestry for the Royal Society, London.

Leeuwenhoek, A. van., 1705. Philosophical Transactions 24, 1843–1855.

Natividade, J.V., 1950. Subericultura. Ministério da Economia, Direcção Geral dos Serviços Florestais e Aquícolas, Lisboa.

Pereira, H., Graça, J., Baptista, C., 1992. The effect of growth rate on the structure and compressive properties of cork from *Quercus suber* L. IAWA Bulletin n.s. 13, 389–396.

Pereira, H., Lopes, F., Graça, J., 1996. The evaluation of the quality of cork planks by image analysis. Holzforschung 50, 111–115.

Pereira, H., Rosa, M.E., Fortes, M.A., 1987. The cellular structure of cork from *Quercus suber* L. IAWA Bulletin n.s. 8, 213–218.

Rosa, M.E., Matos, A.P., Fortes, M.E., Pereira, H., 1991. Algumas características da cortiça verde. Actas do 5° Encontro Nacional da Sociedade Portuguesa de Materiais. Sociedade Portuguesa de Materiais, Lisboa, Vol. 2, pp. 737–746.

Sitte, P., 1962. zum Feinbau der Suberinschichten in Flaschenkork. Protoplasma 54, 555–559.

Von Hohnel, F., 1878. Uber den Kork und verkorkte Gewebe uberhaupt. Sitzungsberichte der kaiserlichen Akademie der Wissenschaften, Mathematisch-Naturwissenschaftliche Classe. LXXVI Band. I Abteilung. Jahrgang 1877, Heft I bis V, 507–602, K. K. Hof- und Staatsdruckerei, Wien.

Wattendorf, J., 1980. Cutinisierte und suberisierte zellwande: erschutzhullen der hoheren planzen. Biologie in unserer Zeit 10, 81–90.

Wisselingh, C. von, 1925. Die Zellmembran. Linsbauers Hanbuch der Pflanzenanatomie, Bd.III-2, Linsbauer K, Berlin.

Chapter 3

The chemical composition of cork

The properties of a material depend on the chemical characteristics of its components, their relative amount and distribution in the solid. In cork, as in other cellular materials, the chemical components are located in the cell faces and cell edges, making up a three-dimensional network of a solid matrix that encircles the hollow air-filled cells. Many of the specific properties of cork, for instance the chemical and biological inertness, and the durability are in direct relation with its chemical composition, while other properties such as the mechanical behaviour and interaction with fluids are the result of both the structural features at the cell level and the cell wall chemical structure.

The composition of cork is different from other plant tissues, namely from wood, in accordance with the role that this tissue plays in the tree: cork establishes a protective barrier between the living and physiologically active tissues of the tree stem and the outside environment. It is thought that cork restricts the loss of water, controls the gas transfer and does not allow passage of large molecules and micro-organisms. These functions are met by the presence of the main component of cork cell walls, suberin, a macromolecular compound, or better, a family of compounds that are found associated to plant membranes or layers that have to function as sealants towards fluid exchange and biotic aggression (i.e. in leaves and roots, and in wound tissue).

The plant cell wall is chemically made up of two types of components: (a) the structural components, which are macromolecules of polymeric nature, that build up the cell wall and define its structure; they are insoluble and cannot be removed from the cell wall without profoundly affecting the cell structure and properties; and (b) the non-structural components, either low-molecular organic compounds that may be solubilised by appropriate solvents, the so-called extractives, or inorganic minerals that make up the ash residue after total incineration (Fig. 3.1).

In cork, the structural components of cell wall are by order of relative importance suberin, lignin and the polysaccharides cellulose and hemicelluloses; the extractives are present in a significant amount and include mostly lipid and phenolic substances. For

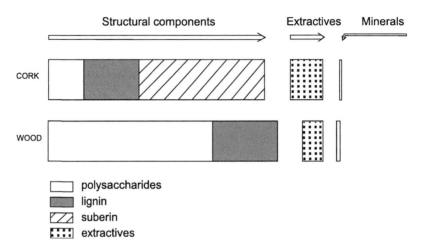

Figure 3.1. Schematic representation of the chemical composition of cork and wood of the cork oak.

comparison, the wood shows as structural components cellulose, hemicelluloses and lignin, and usually a smaller proportion of non-structural substances.

The properties of cork are related with its chemical composition and with the macromolecular structure of its components. Cork attracted the early chemical investigation, and the first studies on cork chemical composition were made, in 1787 by Brugnatelli and later on by Chevreul who published two monographs in 1807 and 1815 and named the main component of cork suberin. Cork chemistry is still a matter under investigation and information is missing regarding many aspects of the polymeric structure of cork components, of their three-dimensional arrangement and biosynthesis.

This chapter summarises the present knowledge on the chemistry of the structural components and of extractives of cork, and suggests some hypothesis for their cell wall topochemistry. The summative composition of cork and the results available regarding its variability are also presented.

3.1. Suberin

Suberin is the main structural component of the cork cells of *Quercus suber* periderm, where it amounts to about 50% of the total material. Suberin is the characteristic chemical component of cork cells in the phellem of tree barks and of other tissues found in underground plant parts (e.g. epidermis, endodermis, exodermis, root phellem and tuber phellem). It is chemically and functionally closely related to cutin, a component of leaf epidermis. Suberin has been studied in several plant materials, mainly in underground periderms, i.e. of potato, as well as in the bark of several tree species (i.e. of birch). It was Chevreul (1807, 1815) who first stated that the acid mixture obtained by chemical degradation of extractive-free cork, which he named suberin, was its main component and the one responsible for the material's properties.

The monomeric composition of suberin is quite well known at present and it is clear that it varies to some extent between plant species. The basic principles underlying the linkages between monomers are also known, but the molecular and supramolecular structures are still substantially unknown and are a matter of hypothetical reasoning.

There is also some controversy about the chemical delimitation of the term "suberin". Classically, suberin (as also cutin) was defined as an aliphatic substance that was found to be a polyester with long-chain fatty acids, and this is still the usual assumption when referring to suberin. For instance, staining with lipid contrasting dyes is the usual histochemical method to identify suberin in cell walls and its quantification is made on the basis of the mass of the aliphatic residues obtained by depolymerisation procedures that cleave the inter-monomeric ester bonds. However, recognition that the breakdown of the macromolecular structure of suberin is accompanied by the simultaneous release of aromatic residues led to the consideration that in addition to its "aliphatic domain" suberin also has an "aromatic domain", and some researchers refer to "aliphatic suberin" and "aromatic suberin" (Bernards and Lewis, 1998; Bernards, 2002). Concurrently others argue that the term suberin should preferably be restricted to the largely aliphatic macromolecule, albeit the occurrence of linked aromatic moieties (Graça and Pereira, 2000b). This discussion is also linked with the controversy regarding the occurrence of lignin in suberised cells, as discussed further on. The fact that research on the chemical composition of suberised cells has been concentrated on a few species (mainly on the potato periderm) and only some tree barks have been studied has contributed to the sometimes conflicting interpretation of results or of their generalisation.

The option of using the term suberin for the aliphatic structural component of the cork cell wall is followed here, and quantification of suberin is based on the aliphatic compounds released by depolymerisation, although the released associated aromatic moieties are also included. In the case of cork from *Quercus suber* bark, the aromatic moieties solubilised by ester cleaving only account to about 1% of all monomers (Graça and Pereira, 2000a). Lignin will be dealt within a separate section (Section 3.2).

As a structural component of the cork cell wall, suberin cannot be removed without impairing the wall integrity and the cell form. In experiments that were made on cork samples from which suberin was partially removed (Pereira and Marques, 1988), it was observed that the cellular structure was substantially destroyed (Fig. 3.2). Suberin is removed after breakdown of the inter-monomeric ester links and solubilisation of the obtained residues. Reactions such as alkaline hydrolysis (NaOH or KOH), alkaline trans-esterification by methanolysis with sodium methoxide and hydrogenolysis with LiAlH$_4$ release fragments in the form, respectively, of fatty acids, methyl esters of fatty acids and fatty alcohols. Gas chromatography and mass spectrometry are subsequently used for separation and identification of monomers.

Other techniques such as Fourier transformed infrared (FT-IR) spectroscopy and nuclear magnetic resonance (^{13}C NMR and ^{1}H NMR) have also been applied to solid cork to characterise the suberin *in situ* (Neto et al., 1995; Gil et al., 1997).

3.1.1. Monomeric composition

The monomers that are obtained by depolymerisation of suberin are fatty acids, fatty alcohols and glycerol. The long-chain monomers are linear, with lengths ranging from 16 to 26 carbons, and include α-acids, α,ω-diacids and ω-hydroxyacids. Some of the hydroxyacids are functionalised at mid-chain by insaturation, vicinal di-hydroxy or epoxide groups.

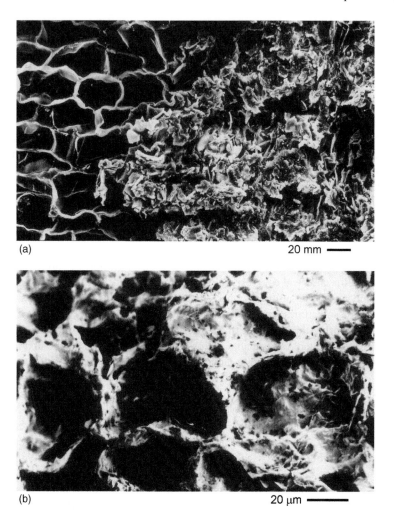

(a)   20 mm ——

(b)   20 μm ——

Figure 3.2. SEM microphotographs from a small cube of cork from which suberin was partially removed from the outer layers: (a) radial section cut through the sample showing that the external region from where suberin was removed has lost its cellular structure and has densified while the unattacked interior part retains the cellular structure; (b) cell wall degradation in one partially desuberinised cell.

The composition of suberin is shown in Table 3.1 in mass percent (Graça and Pereira, 2000a) and calculated also as molar percent. These results were obtained from carefully prepared cork tissue, from which the lenticular channels and lignified inclusions were eliminated, and subjected to exhaustive solvent extraction using dichloromethane, ethanol, water and methanol in sequence. The depolymerisation was carried out by trans-esterification using dry sodium methoxide in methanol and analysis of the methanolic extract was made by gas chromatography (GC-FID) after trimethylsilylation (TMS) derivatisation using internal standards for quantification. Although several studies dealing with the monomeric composition of cork suberin exist, from the first works of Arno et al.

Table 3.1. Monomeric composition of suberin from *Quercus suber* cork (Graça and Pereira, 2000).

Chemical classes and compounds	Formula	Mass %	Mol %
Glycerol	$CH_2OHCHOHCH_2OH$	14.2	40.8
1-Alkanols	$CH_3(CH_2)_nCH_2OH$	1.1	0.8
Alkanoic acids	$CH_3(CH_2)_nCOOH$	1.1	0.7
Saturated diacids	$COOH(CH_2)_nCOOH$	8.7	
Hexadecanedioic acid	$COOH(CH_2)_{14}COOH$	2.0	1.8
Octadecanedioic acid	$COOH(CH_2)_{16}COOH$	0.5	0.4
Eicosanedioic acid	$COOH(CH_2)_{18}COOH$	1.0	0.8
Docosanedioic acid	$COOH(CH_2)_{20}COOH$	4.5	3.2
Tetracosanedioic acid	$COOH(CH_2)_{22}COOH$	0.7	0.5
Substituted diacids		36.8	
8-Octadecenedioic acid	$COOH(CH_2)_7CH=CH(CH_2)_7COOH$	6.2	5.3
9-Epoxioctadecanedioic acid	$COOH(CH_2)_7CHOCH(CH_2)_7COOH$	22.9	18.5
9,10-Dihydroxyoctadecanedioic acid	$COOH(CH_2)_7CHOHCHOH(CH_2)_7COOH$	7.7	5.9
Saturated ω-hydroxyacids	$COOH (CH_2)_nCOOH$	11.4	
16-Hydroxyhexadecanoic acid	$CH_2OH(CH_2)_{14}COOH$	0.4	0.4
18-Hydroxyoctadecanoic acid	$CH_2OH(CH_2)_{16}COOH$	0.1	0.1
20-Hydroxyeicodecanoic acid	$CH_2OH(CH_2)_{18}COOH$	0.5	0.4
22-Hydroxydocosanoic acid	$CH_2OH(CH_2)_{20}COOH$	7.9	5.9
24-Hydroxytetracosanoic acid	$CH_2OH(CH_2)_{22}COOH$	2.4	1.7
26-Hydroxyhexacosanoic acid	$CH_2OH(CH_2)_{24}COOH$	0.1	0.1
Substituted ω-hydroxyacids	$COOH(CH_2)_nCOOH$	14.9	
18-Hydroxy-9-octadecenoic acid	$CH_2OH(CH_2)_7CH=CH(CH_2)_7COOH$	5.4	4.7
9-Epoxi-18-hydroxyoctadecanoic acid	$CH_2OH(CH_2)_7CHOCH(CH_2)_7COOH$	7.3	6.0
9,10,18-Trihydroxyoctadecanoic acid	$CH_2OH(CH_2)_7CHOHCHOH(CH_2)_7COOH$	2.2	1.7
Ferulic acid		0.5	0.6
Others and unidentified		12.0	
Total		100.0	100.0

(1981), Ekman (1983) and Holloway (1972a,b, 1983) to more recent results reported by Garcia-Vallejo et al. (1997), Bento et al. (1998) and Lopes et al. (2000), it is in the work of Graça and Pereira (2000a) that more attention was given to analyse only the suberised cork tissue and to quantify the monomers using standards and taking into account their chromatographic response factors. The quantification technique made in most studies is based only on the peak areas of the trimethylsylilated derivatives of the compounds solubilised as methyl esters chromatograms. Nevertheless, the results of suberin composition given by the different authors deviate more in terms of the relative amounts of the single compounds than on their presence, with the exception of glycerol.

Glycerol has received little attention from modern researchers working on suberins. However, it is known since 1884 when Kugler determined 2.7% glycerol in the saponified extract of pre-extracted cork and concluded that suberin was a lipid in strict sense. The presence of glycerol in cork was reported later by Ribas and Blasco (1940a,b) as 6–7% of cork, by Hergert and Kurth (1952) as 10%, by Parameswaran et al. (1981) as 3% and more recently by Rosa and Pereira (1994) as 4.7% and by Graça and Pereira (1997) as 5.2%. Glycerol was also determined in the depolymerisation products of other suberised materials: 5.8% in the cork of *Pseudotsuga menziesii* (Hergert and Kurth, 1952), 1% in green cotton fibres and 1.2% in potato periderm (Schmutz et al., 1993). However most studies on suberin composition in different plant organs have ignored glycerol as a product of depolymerisation and therefore missed its determining role as a building block of the suberin molecule as shown by Graça and Pereira (1997).

In the results reported in Table 3.1 for the monomers released from suberin after total depolymerisation by methanolysis, glycerol amounts to 14.2% of the solubilised products, corresponding to 8.5% of extractive-free cork (Graça and Pereira, 2000a). The calculation in a molar percent gives glycerol its full importance as a monomer of suberin: 40.8% of molecules released by methanolysis are glycerol.

The composition of the long-chain monomers shows that carboxylic acids represent 1.1%, ω-hydroxyacids 26.3% and α,ω-diacids 45.5% (mass %). Alcohols are also present corresponding to 1.1% of the solubilised products. The most abundant monomers are the 9-epoxyoctadecanedioic acid (22.9%), the 22-hydroxydocosanoic acid (7.9%), the 9,10-dihydroxyoctadecanodioic acid (7.7%) and the 9-epoxy-18-hydroxyoctadecanoic acid (7.3%). In terms of chain length, most fatty acids have 18 and 22 carbons, representing respectively 56.8 and 12.4% of the total monomers.

Figure 3.3 schematically depicts the monomeric composition of cork suberin by major chemical families in mass and in molar percent. The number of functional groups per 100 monomeric moles is also indicated for the carboxylic groups and for the glycerydic and primary hydroxyl groups. It is clear that the monomeric units in cork suberin are predominantly constituted by glycerol and α,ω-diacids, together making up 77% of all molecules, and with an important contribution of ω-hydroxyacids (21%). It results that the functional groups available for monomer assembly are distributed for a 100-monomer unit as 94 carboxylic groups and 146 hydroxyls (of which 124 from glycerol).

3.1.2. Monomer assembly

Suberin is insoluble in organic solvents. *In situ* almost all carboxylic groups of the monomers are esterified (Agullo and Seoane, 1982) and all primary hydroxyls in ω-hydroxyacids are also esterified (Rodriguez-Miguenes and Ribas-Marques, 1972; Agullo and Seoane, 1981). As a consequence of these facts, the chemical bonding that assembles the monomers in suberin is an ester linkage between carboxylic and hydroxyl groups.

Graça and Pereira (1997) have shown that the basis of the macromolecular development of suberin is the glycerydic link between fatty acids and glycerol: in mild depolymerised cork suberin, they could identify glyceryl-acyl dimers of the different fatty acids with glycerol, corresponding to esterification of the 1 and 2 positions of glycerol with carboxylic acids, ω-hydroxyacids and α,ω-diacids (Fig. 3.4). Further, they identified acyl-glyceryl dimers and glyceryl-acyl-glyceryl trimers from a mild suberin methanolysis of potato periderm (Graça and Pereira, 2000b,c), also exemplified in Figure 3.4.

It was also found using size exclusion chromatography and MALDI-MS that the depolymerisation solutions contained also other oligomeric entities in addition to monomeric structures (Bento et al., 2001a,b). Their relative amount is higher when the methanolysis is

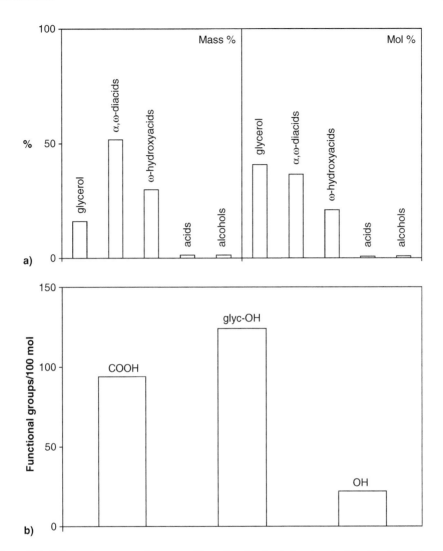

Figure 3.3. Schematic monomeric composition of cork suberin by chemical family classes in mass proportion, molar proportion and in number of hydroxyl and carboxyl groups per 100 molecules.

made in milder conditions. Figure 3.5 shows the high pressure size exclusion chromatogram (HP-SEC) of the eluted extract using 0.5% NaOMe corresponding to 80% of total suberin solubilisation, where the presence of suberinic acids (peak at 400 Da) and of fragments with 2, 3 and 4-5 suberinic acids are detected. The tentative assignment of 2-suberinic acid containing fragments identified with MALDI-MS yields the hypothetical dimers and trimers shown in Table 3.2. The ester bond between two suberinic acids (ω-hydroxyacid to ω-hydroxyacid, or ω-hydroxyacid to diacid, or ω-hydroxyacid to acid) may therefore be present in the suberin molecule, as shown to appear in tomato peel cutin (Osman et al., 1995).

The determination of molecular weights of suberin extracts using vapour osmometry gave M_n average values in the range of 528–968 g mol^{-1} (Lopes et al., 2000), which are consistent

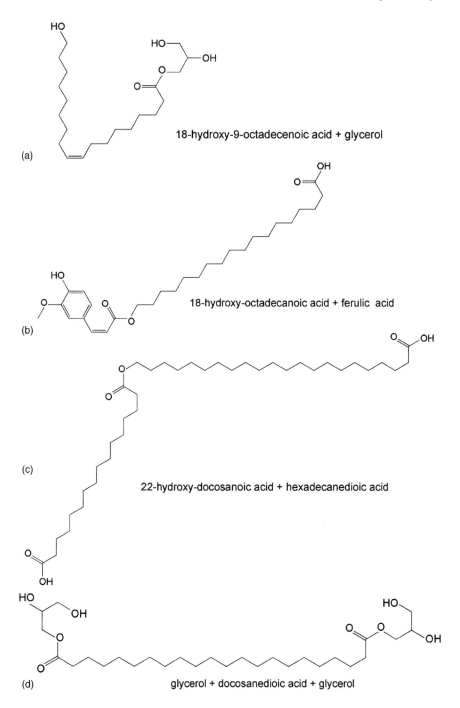

Figure 3.4. Examples of dimers and trimers obtained from suberin using a mild depolymerisation: (a) glycerol–fatty acid (monoacylglycerols); (b) ferulic acid–hydroxy fatty acid (hydroxyacid feruloyl ester); (c) acyl-acyl; (d) glyceryl-acyl-glyceryl.

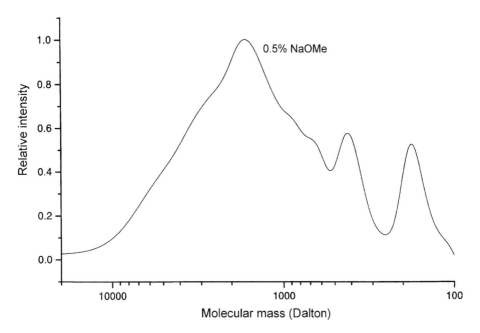

Figure 3.5. HPSEC chromatogram of an extract obtained from cork methanolysis with 0.5% NaOMe concentration as relative mass abundances (calibration of molecular masses with polystyrenes) (Bento et al., 2001b).

with the presence of dimeric/trimeric soluble suberinic residues. The extraction of a dioxane: water soluble fraction of suberin from enzymatically treated cork with cellulose, hemicellulase and pectinase allowed to obtain a polymeric suberin fraction (*ca.* 10% of the total suberin) with a M_n of 2050 g mol^{-1} which proved to be an ester-linked polymer of saturated and unsaturated fatty acids without the presence of aromatic units (Rocha et al., 2001).

Another approach to the structure of suberin is given by the balance of the functional groups that are present in the monomers (hydroxylic *versus* carboxylic groups). The esterification of the hydroxyls from glycerol (about 120/100 monomers) and from ω-hydroxyacid (about 20/100 monomers) requires more than the carboxylic acids available from diacids and ω-hydroxyacids (about 94/100 monomers). Even if the calculations of the number of functional groups are not totally correct due to the uncertainties in the quantification of monomers after depolymerisation, it is nevertheless clear that esterification to other non-lipidic molecular entities has to occur, i.e. to aromatics or to saccharides, in a significative extent (the additional number of carboxylic groups needed for the complete esterification should be around 40/100 monomers). This should be the explanation for the difficulties that are encountered in the chemical analysis of suberised materials, e.g. for determination of polysaccharides and lignin, requiring the previous depolymerisation by ester bond cleavage.

3.1.3. Associated aromatic components

Ferulic acid (4-hydroxy-2-metoxybenzoic acid) is also found in the solution of aliphatic depolymerised products. In Table 3.1 ferulic acid corresponds to 0.5% of all monomers,

Table 3.2. Possible structures present in methanolic extracts obtained with 0.05% NaOMe of cork suberin and corresponding to the main ions found in the region of m/z 540–790 MALDI-MS spectrum (Bento et al., 2001b).

[M+Na]$^+$	Tentative assignment	Molecular Formula	Δm
659.65208	Glycerol+ωOH18:1+FA18:1	$C_{39}O_6H_{72}$	0.130
	ωOH22:0+ FA18:0	$C_{41}O_4H_{80}$	0.057
	ωOH18:0+ FA22:0	$C_{41}O_4H_{80}$	0.057
675.65187	Glycerol+ωOH18:1+ ωOH18:1	$C_{39}O_7H_{72}$	0.134
	ωOH22:0+ ωOH18:0	$C_{41}O_5H_{80}$	0.062
	ωOH20:0+ ωOH20:0	$C_{41}O_5H_{80}$	0.062
	ωOH22:0+ DIA16:0	$C_{40}O_6H_{76}$	0.098
	ωOH20:0+ DIA18:0	$C_{40}O_6H_{76}$	0.098
693.66593	Glycerol+9,10,18-trihydroxyoctadecanoic acid +FA18:1	$C_{39}O_7H_{72}$	0.138
	10-methoxy-9,18-dihydroxyoctadecanoic acid + ωOH20:0	$C_{41}O_5H_{80}$	0.101
	ωOH18:0+10-methoxy-9-hydroxyoctadecanedioc acid	$C_{41}O_5H_{80}$	0.138
753.72829	Glycerol+ωOH20:0+10-methoxy-9, 18-dihydroxyoctadecanoic acid	$C_{42}O_9H_{82}$	0.143
	Glycerol+ωOH18:0+10-methoxy- 9-hydroxyoctadecanedioc acid	$C_{41}O_{10}H_{78}$	0.179
	Glycerol+DIA18:0+10-methoxy-9, 18-dihydroxyoctadecanoic acid	$C_{43}O_8H_{86}$	0.106

FA Alkanoic acids; DIA Diacids; ωOH ω-Hydroxyacids.

but other authors have also reported its presence in the product of suberin depolymerisation in amounts varying between 1.3 and 1.5% (Graça and Pereira, 1997; Lopes et al., 2000) and 5–8% of the soluble monomers (Bento et al., 1998, 2001a,b; Conde et al., 1999). Experimental conditions will certainly play an important role in this quantification and comparisons should be cautious. Other compounds (i.e. triterpenes and simple phenolics) are also present in very small amounts.

In the products obtained with a mild suberin methanolysis, dimeric structures between ferulic acid and the ω-hydroxyacids were found including the feruloyl esters of the 22-hydroxydocosanoic acid and of the 18C monounsaturated ω-hydroxyacid in higher amounts (Graça and Pereira, 1998). Further dimers between ferulic acids and alkanols, ω-hydroxyacids and glycerol were found in the suberin of potato periderm and *Pseudotsuga* bark (Graça and Pereira, 1999, 2000b). Table 3.3 shows the structure of these compounds.

Ferulic acid is known to act as the linking bridge between cell wall confining polymers of different structure (Yamamoto et al., 1989; Negrel et al., 1996). In wound potato it was found that suberin-associated polymeric aromatics could be a ferulic acid-based structure with lignin-like bonds (Bernards et al., 1995). The evidence for cork oak's cork is that ester-linked ferulic acid will play a lesser significant role to suberin due to the substantial smaller amounts of ferulic acid found in the methanolysis solution of suberised cells in comparison to those obtained from other plant origins using the same methodological conditions: <1% in cork oak and *Pseudotsuga* barks and 7.6% in potato periderm (Graça and Pereira, 2000a,b).

Table 3.3. Structure of dimers of glycerol with ω-hydroxyacids, alkanols and ferulic acid (Graça and Pereira, 1999, 2000b).

Dimers

Glyceryl-acyl
Monoacylglyceridic esters of alkanoic acids
1-Monodocosanoylglycerol
1-Monotetracosanoylglycerol
Monoacylglyceridic esters of ω-hydroxi acids
1-Mono (18-hydroxyoctadec-9-enoyl)glycerol
2-Mono (18-hydroxyoctadec-9-enoyl)glycerol
1-Mono (22-hydroxydocosanoyl)glycerol
2-Mono (22-hydroxydocosanoyl)glycerol
1-Mono (23-hydroxytricosanoyl)glycerol
2-Mono (23-hydroxytricosanoyl)glycerol
1-Mono (24-hydroxytetracosanoyl)glycerol
2-Mono (24-hydroxytetracosanoyl)glycerol
Monoacylglyceridic esters of α,ω- diacids
1-Mono (hexadecan-16-oic-1-oyl)glycerol
1-Mono (octadecan-18-oic-1-oyl)glycerol
1-Mono (octadec-9-eno-18-oic-1-oyl)glycerol
2-Mono (octadec-9-eno-18-oic-1-oyl)glycerol
1-Mono (9,10-dihydroxyoctadecan-18-oic-1-oyl)glycerol
2-Mono (9,10-dihydroxyoctadecan-18-oic-1-oyl)glycerol
1-Mono (9(10)-chloro-10(9)-hydroxyoctadecan-18-oic-1-oyl)glycerol
1-Mono (eicosan-20-oic-1-oyl)glycerol
1-Mono (docosan-22-oic-1-oyl)glycerol
2-Mono (docosan-22-oic-1-oyl)glycerol
1-Mono (tetracosan-24-oic-1-oyl)glycerol
Aryl-acyl
Feruloyl esters of ω-hydroxyacids
16-*O*-feruloyloxyhexadecanoic acid
18-*O*-feruloyloxyoctadec-9-enoic acid
18-*O*-feruloyloxyoctadecanoic acid
18-Feruloyl-9,10-dihydroxyoctadecanoic acid
20-*O*-feruloyloxyeicosanoic acid
22-*O*-feruloyloxydocosanoic acid
24-*O*-feruloyloxytetracosanoic acid

3.1.4. *Molecular structure*

The molecular structure of suberin has been the subject of several conjectures. One hypothetical model was proposed by Kolattukudy (1977), where suberin was considered as an
aliphatic–aromatic entity, integrating all the suberised cell wall material excluding polysaccharides and extractives. It was developed mainly on the basis of the results of research
on potato periderm. The model hypothesised on the inter-esterification of the fatty acids
and hydroxyacids and included their ester bonding to ferulic acid and aromatics, as well
as to polysaccharides. Glycerol was not included. This model was extensively quoted in
subsequent studies on suberin. However, it did miss the role of glycerol as a building unit

of the polymer and should no longer be used taken into account the new knowledge gained.

It is now clear that glycerol is the bridge between the monomeric units and the basis for the three-dimensional development of the polymer with a structure that should include the following moieties:

(i) a glyceryl-acyl-glyceryl structure, where the α,ω-diacids are esterified to two glycerol molecules;
(ii) a probable glyceryl-acyl-acyl-glyceryl structure with the esterification between a ω-hydroxyacid and a α,ω-diacid;
(iii) a glyceryl-acyl-feruloyl structure with an ω-hydroxyacid esterified to glycerol and the ferulic acid.

A model was proposed by Bernards in 2001, after a review of suberin chemical composition. This model separates the macromolecule spatially and chemically in two domains: the suberin polyaliphatic domain and the suberin polyaromatic domain. The former includes the glyceridic esterification of the fatty acids, diacids and ω-hydroxyacids, as proposed by Graça and Pereira (1997) and proposes a bilayer arrangement; the latter is aromatic based on the polymerisation of ferulic acid and hydroxycinnamic acids linked by esterification to the aliphatic part. Some ferulate esters of fatty alcohols are also intercalated in the aliphatic polymeric network. Figure 3.6 reproduces this model.

In the case of suberin from cork of *Quercus suber*, this model does not fit well with the available knowledge of its monomeric composition and on the proportion of functional groups discussed above. The aliphatic chains may be arranged in a rather orderly lattice where interaction between aliphatic chains through H-bonding may bring additional supramolecular stability. The development of the aliphatic structure perpendicularly to the lamellae and the cell wall should have a length corresponding to about the chain length of two suberinic acids, with about 4 nm (the distance between two contiguous carbons in one chain is 0.126 nm), corresponding to the light bands seen in the TEM photomicrographs of the suberin layer of the cork cell wall. The width of the light bands in the suberin layer of cork is rather constant at about 3 nm (Sitte, 1962). A study involving the manipulation of chain length of the suberin monomers and the accurate measurement of the lamellae thickness showed that the width of the light lucent layer ranged between 3.4 and 4.4 nm and gave support to this type of structure (Schmutz et al., 1996). Other measurements of the lamellate region of suberised cells in root exodermis (Olesen, 1978) showed 8–10 nm thick translucent lamellae separated by 15–20 nm denser matrix material.

However, most of the aliphatic monomers of cork suberin are functionalised at mi-chain and this adds complexity to the spatial development of the macromolecule due to stereochemical considerations. This is clearly seen in Fig. 3.7, which is proposed as a tentative model for a suberinic oligomer. A total of seven glycerol monomers are represented esterified to 15 different fatty acids in proportions similar to those found in cork, and two ferulic acid moieties were included. The 3D representation of this model structure shows the relevance of the actual chemical features of the monomers in the macromolecular development. The present knowledge is insufficient to further define its molecular architecture, although an overall strip configuration seems probable.

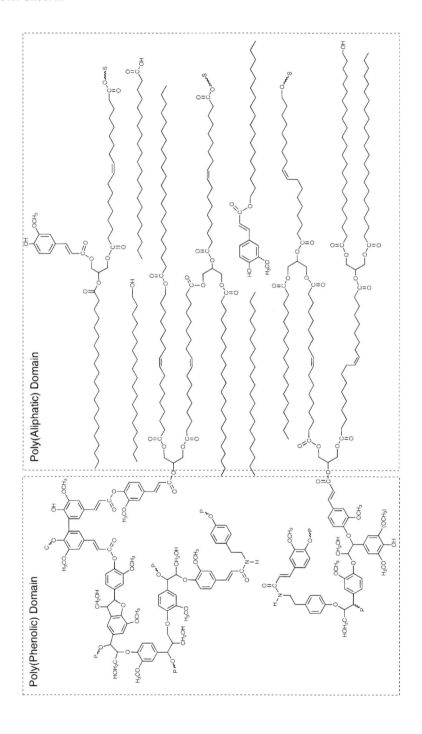

Figure 3.6. Model proposed for potato periderm suberin by Bernards (modified from Bernards, 2002).

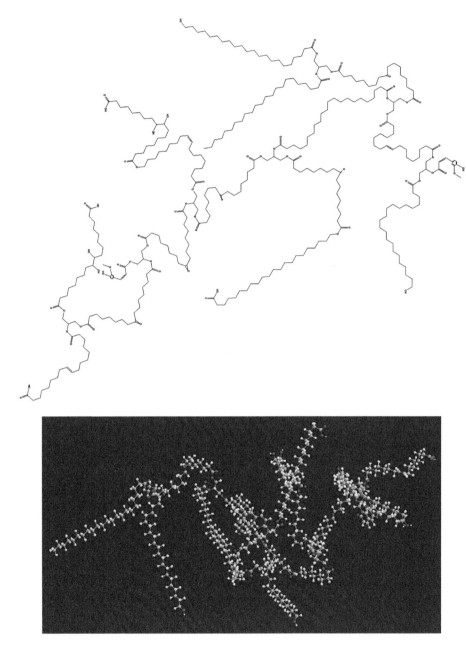

Figure 3.7. Tentative model for a part of the suberin polymer in cork from *Quercus suber* L. showing the chemical structure (top figure) and its three-dimensional representation (bottom figure).

Numerous bonding between hydroxyl groups of glycerol and of fatty alcohols have to occur at the boundary of this aliphatic strips to acid groups of phenolic nature (i.e. to lignin) and to acid monomers of hemicelluloses and pectins.

3.2. Lignin

Lignin is a cross-linked polymer of aromatic nature. It is the second most important structural component of the cork cells where it amounts to about 20–25% of the total material. Research on the chemical structure of lignin in suberised cells is scarce and the first isolation and characterisation of lignin from cork was made only recently (Marques et al., 1996). Therefore, many aspects of the lignin chemistry that are presented here are derived from the knowledge obtained from wood research. In wood, lignin corresponds to about 20–35% of the cell wall material where it is the most complex structural component.

Lignin is heterogeneous and several chemical features, namely the *in situ* lignin structure, are still under investigation. As a result of its chemical structure, lignin is a rigid and hard polymer with strong covalent bonds distributed as a 3D-network, responsible for the stiffening of cell wall and for the wood resistance to compression. It is mostly hydrophobic and its water absorption is low.

In cork, lignin also offers the mechanical support and rigidity to the cell walls. If lignin is selectively removed from the cork, the cells collapse totally. Figure 3.8 shows micrographs taken from a sample where the external part was delignified in which it can be seen how the cells flatten down without the presence of lignin.

3.2.1. Monomeric composition and bonding

Lignin is a macromolecule formed by the polymerisation of three phenylpropane monomers (C9 units), the *p*-hydroxy-cinnamyl alcohols, which differ in the degree of methoxyl substitution in C3 and C5: *p*-coumaryl alcohol, coniferyl alcohol and sinapyl alcohol (Fig. 3.9). The aromatic rings of these alcohols are named respectively *p*-hydroxyphenyl (H), guaiacyl (G) and syringyl (S) on which is based the designation of the different chemical types of lignins.

The proportion of monomers participating in the construction of the macromolecule depends on the type of species and cells and also on its location in the cell wall. For instance, hardwoods have a lignin made up mainly of coniferyl and sinapyl alcohols (guaiacyl-syringyl lignin, GS-lignin) and softwoods of coniferyl alcohol (guaiacyl lignin, G-lignin).

The polymerisation initiates by the formation of phenoxy radicals through an enzymatic dehydrogenation of the alcohol group. These are resonance-stabilised structures with radical character not only on the phenolic oxygen atom, but also on the ring carbons 1, 3 and 5, and on the β carbon of the aliphatic chain. The reaction of these radicals occurs by random coupling to form dimeric structures (dilignols) and bonding may occur at various positions as ether and C–C bonds of various types, such as β-O-4, α-O-4, β–β, β-5, 5-5, 4-O-5, β-1 (Fig. 3.9). Probability of coupling depends on the reactivity of the various positions, with β-O-4 as the most frequent type of linkage in the lignin molecule. The polymerisation proceeds further by the formation of a dilignol radical that may react either with monomeric or other dilignol radicals, yielding tri- or tetralignols. The construction of the molecule continues by this random coupling of monolignols and oligolignols leading to a cross-linked three-dimensionally spreading macrostructure.

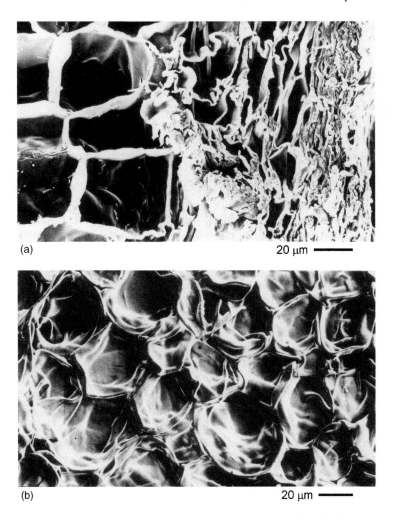

Figure 3.8. SEM micrographs taken from a small cork cube that was delignified in its external part. (a) Radial section cut through the sample where the delignified external part with collapsed cells can be compared with the internal unaffected cellular structure; (b) Tangential section of the external surface of the sample.

The first studies on the chemical characterisation of cork lignin were started by Marques et al. (1994) based on isolated material. They applied the technique which has been extensively used to isolate lignin from wood requiring a long duration ball milling of extractive-free material and the subsequent extraction with dioxane:water, featuring what is called milled wood lignin (MWL) or Bjorkman lignin (Bjorkman, 1956). The process was difficult when applied to cork: the yield of milled cork lignin (MCL) was less than 2% (7–10% is usual for wood lignin), and the isolated material was accompanied by an appreciable amount of aliphatic material from suberin and by pentosans. The isolation was improved by making a methanolysis treatment to remove the suberin previous to the ball milling. In these conditions an almost pure lignin could be obtained with a yield of

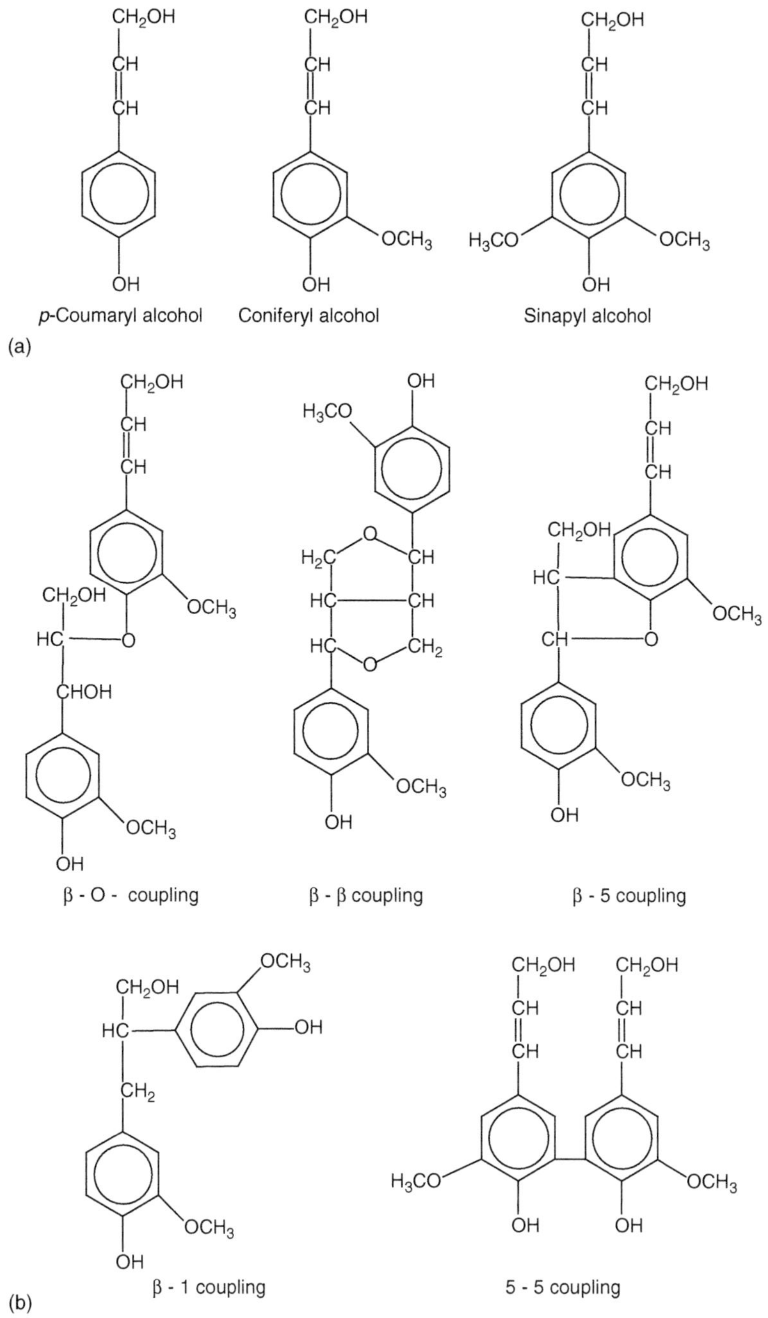

Figure 3.9. Structure of monomer precursors for the enzymatic synthesis of lignin (p-coumaryl alcohol, coniferyl alcohol and sinapyl alcohol) and main intermonomeric couplings (β-O-4, α-O-4, β-β, β-5, 5-5, 4-O-5, β-1).

4.1% of extractive-free cork (11.9% on the transesterified cork) that could be further puri-
fied by separating a small amount of an ether soluble fraction (corresponding to 0.2% of
extractive-free cork). During the procedure, an aqueous soluble lignin–carbohydrate
complex (LCC) was also isolated (0.3% of extractive-free cork). The isolated lignins were
characterised by FT-IR spectroscopy, by analytical pyrolysis and elemental analysis and
chemically characterised by thioacidolysis and permanganate oxidation (Marques et al.,
1996, 1999).

The results of the chemical analysis are shown in Table 3.4. Cork lignin is a G-type
lignin containing 95% guaiacyl units (G), 2% 4-hydroxyphenyl units (H) and 3%
syringyl units (S) with a methoxyl content of 13.95%. The proportion of units with phenolic hydroxyls is 27% and those with β-aryl ether bonds 23%. In comparison with a
G-lignin from wood, cork lignin is more condensed with more biphenyl (5-5), phenyl-
coumaric (β-5) and 4-O-5 linked units.

The isolated cork lignin fulfils the chemical requirements for what is considered a
lignin, namely the methoxyl content and the ratio of oxygenation (O:C) and shows the
typical characteristics (namely derived from FT-IR and analytical pyrolysis) of a G-lignin.
A recent isolation and characterisation of a MCL extracted from the phellem tissue of the
bark of *Pseudotsuga* also showed a G-lignin (Marques et al., 2006).

Fractionation of cork using an ethanol organolsolv treatment was also used for the
isolation of lignin and its characterisation confirmed the chemical features and
monomeric composition of the MCL (Pascoal Neto et al., 1996).

A small quantity of suberinic acids is ester linked to the MCL, corresponding to about
2%. The small fraction of MCL that was soluble in ether contained suberinic acids in
appreciable amount: α,ω-diacids, ω-hydroxyacids and acids, in a proportion in agreement
with the suberin composition in cork.

Table 3.4. Elemental analysis, methoxyl content, monomeric composition (H, G, S), molecular
weight and content of non-aromatic units (carbohydrate and suberinic units) of milled cork lignin
(MCL) obtained from desuberinised extractive-free cork and of its ether soluble fraction (MCL_{ether})
(Marques et al., 1996).

	MCL	MCL_{ether}
Elemental analysis, % of sample		
C	62.82	69.03
H	6.04	8.72
N	0.71	0.16
OMe	13.95	13.19
Lignin composition, % of monomers		
H	0.8	1.8
G	97.6	95.4
S	1.6	2.8
Molecular weight		
Mw	9210	3360
Mn	3070	1300
Mw/Mn	3.0	2.6
Carbohydrates, % of sample	5	17
Suberinic acids, % of sample	2	28

Bonding of lignin to hemicelluloses occurs through the so-called LCCs. In cork, the isolated LCC contained 8.3% arabinose and 0.7% xylose. It is known that the chemical bonding between lignin and hemicelluloses occurs mostly as benzyl ester and ether linkages as well as phenyl glycosidic linkages.

3.2.2. Macromolecular structure

The lignin molecule is a branched and complex structure where the phenylpropanoid units are linked by various types of covalent bonds (C–O and C–C) and have different functional groups: aromatic and aliphatic hydroxyls, benzyl alcohol and ether groups, carbonyl and methoxyl groups. The methoxyl groups are a characteristic feature of lignins, since they derive from the initial building units.

The size of the macromolecule is a matter of discussion and determinations depend largely on the method of isolation. Lignins are polydisperse polymers with average molecular weights (M_w) of a few thousands to more than 80 000 have been reported, corresponding to molecules with just some 20 units up to more than 400 units.

In the case of cork lignin, the average molecular formula has been calculated based on elemental analysis and determination of methoxyl groups as $C_9H_{8.74}O_{2.82}(OCH_3)_{0.85}$ (Marques et al., 1996). The molecular weight of MCL has a monomodal distribution pattern with a maximum at around 7000–8000 Da (corresponding to a polymerisation degree of approximately 40) and two shoulders at 3000 and 1000 Da (Fig. 3.10).

The visualisation of the structure of the lignin macromolecule is facilitated by the use of models that summarise the main linkages and functional groups that occur in the polymer. Figure 3.11 shows a model for cork lignin designed by Marques (1998) taking as a

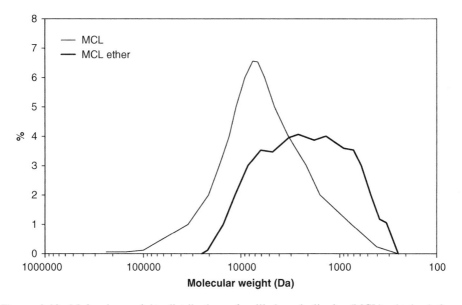

Figure 3.10. Molecular weight distribution of milled cork lignin (MCL) obtained from desuberinised extractive-free cork and of its ether soluble fraction (MCL_{ether}).

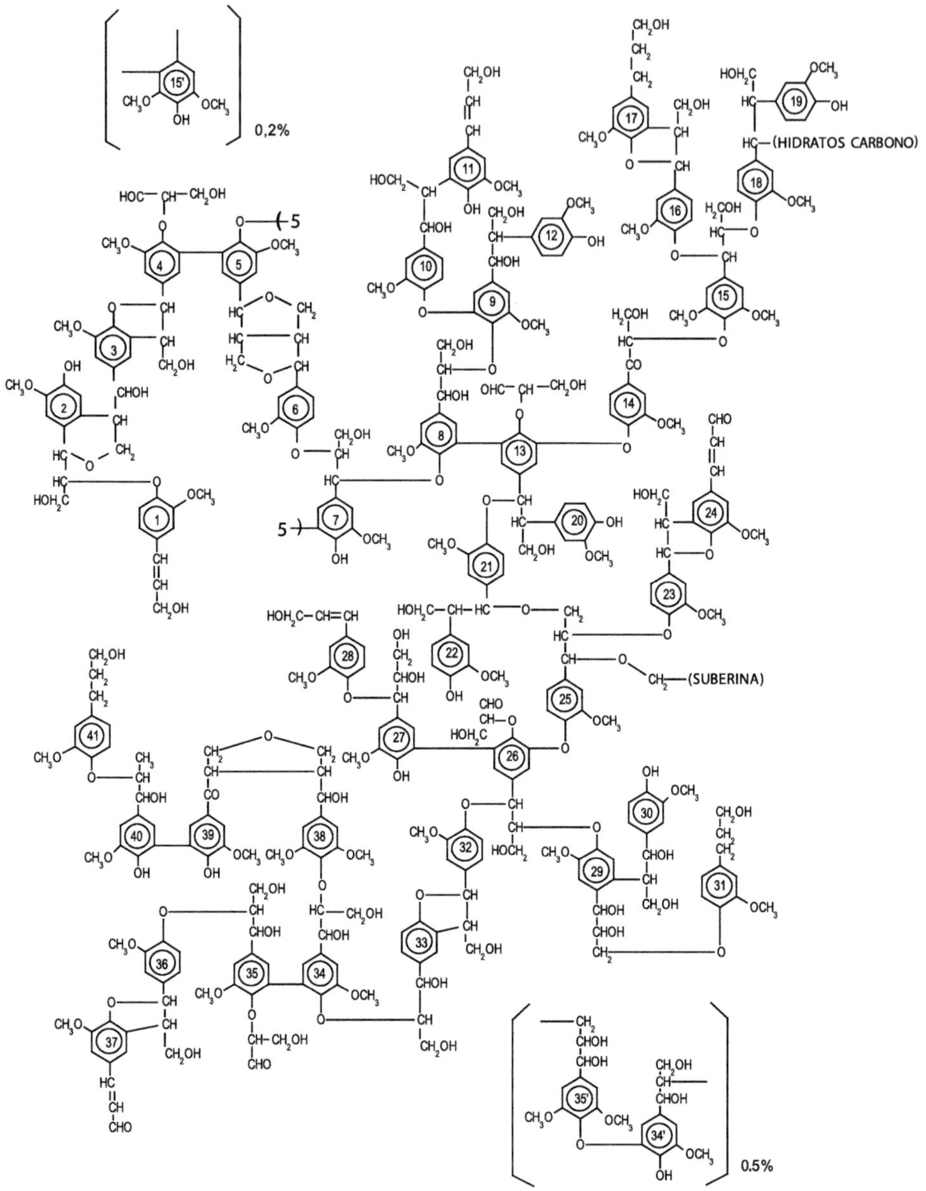

Figure 3.11. Model proposed for the lignin of *Quercus suber* cork (Marques, 1998).

reference a model for spruce wood lignin presented by Sakakibara (1991) and including the available information on monomer bonding. Its three-dimensional representation shows lignin to be amorphous without an organised supramolecular structure and approximately isotropic. However, many aspects regarding the inter-monomeric bonding and the presence and location of functional groups are not known and this representation must be taken only as a work tool.

It is also important to stress that there are covalent links between lignin units and suberin, as well as to hemicelluloses.

3.2.3. Non-lignin aromatics

Several authors have referred to the presence of an aromatic domain in association with suberin, but considered it to be a non-lignin matrix of covalently linked cinnamic acids (Bernards and Lewis, 1998) where feruloyl tyramine (Negrel et al., 1996) was also found. Most evidence on the phenolic analysis of suberised tissues such as potato periderm has shown significant amounts of hydroxycinnamic acids and a proportion of monolignols smaller than expected from a lignified tissue and more cross-linked than in wood lignin, thereby considered a non-lignin polyphenolic domain (Bernards, 2002). However, it can be questioned if this polyphenolic domain associated to suberin in potato wound periderm can be transferred to other plant species, namely to woody ones (Bernards and Razem, 2001).

In the case of cork, methanolysis releases in addition to the suberinic acids also small amounts of ferulic acid (<1%) and acyl ferulates were also isolated from the methanolic extract (Graça and Pereira, 1998). The linkage of ferulic acid units to the macromolecule (either suberin, or non-lignin aromatics) should include ester bonds only to a limited extent. However, higher amounts of ferulic acid are obtained when using stronger reactive conditions for suberin depolymerisation, e.g. amounting to 5–9% of suberin polymers (Garcia-Vallejo et al., 1997) or to 6% (Bento et al., 1998). Cork and MCL also include a small amount of N as shown by elemental composition (Table 3.4) which is compatible to the presence of hydroxycinnamoyl amide links in the polymer.

The total amount of aromatics in cork was estimated indirectly from the results of analytical pyrolysis as approximately 40% of the extractive-free cork (Marques et al., 1994). A comparative analysis of the aromatic monomeric composition of the isolated MCL with the result of its direct determination on cork shows that cork contains in addition to lignin other aromatic compounds with a different composition. The non-lignin aromatics were estimated to amount to about half of the total aromatics in cork and to include 25% H units and 75% G units, while showing esterification to suberinic acids and hemicelluloses (Marques et al., 1999).

In spite of the recent evidence gained on the characterisation of the aromatic components of suberised tissues, the chemical characterisation of the non-lignin aromatic content of cork cell walls is still more a matter of speculative reasoning than of scientific evidence. The same applies to the bonding between the different components of the cork cell wall: lignin, non-lignin-aromatics, suberin and hemicelluloses.

3.3. Polysaccharides

The structural polysaccharides of plant cell walls are cellulose and hemicelluloses. In cork the total polysaccharides represent approximately 20%, much less than in wood where cellulose and hemicelluloses correspond to 70–80% of the structural components of the cell wall.

Table 3.5. Monosaccharide composition after acid hydrolysis of *Quercus suber* virgin and reproduction cork (Pereira, 1988).

Monosaccharides	% of total monosaccharides	
	Virgin cork	Reproduction cork
Glucose	50.7 (6.4)	45.4 (6.2)
Xylose	34.0 (5.1)	32.3 (5.5)
Arabinose	6.4 (0.8)	13.2 (2.3)
Galactose	3.6 (0.9)	5.1 (2.6)
Mannose	3.7 (0.7)	3.2 (1.1)
Rhamnose	1.7 (0.4)	0.8 (0.1)

Note: Mean of 10 trees and standard deviation.

Holocellulose, comprising the cellulose and hemicelluloses, can be isolated from extractive-free and desuberinised cork as a white solid using acid sodium chlorite with a yield of 26% of cork (Conde et al., 1998), but Asensio (1987c) obtained a holocellulose yield of only 13%. Upon total hydrolysis, the extractive and suberin-free cork yields neutral sugars and uronic acids. Glucose and xylose correspond to 50 and 35% of all neutral sugars and are accompanied by smaller amounts of arabinose, mannose, galactose and rhamnose (Table 3.5). The uronic acid content of cork polysaccharides is approximately 12% (Rocha et al., 2004).

3.3.1. Cellulose

Cellulose is the main component of wood where it represents about half of the cell wall material. It is therefore an important component and its chemical characteristics and supramolecular structure are responsible for many of the wood properties.

Cellulose is a linear polymer of β-glucose molecules linked together by β-(1-4) glycosidic bonds in a long chain containing several thousands of anhydroglucose units (Fig. 3.12). The cellulose chains are aligned parallel to each other, building up a densely packed structure featuring intramolecular H-bonds between adjacent monomeric units and intermolecular H-bonds between adjacent chains (Fig. 3.13). The supramolecular structure of cellulose is characterised by a highly ordered arrangement with densely packed molecules, building up a fibrous-like rod structure named microfibril. This regular arrangement of the cellulose molecules forms what is called the crystalline cellulose, which is surrounded by less ordered cellulosic regions, the amorphous cellulose, and by hemicelluloses.

Cellulose is insoluble and interaction with water and reaction only occur under moderate conditions in the amorphous regions of cellulose. Cellulose is strongly anisotropic, with the chain length direction as the more resistant one. In wood anisotropy of several properties (i.e. mechanical) are directly related to the topochemistry of the cellulose molecules. For instance, tensile strength of wood is larger in the axial direction, corresponding to the major cell alignment and within-the-cell-wall orientation of the cellulose microfibrils.

In the cork cells, cellulose is located in the primary wall and in the tertiary wall. Its presence in the cell wall is known since 1927 by the work of Zetsche and Rosenthal. However, very little is known on cork cellulose and details on polymerisation degree, cristallinity and chain orientation are not known. The cellulose content of cork has been estimated at approximately 10% based on the neutral sugar composition of hydrolysates (Pereira, 1988).

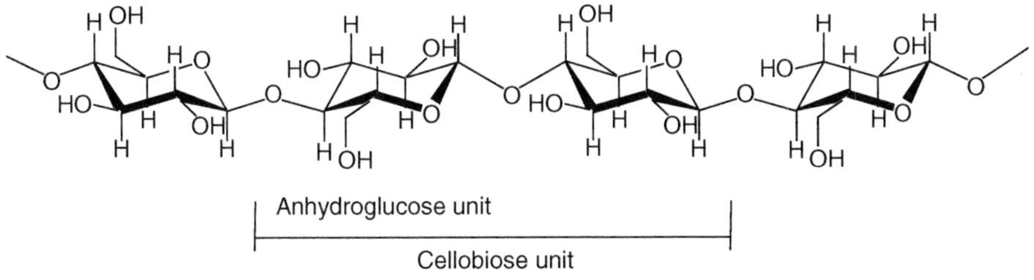

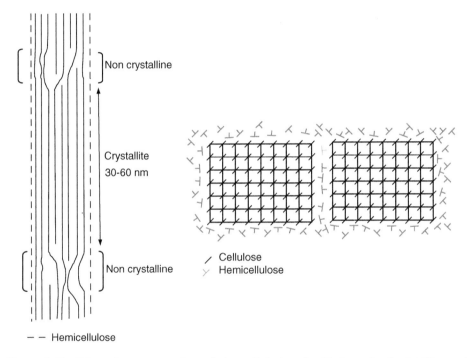

Anhydroglucose unit

Cellobiose unit

Figure 3.12. Structure of one cellulose chain.

Non crystalline

Crystallite
30-60 nm

Non crystalline

/ Cellulose
Y Hemicellulose

− − Hemicellulose

Figure 3.13. Schematic representation of the cellulosic microfibrils and of cristalline and amorphous regions.

The small amount of cellulose in the cork cell wall makes that it will not have a determining role in cork properties, in opposition to what happens in wood. In cork this role is played by suberin.

3.3.2. Hemicelluloses

Hemicelluloses are polysaccharides that comprise various compounds of different chemical composition and molecular structure. They are heteropolymers, usually classified according to the main types of sugar residues that are present. Composition and content of hemicelluloses vary with species and cell type. The most important hemicelluloses in hardwoods are of the xylan family, mainly the 4-*O*-methylglucuronoxylans.

The monosaccharides that are found in hemicelluloses include pentoses (β-D-xylose, α-L-arabinose), hexoses (β-D-mannose, β-D-glucose, α-D-galactose), uronic acids (β-D-glucuronic acid, α-D-4-*O*-methylglucuronic acid and α-D-galacturonic acid) (Fig. 3.14). To a smaller extent may appear α-L-rhamnose and α-L-fucose. Compared with cellulose, hemicelluloses are smaller branched polymers, amorphous and more reactive.

The hemicelluloses of cork were studied by Asensio (1987a,b, 1988a,b) who isolated and characterised three different xylan-based hemicelluloses: 4-*O*-methylglucuronoxylan, arabino-4-*O*-methylglucuronoxylan and 4-*O*-methylglucurono-arabinogalactoglucoxylan. The hemicelluloses were isolated using alkaline solutions from holocellulose prepared from extractive-free cork after desuberinisation by methanolysis. Their structural features are as follows:

• the *4-O-methylglucuronoxylan* has a backbone of β-D-xylopyranosyl units linked by β-(1-4) glycosidic bonds with units of 4-*O*-methyl-α-D-glucopyranosyluronic acid

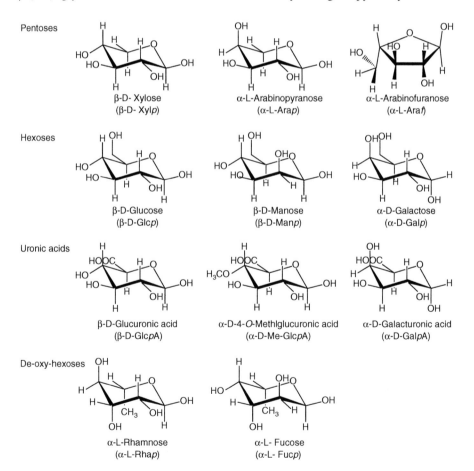

Figure 3.14. Monosaccharides that may be present as monomeric units in hemicelluloses.

attached by α-(1-2) glycosidic bonds on average to every 16 xylosyl residue, corresponding to a molar proportion of xylose to 4-*O*-methylglucuronic acid units of 94:6. This hemicellulose was solubilised with 10% NaOH and precipitated by acidification at pH 5, with a yield of 34.6% of the cork holocellulose;

- the *arabino-4-O-methylglucuronoxylan* has a linear backbone of β-D-xylopyranosyl units linked by β-(1-4) glycosidic bonds with side chains units attached by α-(1-2) glycosidic bonds of 4-*O*-methyl-α-D-glucopyranosyluronic acid units to every 15 xylosyl units, of β-D-xylopyranosyl to every 13 xylose units, and of α-L-arabinofuranosyl units to every 56 xylosyl units of the chain. The molar ratios of xylose, 4-*O*-methylglucuronic acid and arabinose were 170:13:3; this hemicellulose was isolated from the acid soluble fraction of the 10% NaOH solubilised material after precipitation with ethanol;

- the *4-O-methylglucurono-arabinogalactoglucoxylan* was isolated after solubilisation with 4% OHNa is a highly branched polymer with a backbone including β-(1-4) and β-(1-3) linked β-D-glucopyranosyl units, β-(1-4) linked β-D-xylopyranosyl units, and β-(1-4) and β-(1-6) linked galactopyranosyl residues. The side chains to this backbone include: one α-(1-4) linked 4-*O*-methylglucuronosyluronic acid unit to every 5 xylosepyranosyl residue in the main chain; one α-(1-3) linked 3-*O*-(xylopyranosyl-arabinofuranosyl) group to every 10 xylosepyranosyl residue in the main chain; one α-(1-6) linked arabinofuranosyl unit to every 2 glucopyranosyl residue in the main chain; one (1-3) linked galactopyranosyl unit to every 6 galactopyranosyl residues in the main chain. Some α-rhamnosyl groups are also present as non-reducing end-groups. The molar ratios of xylose, arabinose, glucose, galactose, 4-*O*-methylglucuronic acid and rhamnose were 17:12:12:6:4:1; this hemicellulose was isolated from the soluble fraction using 4% NaOH.

Figure 3.15 schematically represents the structure of these three hemicelluloses from cork, based on the results of Asensio (1987a,b, 1988a,b).

There are no studies on the molecular size of cork hemicelluloses. A parallel with wood 4-*O*-methylglucuronoxylans would indicate a number-average degree of polymerisation in the range of 50–200.

As regards the supramolecular structure, the xylans in the cell wall are amorphous and the irregular presence of side groups and branching along the linear chain do not allow a strong intermolecular association by H-bonds. The highly branched 4-*O*-methylglucurono-arabinogalactoglucoxylan may be partially soluble, as shown by a long duration extraction of cork with 12% ethanolic solution during 6 months (Rocha et al., 2004) that could remove 1.4% of cork as polysaccharides with the following composition: arabinose 10%, xylose 53%, mannose 3%, galactose 4%, glucose 17%, uronic acids 13%.

The presence of hydroxyl groups allow H-bonding, for instance to available positions in neighbouring cellulose molecules, and also chemical bonding with lignin and other phenolics, as well as with the fatty acids of suberin.

3.3.3. Pectins

Cork contains pectins that are located in the middle lamella in amounts corresponding to approximately 1.5% of cork (Rocha et al., 2000). The pectins are slightly branched polymers of hexuronic acid units with side chains of arabinose residues (corresponding

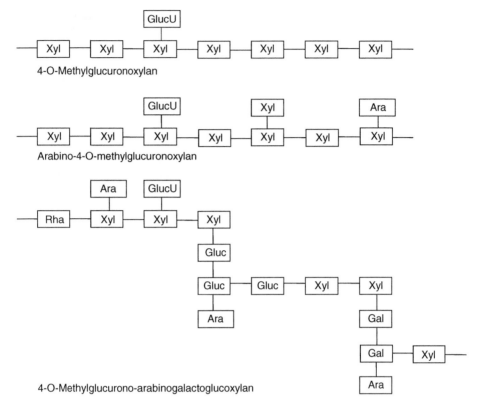

Figure 3.15. Model structure from the three xylans of *Quercus suber* cork.

to 5% of the total monomers). Rhamnose, fucose, xylose, mannose, galactose and glucose are present in trace amounts.

3.4. Extractives

Extractives are low or medium molecular weight molecules that may be removed from the
cells by solvent extraction without affecting the cellular structure of the material and its
mechanical properties. In wood extractives represent a small proportion of the material,
as a rule under 10%. In cork the amount of extractives is higher and very variable, from
8 to 24% with average values in the range of 14–18% of cork (Table 3.6).

   The extractives may include hundreds of different molecules. They can be classified in
several manners, for instance based in their different polarity or solvent in which they are
solubilised, or organised by chemical families.

   The extractives of cork include *n*-alkanes, *n*-alkanols, waxes, triterpenes, fatty acids,
glycerides, sterols, phenols and polyphenols. Usually they are classified in two groups:
aliphatics that are solubilised with low-polarity solvents (e.g. hexane, dichloromethane,
chloroform) and phenolics extracted by polar solvents (e.g. ethanol and water).

Aliphatics represent about one-third of extractives, corresponding to 5–8% of cork. The main components are triterpenes (e.g. cerin, friedelin, betulin and betulinic acid as well as sterols), corresponding to half of the aliphatics, and *n*-alkanes (from C16 to C34), *n*-alkanols (from C20 to C26) and fatty acids (monoacids, diacids and hydroxy-acids).

Phenolic compounds represent 6–9% of cork. They include simple phenols (phenols, benzoic acids and cinnamic acids) and polymeric phenols (tannins), corresponding to more than 90% of the total.

3.4.1. Long-chain aliphatics

The non-polar extractives of cork include aliphatics like long-chain *n*-alkanes, *n*-alkanols, and fatty acids usually combined in a common designation of cork waxes. The alcohols represent 2.5% and the fatty acids 1.1% of cork (Conde et al., 1999).

The fatty alcohols comprise all the even members from C16 to C26, with traces of the intermediate odd members and some unsaturated groups; docosanol and tetracosanol are the main compounds (1.4 and 0.7% of cork, respectively).

The fatty acids include the saturated even series from C14 to C24 and also the C15, C17, C21 acids, accompanied by a large amount of unsaturated acids and some ω-hydroxy-acids: most abundant were the docosanoic and hexadecanoic acids and the C18-acids 9-octadecenoic, 9,12-octadecadienoic and octadecanoic acids.

3.4.2. Terpenoid extractives

Terpenoids can be seen as structural multiples of isoprene (2-methyl-1,3-butadiene), e.g. triterpenoids have six isoprenic units. Triterpenoids are of widespread presence in hardwoods. They are based on many skeletal types, most of them pentacyclic. A tetra-cyclic group of triterpenoids is the steroids, which includes β-sitosterol, the major sterol found in woods and in many other plant tissues.

Cork contains triterpenes from the families of friedelan, lupane, and sitostane, including mainly cerin and friedelin (Pereira, 1979; Caldas et al., 1985; Conde et al., 1999) that are solubilised by solvents of low polarity. With the compositional data given

Table 3.6. Extractives removed from virgin and reproduction cork by successive extraction with solvents of increasing polarity (Pereira, 1988).

Extractives	% of cork	
	Virgin cork	Reproduction cork
Total	16.9 (2.5)	14.2 (1.1)
Dichloromethane	7.9 (1.7)	5.4 (0.5)
Ethanol	5.8 (1.2)	4.8 (0.7)
Water	3.1 (0.5)	4.0 (0.7)

Note: Mean of 10 trees and standard deviation.

for a chloroform extraction of cork (Conde et al., 1999), the following calculation of terpenic content in cork may be made: the solubilised triterpenes correspond to 3.4% of cork, including mainly friedelin (1% of cork), betulin (0.5%) and cerin (0.4%). Betulinic acid and 3-α-hydroxyfriedelan-2-one are also found in the chloroform extract of cork (Castola et al., 2002).

Cork also contains sterols, mainly β-sitosterol and sitost-4-en-3-one in significative amounts. A CO_2 supercritical extraction seems to be more selective towards sterols than a chloroform extraction and a sitost-4-en-3-one yield of 1% of cork was reported (Castola et al., 2005).

Figure 3.16 shows the chemical structures of the main components of the triterpenes present in cork extractives.

3.4.3. Phenolic extractives

A major group of extractives in wood and other lignocellulosics includes compounds with phenol units in their structure. Many of these phenolic extractives are found as glycosides, linked to glucose and other sugars. Some of the phenolic compounds, namely the group of tannins, can reach relatively high molecular weights.

One important group of phenolic extractives are the flavonoids. Flavonoids are C6–C3–C6 three-ring structures, and the structure of the central ring defines different classes of flavonoids. Flavonoids exist as such, as glycosides and also in oligomeric and polymeric form. Flavanes like catechin (flavan-3-ol) and leucocyanidin (flavan-3,4-diol) condense in dimeric forms, as biflavonoids (proanthocyanidins), or in higher degree to form polyflavonoids known as condensed tannins. The other type of tannins present in wood are the hydrolysable tannins, named as such because they are hydrolysed to monomers with

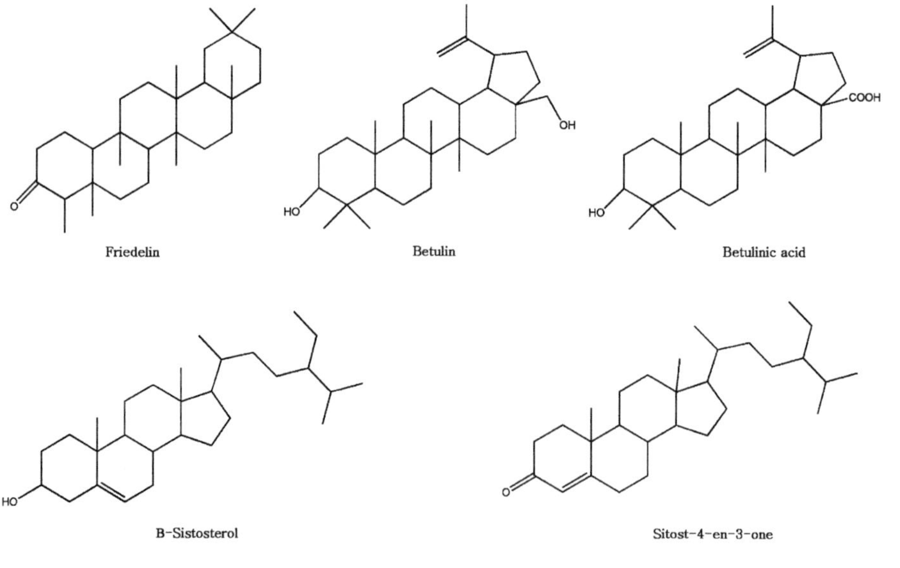

Figure 3.16. Chemical structures of the main components of the triterpenes present in cork extractives.

acids. They are esters of gallic acid and of its dimers, digallic and ellagic acids, with sugars, usually glucose. Figure 3.17 shows the chemical structures of some of these flavonoids.

In cork ethanolic extracts simple phenolic molecules were identified that include benzoic and cinnnamic acid derivatives, as well as vanillin, syringaldehyde and acetovanillone (Mazzoleni et al., 1998). The following low molecular weight polyphenols were identified in cork extracts (Conde et al., 1997): gallic, protocatechic, vanillinic, caffeic, ferulic and ellagic acids; protocatechuic, vanillic, coniferylic and sinapic aldehydes and aesculetin and scopoletin. Ellagic acid is the main component.

In the polyphenolic extractives of cork, the hydrolysables tannins are the most important. Ellagitannins, roburins A and E, grandinin, vescalagin and castalagin were identified (Cadahia et al., 1998).

3.5. Inorganic components

The inorganic components of lignocellulosic materials are usually included in what is called the ash content, representing the solid residue after total combustion. The ash content is usually below 1% in wood from temperate regions but much higher in barks.

Comparatively cork has rather low ash content, corresponding to values between 1 and 2%. Calcium is the most important element, with more than 60% of the total. Phosphorous, sodium, potassium and magnesium are also present with values of some significance while other minerals are identified in very small amounts.

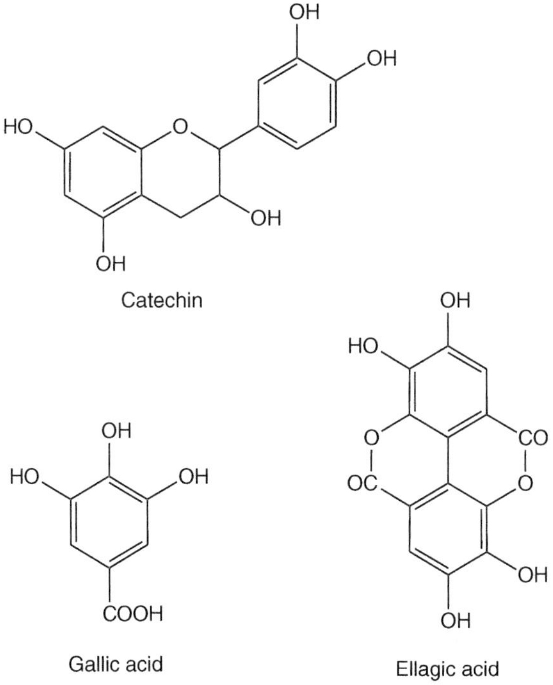

Figure 3.17. Chemical structures of simple phenolics and of flavonoids present in cork.

The mineral composition of reproduction cork was determined as follows, in percentage of cork dry mass (Mata et al., 1986): Na 2.91×10^{-2}, K 2.31×10^{-1}, Ca 6.25×10^{-1}, Mg 2.49×10^{-2}, Mn 7.03×10^{-3}, Fe 0.87×10^{-2}, Cu 0.73×10^{-3}, Zn 0.77×10^{-3}, P 2.46×10^{-2}.

3.6. Chemical composition of cork

The chemical composition of cork is substantially different from other plant materials, namely from wood and from the other tissues in barks. Chemically the composition of cork is dominated by the presence of suberin as the main structural cell wall component (Fig. 3.1) and by its close association with the other components, mainly the lignin and the non-lignin aromatics. This results in a difficult methodological approach to comprehensively cover cork in terms of its chemical components. In addition to the general difficulties encountered in the determination of the summative chemical composition of lignocellulosics, the work becomes more complex in the case of cork due to the lack of a standard approach to analytical procedures and to the uncertainties in chemical definition of the macromolecules, as discussed before. Therefore, results obtained in the various studies from different authors may differ to a considerable extent and attention to experimental details is mandatory.

Klauber (1920) presented the following chemical composition for cork (% dry mass): suberin 58%, lignin 12%, cellulose 22%, cerin 2%, other components 1% and water 5%. Subsequent results report variable suberin contents from 33 to 50%.

Table 3.7 summarises the average results and their range obtained for the determination of the summative chemical composition of virgin cork determined in a total of 40 trees. It includes also the comparison of the chemical composition of virgin and reproduction cork in 10 trees. On average, extractives amount to 17%, suberin to 45%, lignin to 25% and cellulose and hemicelluloses to 20%. However, some remarks must be made regarding the implication of the methods used on these results (Pereira, 1988). The suberin value included in these determinations refers only to the long-chain fatty components, excluding glycerol, which was the usual practice in cork chemical analysis, meaning that the total suberin content should be higher. Lignin was determined as the sum of the residue after acid hydrolysis (klason lignin) and of the acid soluble aromatics, and therefore it contains the lignin and the non-lignin aromatics.

Table 3.7. Chemical composition of virgin cork and of reproduction cork from *Quercus suber* in different trees and locations in Portugal (Pereira, 1988).

	Four sites (40 trees)	One site (10 trees)	
	Virgin cork	Virgin cork	Reproduction cork
Ash	0.7 (0.2)	0.9 (0.2)	1.2 (0.2)
Extractives	15.3 (1.7)	16.9 (2.5)	14.2 (1.1)
Suberin[a]	38.6 (4.2)	35.2 (3.1)	39.4 (1.7)
Lignin	21.7 (0.9)	22.4 (1.1)	23.0 (0.8)
Polysaccharides	18.2 (2.5)	21.3 (2.4)	19.9 (2.6)

[a]Suberin only includes fatty acids (without glycerol).

More recent determinations of cork chemical composition from samples with different geographical origins in Portugal (with a total of 58 cork samples) confirm the average values already referred. The same occurred when analysing more intensively different trees in two sites (20 trees/site). Table 3.8 summarises the variability of results found for the different sites and the between-tree difference in each site. The following results were obtained: total extractives 15.9% (range 10.7–29.7%), suberin 41.2% (range 22.1–51.7%) and lignin 20.4% (range 15.6–27.6%). In spite of a broad range of values, the coefficient of variation of the mean was rather low: 12% for lignin, 15% for suberin and 23% for the extractives.

A study on the chemical variation of cork from different provenances in Spain showed the following results (Conde et al., 1998): total extractives 18.4% (range 16.5–21.0%), lignin 22.7% (range 21.5–24.3%) holocellulose 25.9% (15.3–34.3%); the suberin content determined in this work due to experimental procedure cannot be used for comparison.

In spite of the range of values found for each component, it is clear that cork chemical composition is marked by a high suberisation degree (potato periderm has about 17% of suberin on a total dry basis, Graça and Pereira, 2000b) and by a lignin content similar to the values found in hardwoods. Cellulose and hemicelluloses represent a small fraction of the cork cell wall (in wood they correspond to about 70%). The composition of the polysaccharides in terms of their neutral monosaccharides was shown in Table 3.5. There is also some between-sample variation, but overall glucose represents about half of the sugars, and xylose more than one-third. Taking into account the chemical composition of cork hemicelluloses, it is clear that cellulose corresponds to approximately half of the cork polysaccharides, amounting to nearly 10% of cork.

3.7. Topochemistry of cell walls

A more complete chemical knowledge of cork requires in addition to its composition and the chemical features of the individual components, also the insight of where within the

Table 3.8. Chemical composition of reproduction cork from *Quercus suber* in different trees and locations in Portugal.

	Extractives				Suberin[a]	Lignin
	Cl_2CH_2	EtOH	Water	Total		
Site A						
Mean	5.73	5.59	5.38	16.70	42.12	20.77
Standard deviation	0.73	2.57	2.26	4.04	3.79	2.69
Coefficient of variation	12.7	50.0	42.0	24.2	9.0	13.0
Maximum	7.39	13.24	11.19	28.86	48.84	26.59
Minimum	4.67	2.48	2.97	12.27	33.86	16.10
Site B						
Mean	5.80	7.68	4.26	17.73	38.35	n.d.
Standard deviation	0.84	4.85	1.76	5.77	6.57	n.d.
Coefficient of variation	14.5	63.2	41.3	32.5	17.1	n.d.
Maximum	6.70	22.03	8.47	32.89	46.85	n.d.
Minimum	2.64	1.69	0.95	8.58	22.95	n.d.

[a]Includes only the long-chain fatty acids (without glycerol).

cell wall they are distributed, the so-called cell wall topochemistry, and the eventual inter-action between them.

The gaps in knowledge regarding the structural components of cork have been discussed and it is clear that they lie mainly in their macromolecular characteristics. The fact that most evidence is based on the analysis of very small moieties obtained from the depolymerisation of the macromolecule that destroys its in-situ structure is one of the reasons. Also, the com-plexity of whole-material analysis using holistic approaches, such as ^{13}C NMR or ^1H NMR, FT-IR, contribute to a still speculative approach to cell wall chemical architecture.

An important aspect that has to be taken into consideration when trying to design the chemical assembly in cork cell walls is the absolute proportion of the components. Although there is natural variation between samples and the proportion of components lie within a range as previously referred, three points are clear: (i) suberin is the main con-struction element of the cell wall (with more than half of its solid material); (ii) lignin, including other macromolecular polyphenolics, represents about one-third of the cell wall material; (iii) cellulose has a lesser role in the construction of the cell wall, and its pro-portion should be less than 10%; hemicelluloses also have a similar proportion.

Suberin is a flexible polymer made up of long-chain aliphatic chains linked through glyceridic anchor points. The macromolecular structure hypothesised in Figure 3.7 shows an approximate bilayer arrangement with a width represented by about 4–6 nm (a straight C20 chain would have a length of 2.5 nm). The suberin carbons show two different mobil-ities: most have a higher mobility (about 85% of the total, resonating at 30 ppm) and a smaller proportion are more restricted (about 15% of the total, resonating at 33 ppm) (Lopes et al., 2000a). We can interpret these results as referring to the long-chain CH_2 car-bons and the glyceridic restricted carbons, respectively.

Lignins present a more or less spherical appearance under the microscope with globu-lar particles in the range of 10–100 nm with the aromatic rings giving bulk and rigidity to the structure. Cork lignin is considerably condensed but the aliphatic short chains allow some flexibility to adapt to various conformations and packing arrangements, namely in conjunction to the occurrence of non-lignin, cinnamic acid based moieties.

It is proved that there are some links between aromatic units and hemicelluloses (LCCs), as well as the possibility of aromatic–suberinic ester links as given by the presence of feru-loyl acyl esters (Fig. 3.4). In the cell wall, the suberin, lignin and hemicellulosic fractions build up a spatially strong and chemical resistant assembly, and their fractioning is particu-larly difficult whatever the method used. All the standard analytical procedures that have been developed for wood encounter obstacles when they are applied to cork and yields are in general poor, as it was evident from the studies on cork lignin (Marques et al., 1994, 1996). The substantial disruption of the cell wall structure by removing the suberin is usu-ally required but even in this case residues of the other components are still present.

This clearly rules out past assumptions that lignin in cork is restricted to the middle lamella or that the secondary wall is made up only of suberin in association with the aliphatic extractives.

On the other side, the analysis of solid cork has shown that the suberin carbons and protons are spatially segregated from the aromatic ones and the pool of aromatic and poly-sacharidic signals has been considered as more spatially interlinked (Lopes et al., 2000a).

One important aspect when trying to put together the cell wall chemical assembly in cork is also the microscopic evidence. The secondary wall of cork has a lamellated structure with translucent layers of constant thickness of about 4–5 nm and electron dense dark layers of more variable width. This is compatible with a structure where the suberinic bilayers alternate with the dark layers of lignin and aromatics with bonding occurring at the boundary between the suberin and cinammic acid moieties. In the middle lamella lignin, hemicellulose and pectins make up a composite matrix, while the tertiary wall includes the cellulose and hemicelluloses.

The construction of the cork cell wall may be schematically drawn as in Figure 3.18. The lignin matrix of the middle lamella cements the cells together and rigidifies the overall structure. The lignin and aromatics layers between the suberin lamella also give some structural rigidity to the cell walls, quite in a way of a scaffolding that supports the flexible aliphatic component. The cellulosic tertiary wall makes the lining to the lumen side of the cells.

Some evidence that this structure is in accordance to the material behaviour may be obtained when chemical selective handling of cork is made. The mechanical support given by lignin was already shown in Figure 3.8, representing the collapsed cellular structure upon removal of lignin. A more detailed pictorial representation of the effects of cork delignification (Fig. 3.19) clearly shows that it is not only cellular separation (due to removal of lignin in the middle lamella) that occurs, but also the collapse of the cell walls that become a completely flexible layer that is able to adapt to various forms.

The removal of suberin leaves a structure where the cellular arrangement still can be recognised but where the cell walls appear as made up of a porous agglomerate of debris (Fig. 3.20a). In regions where only a partial suberin removal was obtained, ribbon-like structures were observed that conform to the spatial development of the suberin macromolecule as discussed previously (Fig. 3.20b). After delignification, the drying of the suberin cell walls may originate regularly arranged and parallel lying fracture lines that are compatible with such structural features (Fig. 3.20c). When observing the region of

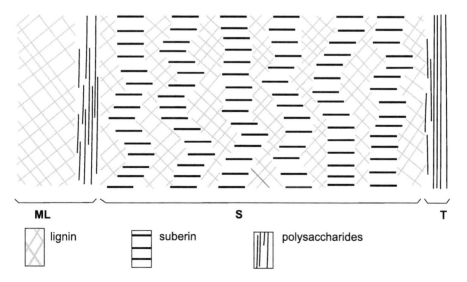

Figure 3.18. Schematic representation of the cork cell wall in *Quercus suber* showing the location of the structural components.

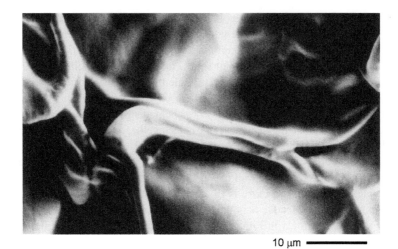

10 μm ━━━━━━

Figure 3.19. SEM photographs of delignified cork showing cellular separation and the collapse of the cell walls.

the phellogen after removal of the cork layer, the fragile cell walls of the newly formed cells can be observed (Fig. 3.21). Many cell walls appear porous and present ribbon like structures. It was also possible to observe the intact tertiary wall that becomes separated from the suberinic secondary wall (Fig. 3.22).

Although speculative, the cork cell wall structure depicted in Figure 3.18 is in agreement with the available knowledge and may be used as a work tool to study the properties and behaviour of cork. A more detailed investigation at the nanometer level of the cell wall structure in cork will certainly bring a new insight into this intriguing and complex cellular solid.

3.8. Methods of chemical analysis

The conventional approach to the chemical characterisation of cork, as of other lignocellulosic materials, is to report its summative chemical composition by giving the contents of inorganic ashes, organic extractives and the structural components. This is a chemical step wise approach that disturbs the *in situ* characteristics of the structural components.

Another more recent approach is to characterise the cork structural components without their removal from the cell wall. Techniques such as FT-IR, ^{13}C-NMR and ^{1}H-NMR are being increasingly used on the cork solid material to analyse its main chemical features regarding functional groups and their chemical environments. Since chemical transformation is reduced to a minimum (usually only involving removal of extractives), the information obtained refers to the macromolecules as they are assembled in the cork cell wall.

3.8.1. Summative analysis

The chemical analysis of cork has adapted the analytical approach used for wood and other lignocellulosics. When performing a summative chemical analysis it is the aim to

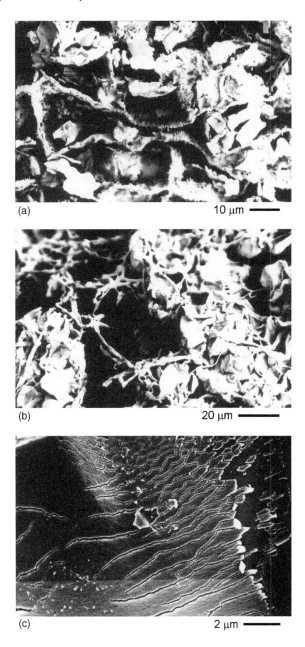

(a)   10 μm

(b)   20 μm

(c)   2 μm

Figure 3.20. SEM photographs of (a) desuberinised cork cells showing the remaining cell walls as a porous agglomerate of debris; (b) ribbon-like structures that were observed in cells with only a partial suberin removal and (c) delignified cork cell walls showing fracture lines after drying.

quantify the structural and non-structural components as accurate as possible in spite of the difficulties of separating and measuring each of the structural components. This is already known in wood chemical analysis, where summative compositions using standardised methods represent frequently only about 95% of the material.

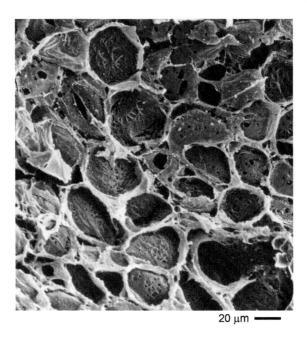

20 µm ━━━

Figure 3.21. SEM photograph of the newly formed cells after separation from the phellogen.

10 µm ━━━

Figure 3.22. SEM photographs of cork cells after delignification showing fragments of the intact tertiary wall after separation from the suberinic secondary wall.

In cork, the determination of the summative chemical composition is more complex due to the occurrence of another structural component (the suberin) and to the special topochemistry of the cell wall. With the analytical procedures used at present a total mass recovery usually lies only slightly over 90%. Tables 3.7 and 3.8 are examples of such summative chemical compositions.

As previously mentioned, the specific chemical protocol may have a direct influence in the result obtained, and therefore it is necessary to pay attention to detail when analysing data, namely for comparative purposes. Although a few proposals for chemical fractioning of cork have been made, i.e. an organosolv ethanol-based extraction (Cordeiro et al., 2002), enzymatic treatments (Rocha et al., 2001) or high-pressure dioxane extraction (Miranda et al., 1996), the most used and solid approach has evolved as an adaptation of the standard methodologies used for lignocellulosic materials with the introduction of suberin removal and determination (Marques and Pereira, 1987).

Figure 3.23 schematically represents the analysis of cork as a flow sheet process and each unit operation is described below. The process starts with the sample preparation (grinding and sieving) since different tissues are present with variable extents that can influence the chemical results.

3.8.1.1. Sample preparation

Cork analysis should be made on granulated material with particle size between 40–60 mesh (850–425 mm). Trituration of the cork samples may use knife mills, but the operation is

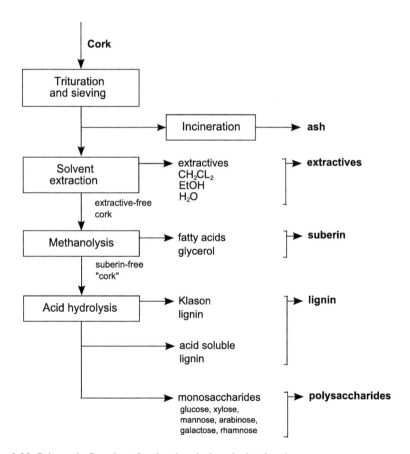

Figure 3.23. Schematic flowsheet for the chemical analysis of cork.

difficult and the yield is low due to the deformability of the cork pieces. It is advised to carry out the operation stepwise, first with an output grid of 2 mm, an intermediate separation of the material below 60 mesh and the successive regrinding of granules above 40 mesh and further separation.

The granulometric classification is important to separate material that is not made up by the cork suberous cells. In fact the cork planks contain woody inclusions and lenticular material of different chemical composition (as detailed in the following item) that have a different particle size distribution after milling and constitute the fine fractions. In the case of analysis of reproduction cork, it is necessary to remove the phloemic layer at the external part of the cork plank, the cork back, before grinding, since it has a different chemical composition (Pereira, 1987).

The sample preparation is a crucial step to obtain a representative material and the guarantee that the results truly refer to the cork tissue. This is of particular importance when making comparative studies, i.e. between provenances, trees or treatments, where the small differences may result from different content of non-suberous material (the so-called quality of cork as detailed in Chapter 7). This applies also to studies on the fine chemistry of the components, where further separation operations may be advisable, such as a density classification using, e.g. an elutriator (Marques et al., 1994).

For the analytical sequence an amount of 10 g of sample is sufficient for duplicate measurements.

3.8.1.2. Ash determination

The determination of ash content is made by incineration of a 2 g sample at $575 + 25°C$ in an oven during 3 h and weighing of the residue.

3.8.1.3. Solvent extraction

The objective of this analysis step is to completely remove all the non-structural components of cork, allowing their quantification and preparing an extractive-free sample for the subsequent analysis steps. A 3 g sample is adequate for the procedure.

The extractives are removed by a sequential extraction with solvents of increasing polarity with a duration required for an exhaustive removal of soluble compounds. The soxhlet extraction with the following sequence has been often used: a first extraction with dichloromethane during 4 h, followed by ethanol during 6 h and water during 8 h. The solution after each extraction step is evaporated and the solid residue weighed. The duration of extraction should be extended if required for a complete removal and other solvents may be used (for instance chloroform). However, alkaline solutions (i.e. 1% NaOH, as often used in bark chemical determinations) cannot be used since they react with suberin.

3.8.1.4. Methanolysis

The determination of suberin is based on its removal by depolymerisation. Although several reactions may be used to cleave the glyceridic esther bonds, an adequate procedure for cork desuberinisation is to use 3% $NaOCH_3$ in dry methanol. A 1.5 g sample is refluxed in 250 mL methanolic solution during 3 h, filtrated and the residue refluxed again with 100 mL methanol for 15 min. The residue is filtrated and washed with methanol and the methanolic filtrates are combined. The solid residue is dried and weighed. It represents the desuberinised cork sample to use for subsequent analysis. In this analysis procedure it is important to avoid water input and all materials should be dry.

Quantification of suberin may be made in two ways: by determining the mass loss by methanolysis using the solid residue, and by quantifying the fatty aliphatic and the glycerol components of suberin in the methanolic extracts. For this, the combined filtrates are acidified to pH 6, evaporated to dryness and the residue suspended in 50 mL water and extracted with 50 mL chloroform three times: the organic solution layer that contains the fatty aliphatics is dried over Na_2SO_4, filtrated, evaporated and determined as fatty aliphatic suberin; the water solution is used for determination of the glycerol, i.e. by HPLC or with an enzymatic assay. For most cork chemical compositions available in the literature (i.e. Tables 3.7 and 3.8), only the fatty aliphatic components of suberin were determined and the glycerol was not measured.

The composition of suberin may be analysed on the methanolic filtrate obtained after transesterification, by GC after trimethylsilyl derivatisation (Graça and Pereira, 2000a).

3.8.1.5. Acid hydrolysis

The desuberinised cork sample is used for the determination of lignin and polysaccharides following standard procedures based on acid hydrolysis. A 350 mg sample is hydrolysed with 3 mL 72% H_2SO_4 at 30°C for 60 min, diluted with 56 mL of water and autoclaved at 121°C and 1.2 atm. for 60 min. The residue is washed with hot water, dried and weighed as klason (acid insoluble) lignin. The filtrate is used for determination of acid soluble lignin by UV measurement at 200–205 nm. The monosaccharide composition is determined in the filtrate after derivatisation of the neutral sugars to alditol acetates by GC separation and quantification using myo-inositol as internal standard.

The acid sugar monomers are not determined in this procedure and a separate analysis for uronic acids has to be carried out.

3.8.2. FT-IR characterisation

The infra-red spectroscopy gives information on the different functional groups and chemical bonds that are present in a material based on their specific absorbance of vibrational energy. The FT-IR spectroscopy has been increasingly used to detect functional groups in lignocellulosic materials and also to quantify specific components. In cork, FT-IR has been used as a tool to monitor chemical changes during sample manipulation and to ascertain the purity of isolated components.

Figure 3.24 shows a Fourier transformed absorbance spectrum in the mid infrared (FT-IR) for an extractive-free pure cork from which extraneous materials were removed. The spectrum is dominated by the absorbance bands of suberin: two peaks at approximately 2927 and 2854 cm^{-1} corresponding to the asymmetric and symmetric vibration, respectively, of C–H in the olephinic chains and an intense band at 1740 cm^{-1} corresponding to the carbonyl C=O in aliphatic acids and esters. The band at 1162 cm^{-1} can also be

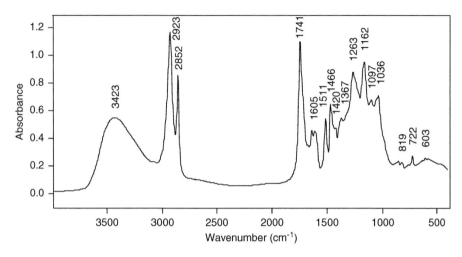

Figure 3.24. FT-IR spectrum of an extractive-free pure cork sample.

attributed to the C–O–C of the ester group in suberin. The intensity of these bands is related with the content of suberin and they practically disappear in fully desuberinised cork samples. Suberin also contributes to the 1263 cm^{-1} peak corresponding to C–O stretch, through the epoxide group and in desuberinised cork samples this peak also decreases significantly.

The bands at approximately 1511, 1466 and 1267 cm^{-1} are typical for a guaiacyl lignin. The peaks at 1087 and 1035 cm^{-1} correspond to the C–O bond and are characteristic of polysaccharides (cellulose and hemicelluloses). They have a relative smaller intensity.

3.8.3. *Nuclear magnetic resonance*

Nuclear magnetic resonance spectroscopy has been applied to cork, suberin and chemically transformed cork including solid-state and liquid ^{13}C NMR and ^1H NMR (Gil et al., 1997, 1999; Lopes et al., 2000a,b, 2001; Rocha et al., 2001).

In most regions of the spectrum there is significant overlap of signals from the different cork components. Table 3.9 shows the ^{13}C NMR assignments of chemical shifts for cork components. The spectra is dominated by the signals at 22–40 ppm that are assigned to suberin aliphatic methylenes which resonates in a spectral region free of carbohydrate and aromatic resonances. Two main distinct $-(CH_2)-$ environments are found in suberin, which have resonance at 30 and 33 ppm: the 33-ppm methylenes are more hindered than the 30-ppm methylenes.

Further peaks of the solid state of cork are: a peak at 21 ppm assigned to the CH_3COOR groups of hemicellluloses, a peak at 56 ppm assigned to the methoxyl groups of lignin and hemicelluloses, peaks at 65–105 ppm mainly assigned to the carbohydrate carbons, peaks at 110–155 ppm assigned to lignin aromatic carbons and the peak at 172 ppm to the suberin ester groups. The extractives composed by waxes and polyphenols contribute to the overall spectrum at 25–50 and 110–145 ppm, respectively.

Table 3.9. Assignment of the ¹³C CP/MAS spectrum of cork (Gil et al., 1997).

Chemical shifts (ppm)	Functional group	Component
20.8	CH_3-COO−	Carbohydrate
25.5	$-(CH_2)_n-$	Suberin
29.9	$-(CH_2)_n-$	Suberin
32.7	$-(CH_2)_n-$	Suberin
55.8	$Ar-OCH_3$, $-OCH_3$	Lignin and /or suberin, carbohydrate
64.6	$-C\gamma H_2O-$, $-CH_2OH$	Lignin and/or suberin, C6: carbohydrate
72.2	−CHOH−	C2, 3,5: carbohydrate
74.1	−CHOH−	C2, 3,5: carbohydrate
	$-C\gamma-OR$, $-Cb-OR$	Lignin and/or suberin
	Ar−CαHOH−	Lignin and/or suberin
81.6	−CHOH	C4: carbohydrate
	−CHOAr−	Lignin and/or suberin
87.5	−CH−O−	Lignin and/or suberin
	−CHOH−, Cβ−OR, Cα-R	C4: carbohydrate, lignin and/or suberin
104.9	−CH−	C1: carbohydrate, G2, S2, S6: lignin
113.3	−CH−	G5: lignin and/or suberin
122.2	−CH−, Cβ	G6: lignin and/or suberin
	Ar−CH≈CHOH−	Lignin and/or suberin
129.9	Ar−CHC−, −CH≈CH−	Lignin and/or suberin
147.8	Ar−CH≈CHCO−	Lignin and/or suberin
	Quaternary aromatic C	G4, S4: lignin and/or suberin
152.6	Quaternary aromatic C	S3, S5: lignin and/or suberin
172.8	−COO−	Suberin and/or lignin, uronic acids
	CH_3−COO−	Carbohydrate

3.9. Conclusions

Cork properties are largely derived from the specificities of its chemical composition and of the chemical structure of its components. The main chemical characteristics of cork may be summarised as follows:

- suberin is the main structural component of the cell wall and the responsible for many of the properties of the material, corresponding to nearly half of the cell wall structural polymers;
- lignin is the second most important structural component and it exists in an amount similar to that occurring in the wood of hardwood species; it is a component that is also determinant for the behaviour of cork as a material;
- cellulose has a minor role in the construction of the cell wall of cork, representing about only 10%, and this is a major difference in relation to lignocellulosic tissues; the content of hemicelluloses is similar;
- cork also contains an appreciable content of extractives that can be solubilised without impairing the materials properties; the extractives include non-polar compounds and polar, mostly phenolic, compounds.

The suberin macromolecule is a matter of research although several features are well established as follows:

- glycerol has a central role in the polymer construction and it is one of the most important monomers in molar percent;
- long-chain fatty α,ω-diacids and ω-hydroxyacids are the main monomeric families, mostly with 18 and 22 carbons and partly including an epoxide functionalisation at mid-chain;
- the polymer is constructed by an ester linking between the OH and COOH functional groups of the monomers, and no free functional groups remain;
- aromatic compounds, i.e. ferulic acid are linked to the polymer.

Cork contains lignin, which is a guaiacyl lignin (G-lignin) very similar to the lignin of wood from softwoods. It is closely associated with suberin in the cell wall, with probable links to suberinic acids. Cork also contains other non-lignin aromatic macromolecular structures of polyphenolic nature, linked to suberinic acids and to hemicelluloses, probably in an appreciable amount, but experimental data for cork are still missing.

The wall of cork cells is a layered construction dominated by a secondary wall that shows in the microscope a characteristic lamellar structure of alternating light and dark layers. It is proposed that suberin makes up successive layers, of rather constant thickness, that are imbedded in a lignin matrix. The between-cell region of the middle lamella is also lignified to a large extent.

The structural roles of lignin and suberin in the cells may be schematised as follows taking into account this topochemistry and their macromolecular characteristics: lignin is a cross-linked isotropic polymer which is responsible for the structural rigidity of the cells and their resistance to compression, as well as to their continuous assembly; suberin is a polymer with subunits of a linear type that are assembled forming ribbon-like structures, therefore being the component that is responsible for the elastic properties of cork and allowing the bending and collapse of cell walls.

The chemical analysis of cork, as well as of other suberous tissues, has to be adapted for the materials chemical characteristics. The proposed analysis is a stepwise procedure involving a thorough extraction with a succession of solvents of different polarity, followed by the depolymerisation of suberin by ester-bond cleavage through methanolysis: the lignin in the residue is determined following acid hydrolysis as well as the hydrolysed monosaccharides. It is also important to refer that cork contains other tissues with different chemical composition, i.e. lenticular filling material and woody inclusions, which requires attention in sample preparation for analysis.

References

Agullo, C., Seoane, E., 1981. Free hydroxyl-groups in the cork suberin. Chemistry & Industry 17, 608–609.

Agullo, C., Seoane, E., 1982. Hydrogenolysis of cork suberin by LiBH4 – free carboxyl groups. Anales de Quimica Serie C – Quimica Organica y Bioquimica 78, 389–393.

Arno, M., Serra, M.C., Seoane, E., 1981. Metanolisis de la suberina del corcho. Identificación y estimación de sus componentes ácidos como ésteres metílicos. Anales de Quimica 77C, 82–86.

Asensio, A., 1987a. Structural studies of the hemicellulose A from the cork of *Quercus suber*. Carbohydrate Research 161, 167–170.

Asensio, A., 1987b. Structural studies of the hemicellulose B fraction from the cork of *Quercus suber*. Carbohydrate Research 165, 134–138.

Asensio, A., 1987c. Polysaccharides from the cork of *Quercus suber*. I. Holocellulose and cellulose. Journal of Natural Products 50, 811–814.

Asensio, A., 1988a. Structural studies of hemicellulose B fraction (B-2) from the cork of *Quercus suber*. Canadian Journal of Chemistry 66, 449–453.

Asensio, A., 1988b. Polysaccharides from the cork of *Quercus suber*. II. Hemicellulose. Journal of Natural Products 51, 488–491.

Bento, M.F., Pereira, H., Cunha, M.A., Moutinho, A.M.C., van der Berg, K.J., Boon, J.J., 1998. Thermally assisted transmethylation gas chromatography-mass spectrometry of suberin components in cork from *Quercus suber* L. Phytochemical Analysis 9, 1–13.

Bento, M.F., Pereira, H., Cunha, M.A., Moutinho, A.M.C., van der Berg, K.J., Boon, J.J., 2001a. A study of variability of suberin composition in cork from *Quercus suber* L. using thermally assisted transmethylation GC-MS. Journal of Analytical and Applied Pyrolysis 54, 45–55.

Bento, M.F., Pereira, H., Cunha, M.A., Moutinho, A.M.C., van der Berg, K.J., Boon, J.J., van den Brink, O., Heeren, R.M.A., 2001b. Fragmentation of suberin and composition of aliphatic monomers released by methanolysis of cork from *Quercus suber* L. analysed by GC-MS SEC and MALDI-MS. Holzforschung 55, 487–493.

Bernards, M.A., 2002. Demystifying suberin. Canadian Journal of Botany 80, 227–240.

Bernards, M.A., Lewis, N.G., 1998. The macromolecular aromatic domain in suberized tissue: a changing paradigm. Phytochemistry 47, 915–933.

Bernards, M.A., Lopez, M.L., Zajicek, J., Lewis, N.G., 1995. Hydroxycinnamic acid-derived polymers constitute the polyaromatic domain of suberin. Journal of Biological Chemistry 270, 7382–7386.

Bernards, M.A., Razem, F.A., 2001. The poly(phenolic) domain of potato suberin: a non-lignin cell wall biopolymer. Phytochemistry 57, 1115–1122.

Bjorkman, A., 1956. Studies on finely divided wood. Part 1. Extraction of lignin with neutral solvents. Svensk Papperstidning 59, 477–485.

Brugnatelli, D., 1787. Elementi di chimica. Tomo II.

Cadahía, E., Conde, E., de Simon, B.F., García-Vallejo, M.C., 1998. Changes in tannic composition of reproduction cork from *Quercus suber* throughout the industrial processing. Journal of Agriculture and Food Chemistry 46, 2332–2336.

Caldas, M.M., Ferreira, J.M.L., Borges, M.A., 1985. Abordagem sobre a caracterização química da cortiça nas várias etapas do processamento industrial. Cortiça 560, 549–560.

Castola, V., Bigelli, A., Rezzi, S., Melloni, J., Gladiali, S., Desjobert, J.M., Casanova, J., 2005. Composition and chemical variability of the triterpene fraction of dichloromethane extracts of cork (*Quercus suber* L.). Industrial Crops and Products 21, 65–69.

Castola, V., Marongiu, B., Bigelli, A., Floris, C., Lai, A., Casanova, J., 2002. Extractives of cork (Quercus suber L.): chemical composition of dichloromethane and supercritical CO_2 extracts. Industrial Crops and Products 15, 15–22.

Chevreul, M., 1807. De l'action de l'acide nitrique sur le liège. Annales de Chimie 92, 323–333.

Chevreul, M., 1815. Mémoire sur le moyen d'analyser plusieures matières végétales et le liège en particulier. Annales de Chimie 96, 141–189.

Conde, E., Cadahía, E., García-Vallejo, M.C., Adrados, J.R.G., 1998. Chemical characterization of reproduction cork from Spanish *Quercus suber*. Journal of Wood Chemistry and Technology 18, 447–469.

Conde, E., Cadahía, E., García-Vallejo, M.C., de Simon, B.F., Adrados, J.R.G., 1997. Low molecular weight polyphenols in cork of *Quercus suber*. Journal of Agriculture and Food Chemistry 45, 2695–2700.

Conde, E., García-Vallejo, M.C., Cadahía, E., 1999. Waxes composition of *Quercus suber* reproduction cork from different Spanish provenances. Wood Science and Technology 33, 270–283.

Cordeiro, N., Neto, C.P., Rocha, J., Belgacem, M.N., Gandini, A., 2002. The organosolv fractionation of cork components. Holzforschung 56, 135–142.

Ekman, R., 1983. The suberin monomers and triterpenoids from the outer bark of *Betula verrucosa* Ehrh. Holzforschung 37, 205–211.

García-Vallejo, M.C., Conde, E., Cadahía, E., Simón, F., 1997. Suberin composition of reproduction cork from *Quercus suber*. Holzforschung 51, 219–224.

Gil, A.M., Lopes, M., Neto, C.P., Rocha, J., 1999. Very high resolution ^{1}H MAS-NMR of a natural polymeric material. Solid State Nuclear Magnetic Resonance 15, 59–67.

Gil, A.M., Lopes, M., Rocha, J., Neto, C.P., 1997. A 13C solid state nuclear magnetic resonance spectroscopic study of cork cell wall structure: the effect of suberin removal. International Journal of Biological Macromolecules 20, 293–305.

Graça, J., 1998. *A suberina da cortiça. Lípidos macromoleculares estruturais das paredes celulares suberificadas da cortiça de Quercus suber L.* Ph.D. Thesis, Instituto Superior de Agronomia, Lisboa.

Graça, J., Pereira, H., 1997. Cork suberin: a glyceryl based polyester. Holzforschung 51, 225–234.

Graça, J., Pereira, H., 1998. Feruloyl esters of ω-hydroxyacids in cork suberin. Journal of Wood Chemistry and Technology 18, 207–217.

Graça, J., Pereira, H., 1999. Glyceryl-acyl and aryl-acyl dimers in *Pseudotsuga menziesii* bark suberin. Holzforschung 53, 397–402.

Graça, J., Pereira, H., 2000a. Methanolysis of bark suberins: analysis of glycerol and acid monomers Phytochemical Analysis 11, 45–51.

Graça, J., Pereira, H., 2000b. Suberin structure in potato periderm: glycerol, long chain monomers and glyceryl and feruloyl dimmers. Journal of Agricultural Food Chemistry 48, 5476–5483.

Graça, J., Pereira, H., 2000c. Diglycerol alkanedioates in suberin: building units of a poly(acylglycerol)polyester. Biomacromolecules 1, 519–522.

Hergert, H., Kurth, E., 1952. The chemical nature of cork from Douglas fir bark. Tappi 35, 59–66.

Holloway, P., 1972a. The composition of suberin from the corks of *Quercus suber* L. and *Betula pendula* Roth. Chemistry and Physics of Lipids 9, 158–170.

Holloway, P., 1972b. The suberin composition of the cork layers from some *Ribes* species. Chemistry and Physics of Lipids 9, 171–179.

Holloway, P., 1983. Some variation in the composition of suberin from cork layers of higher plants. Phytochemistry 22, 495–502.

Klauber, A., 1920. Die Monographie des Korkes. Berlin.

Kolattukudy, P., 1977. Lipid polymers and associated phenols, their chemistry, biosynthesis and role in pathogenesis. Recent Advances in Phytochemistry 77, 185–246.

Lopes, M.H., Barros, A.S., Pascoal Neto, C., Rutledge, D., Delgadillo, L., Gil, A.M., 2001. Variability of cork from Portuguese *Quercus suber* studied by solid-state ^{13}C-NMR and FTIR spectroscopies. Biopolymers 62, 268–277.

Lopes, M., Gil, A., Silvestre, A., Pascoal Neto, C., 2000a. Composition of suberin extracted upon gradual alkaline methanolysis of *Quercus suber* L. cork. Journal of Agriculture and Food Chemistry 48, 383–391.

Lopes, M.H., Pascoal Neto, C., Barros, A.S., Rutledge, D., Delgadillo, L., Gil, A.M., 2000b. Quantitation of aliphatic suberin in *Quercus suber* cork by FTIR spectroscopy and solid-state ^{13}C-NMR spectroscopy. Biopolymers 57, 344–351.

Lopes, M.H., Sarychev, A., Pascoal Neto, C., Gil, A.M., 2000c. Spectral editing of ^{13}C CP/MAS-NMR spectra of complex systems: application to the structural characterisation of cork cell walls. Solid State Nuclear Magnetic Resonance 16, 109–121.

Marques, A., 1998. Isolamento e caracterização estrutural da lenhina da cortiça de *Quercus suber* L. Ph.D. Thesis, Instituto Superior de Agronomia, Lisboa.

Marques, A.V., Pereira, H., 1987. On the determination of suberin and other structural components in cork from *Quercus suber* L. Anais do Instituto Superior de Agronomia (Lisboa) 42, 321–335.

Marques, A.V., Pereira, H., Meier, D., Faix, O., 1994. Quantitative analysis of cork (*Quercus suber* L.) and milled cork lignin by FTIR spectroscopy, analytical pyrolysis and total hydrolysis. Holzforschung 48(suppl.), 43–50.

Marques, A.V., Pereira, H., Meier, D., Faix, O., 1996. Isolation and characterization of a guaiacyl lignin from saponified cork of *Quercus suber* L. Holzforschung 50, 393–400.

Marques, A.V., Pereira, H., Meier, D., Faix, O., 1999. Structural characterization of cork lignin by thioacidolysis and permanganate oxidation. Holzforschung 53, 167–174.

Marques, A.V., Pereira, H., Rodrigues, J., Meier, D., Faix, O., 2006. Isolation and comparative characterization of a Bjorkman lignin from the saponified cork of Douglas-fir bark. Journal of Analytical and Applied Pyrolysis 77, 169–176.

Mata, F., Marques, V., Pereira, H., 1986. Influência da granulometria na determinação de elementos minerais na cortiça. Cortiça 569, 68–72.

Mazzoleni, V., Caldentey, P., Silva, A., 1998. Phenolic compounds in cork used for production of wine stoppers as affected by storage and boiling of cork slabs. American Journal of Enology and Viticulture 49, 6–10.

Miranda, A.M., Machado, A.S.R., Pereira, H., daPonte, M.N., 1996. High-pressure extraction of cork with CO_2 and 1,4-dioxane. In: von Rohr, P.R., Trepp, C. (Eds), High Pressure Chemical Engineering 12, Elsevier Science Publisher, Amsterdam, pp. 417–422.

Negrel, J., Pollet, B., Lapierre, C., 1996. Ether-linked ferulic acid amides in natural and wound periderms of potato tuber. Phytochemistry 43, 1195–1199.

Neto, C.P., Rocha, J., Gil, A., Cordeiro, N., Esculcas, A.P., Rocha, S., Delgadillo, I., Pedrosa de Jesus, J.D., Ferrer Correia, A.J., 1995. 13C solid state nuclear magnetic resonance and Fouriei transform infrared studies of the thermal decomposition of cork. Solid State Nuclear Magnetic Resonance 4, 143–151.

Olesen, P., 1978. Studies on thee physiological sheaths in roots. I. Ultrastructure of the exodermis in *Hoya carnosa* L. Protoplasma 94, 325–340.

Osman, S., Gerard, H., Fett, W., Moreau, R., Dudley, R., 1995. Method for the production and characterization of tomato cutin oligomers. Journal of Agriculture and Food Chemistry 43, 2134–2137.

Parameswaran, N., Liese, W., Gunzerodt, H., 1981. Características do verde da cortiça de *Quercus suber* L. Cortiça 514, 179–184.

Pascoal Neto, C., Cordeiro, N., Seca, A., Domingues, F., Gandini, A., Robert, D., 1996. Isolation and characterization of a lignin-like polymer of cork of *Quercus suber* L. Holzforschung 50, 563–565.

Pereira, H., 1979. Química da cortiça. III. Extracções da cortiça com solventes orgânicos e água. Cortiça 509, 57–59.

Pereira, H., 1987. Composição química da raspa em pranchas de cortiça de produção amadia. Cortiça 587, 231–233.

Pereira, H., 1988. Chemical composition and variability of cork from *Quercus suber* L. Wood Science and Technology 22, 211–218.

Pereira, H., Marques, A.V., 1988. The effect of chemical treatments on the cellular structure of cork. IAWA Bulletin n.s. 9, 337–345.

Ribas, I., Blasco, E., 1940. Investigaciones sobre el corcho.II. Determinacion cuantitativa de la glicerina existente. Anales de la Real Sociedad Espanola de Fisica y Quimica 36B, 248–254.

Rocha, M.S., Coimbra, M.A., Delgadillo, I., 2000. Demonstration of pectic polysaccharides in cork cell wall from *Quercus suber* L. Journal of Agriculture and Food Chemistry 48, 2003–2007.

Rocha, M.S., Coimbra, M.A., Delgadillo, I., 2004. Occurrence of furfuraldehydes during the processing of *Quercus suber* L. cork. Simultaneous determination oof furfural, 5-hydroxymethylfurfural and 5-methylfurfural and their relation with cork polysaccharides. Carbohydrate Polymers 56, 287–293.

Rocha, M.S., Goodfellow, B.J., Delgadillo, I., Neto, C.P., Gil, A.M., 2001. Enzymatic isolation and structural characterization of polymeric suberin from *Quercus suber* L. International Journal of Biological Macromolecules 28, 107–119.

Rodriguez-Miguenes, B., Ribas-Marques, I., 1972. Contribucion a la estrutura quimica de la suberina. Anales de Quimica 68B, 1301–1306.

Rosa, M.E., Pereira, H., 1994. The effect of long term treatment at 100ºC–150ºC on structure, chemical composition and compression behaviour of cork. Holzforschung 48, 226–232.

Sakakibara, A., 1991. Chemistry of lignin. In: David, N., Hon, S., Siraishi, N. (Eds), Wood and Cellulosic Chemistry. Marcel Dekker Inc., New York, pp. 113–175.

Schmutz, A., Buchalla, A., Ryser, U., 1996. Changing the dimensions of suberin lamellae of green cotton fibers with a specific inhibitor of the endoplasmic reticulum-associated fatty acid-elongases. Plant Physiology 110, 403–411.

Schmutz, A., Jenny, T., Amrhein, N., Ryser, U., 1993. Caffeic acid and glycerol are constituents of the suberin layers in green cotton fibers. Planta 189, 453–460.

Sitte, P., 1962. Zum Feinbau der Suberinschichten in Flaschenkork. Protoplasma 54, 555–559.

Yamamoto, E., Bokelman, G.H., Lewis, N.G., 1989. Phenylpropanoid metabolism in cell wall – an overview. ACS Symposium Series 399, 68–80.

Part II
Cork production

Chapter 4
The cork oak

The cork oak (*Quercus suber* L.) is an evergreen oak that is characterised by the presence of a conspicuous thick and furrowed bark with a continuous layer of cork in its outer part. It is this cork bark that gave the cork oak its notoriety and economic importance as a cork producer, as well as its ornamental value in many parks and urban areas around the world.

The cork oak spreads in the western Mediterranean areas of southern Europe and North Africa, mostly integrating multifunctional agro-forestry systems (called montado in Portugal and dehesa in Spain), usually combining the production of cork with cattle grazing, hunting and other non-wood productions. The tree is well adapted to hot and dry summers, and to soils of low fertility.

Given the regions where they are distributed, the cork oak forests are also valued for their ecological role to contain desertification and soil erosion, and for their contribution to biodiversity maintenance. However, the present cork oak systems are not natural systems; instead they are the result of centuries' continuing management and objective-oriented silviculture that have shaped the trees and the cork oak landscape.

The distribution, botanical and ecological aspects of the cork oak are summarised in this chapter following general reviews for the species (Natividade, 1950; Montoya, 1988; Montero and Canellas, 1999), and the tree growth is analysed in relation to its within-the-year rhythm and the between-the-year variation, namely in relation to tree age and climatic factors. The silvicultural practices that are applied to the cork oak stands are detailed in relation to the phases of establishment, juvenile and mature stages, and a description is made of the cork oak dense forests and of the agro-forestry cork oak systems. The cork extraction is explained in detail in the next chapter.

4.1. Geographic distribution

Cork oak spreads in the countries of the western Mediterranean basin, on a total area over 2 million hectares. It extends in the western coastal Mediterranean and adjoining Atlantic areas between 33°N and 45°N latitude. Cork oak forests and woodlands cover nowadays considerable areas in the southern European Iberian peninsula (Portugal and Spain) and in the northern African countries of Morocco, Algeria and Tunisia. Cork oaks are also present in France and Italy.

The cork oak is one of the most recent *Quercus* having originated in the Tertiary (Oligocene–Miocene). It is thought that the centre of dissemination lied in what is now the Tyrrhenian sea and the species travelled north and west along the European coast and through Sicily to the south and west of the African coast, possibly also with a migration from Iberia to North Africa through the Gibraltar strait (Fig. 4.1). Pollen studies suggest that cork oaks survived the last glacial period in southern and coastal areas of Iberia and in North Africa and with a possible post glacial re-colonisation between both areas (Carrión et al., 2000), while chloroplast DNA analysis showed potential glacial refuges in Italy, North Africa and Iberia (Lumaret et al., 2005). At present two genetically distinct groups of cork oak populations were identified corresponding to two distinct geographical areas: a group in the Iberian peninsula and adjoining French regions, showing higher within-population genetic diversity and lower among-population differentiation than the second group in North Africa, Sicily, Sardinia, Corsica and Provence (south-eastern France) (Toumi and Lumaret, 1998).

The area occupied by present cork oak forests and cork oak based agro-forestry systems is not known with exactitude. A total area of cork oaks is often quoted as over 2.4 million hectares, but the real value is probably somewhat lower. Using the existing data for national forest inventories, mostly referring to the 1990s, the following figures can be collected for the cork oak areas: 713 000 ha in Portugal (of which 592 000 ha of pure stands), 475 000 ha in Spain (365 000 ha of pure stands), 68 000 ha in France, 65 000 ha in Italy, 348 000 ha in Morocco and 90 000 ha in Tunisia. For Algeria the value of 440 000 ha is usually referred but no recent assessments have been made and it is likely that the area has decreased to about 230 000 ha. On the other side, higher values have been reported from regional evaluations in Italy, such as 122 000 ha in the region of Sardinia, the main area of cork oaks in the country (Barberis et al., 2003).

When considering the present or potential value of the cork oak forests as producers of cork, it may be misleading to regard only the figures of areas. In fact these values also

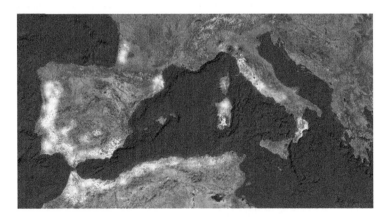

Figure 4.1. Spread of the cork oak around the Mediterranean sea.

include the stands with mixed species as well as dense and small-sized matorral-type formations that contribute less to the production of cork. The two main countries where cork oaks are managed for cork production, Portugal and Spain, are in the process of completing the revisions of their forest inventories and more accurate data will be available soon. The distribution of cork oaks in the Iberian peninsula is shown in Figure 4.2: in Portugal the highest concentration is found south of the river Tagus in the regions of Alentejo and Tagus valley (districts of Setúbal, Évora, Beja, Portalegre and Santarém) and in Spain the cork oaks concentrate in the south-western regions of Andalusia and Estremadura, with a small area in Catalonia.

The annual world production of cork totals about 374 000 tons mostly from the cork oaks of Portugal and Spain, who produce 74% of the total (respectively 51 and 23%). The annual production of cork is rather stable with some oscillations due to climatic or accidental occurrences, i.e. the 2003 forest fires or the 2004 drought in Portugal.

4.2. Botanical description

Cork oak belongs to the order of *Fagales* and the family of *Fagaceae*. It was described by Linneo in 1753.

The cork oaks are polymorphic and show different botanical forms that differ in traits such as tree form, flowering period, form and size of leaves, flowers and fruits. It is thought that the species contains more than 40 varieties that were grouped in four: genuine, subcrinita, macrocarpa and occidentalis (Pereira Coutinho, 1939; Natividade, 1950; Vicioso, 1950). Natural hybrids also occur, namely between *Q. ilex* and *Q. suber*, which have very similar breeding systems (Elena-Rosselló et al., 1992; Belahbib et al., 2001; Boavida et al., 2001). The high levels of morphological and phenotypical diversity should

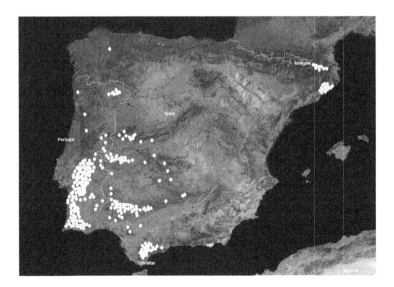

Figure 4.2. Distribution map of cork oaks in Portugal and Spain.

largely result from past introgressive hybridisation. A large variation between trees in the same population is also found in several characteristics, namely of cork quality, and the progeny of one tree may show remarkable differences between individual trees.

The cork oaks are low-spreading trees with a short stem and thick branches. The trees do not attain heights greater than 14–16 m, but open-grown trees may have very large crown dimensions, e.g. 500 m^2 of crown projection in some mature trees with 150–200 years of age, and large stem circumferences. When growing in dense stands, the shape of the tree is strongly influenced by competition, originating trees with narrower crowns and higher stems. Figure 4.3 shows examples of an isolated cork oak and of a densely forested

Figure 4.3. Photographs of cork oaks: (a) isolated tree and (b) dense forest stand (see Colour Plate Section).

area. However, the cork oaks seen nowadays in the managed montado system are some-what differently shaped as a result of early and continuing pruning: most trees have a stem bifurcation at a low height with two or three main boughs set with open angles to the stem, and a circular crown with a flattened top. This tree form is perceived as the symbol for the cork oak landscape and used for cork-related logos, i.e. for the Alentejo region or for the wine stoppers (Fig. 4.4).

The root system is characterised by a dimorphic distribution with a strong and long tap root with thick lateral ramifications that in open-grown trees may show a large horizontal expansion with many superficial roots. The central root may penetrate several meters down in the soil and this explains why in the summer the cork oak is able to extract water from deep lying aquifers to maintain high leaf hydration (Nardini et al., 1999), and there-fore to maintain growth during the period of high water demand when radiation is high. The association of the root system with different miccorhyza is quite frequent.

Buds are dark purple, ovoid and small with about 2 mm length. The leaves are dark green, with a dense white pubescence in the lower face (see Fig. 1.4) and numerous stom-ata, around 430 per mm^2 of abaxial leaf surface with the guard cells protruding from the epidermal plane (Molinas, 1991). The leaf shape is more or less ovoid to oblong with a crinkled or waved margin, and the leaf size varies in the range of 4–7 cm of length and 2–3 cm of width (Fig. 4.5a). Leaf shape and size variation is large both between trees as well as within the canopy.

The initiation of the physiological activity occurs in February/March with bud devel-opment. The shoot elongation and emergence of new leaves start in early spring and con-tinue until June. The leaf duration is approximately 14 months, with a range of 11–18 months (Escudero et al., 1992; Oliveira et al., 1994; Robert et al., 1996; Fialho et al., 2001). Therefore, leaf fall occurs in spring except if some accidental occurrence such as heavy winter rains leads to a more rapid defoliation (Fig. 4.6). This seasonality of leaf falling is the reason why cork oaks do not look their best in spring time, because the crown is still dominated by the fading colour of the previous year leaves. In contrast with other

(a)

(b)

Figure 4.4. Typical cork oak architecture in the managed agro-forestry systems (a) (see Colour Plate Section) and symbol of natural cork wine stoppers (b).

species displaying the bright spring green of new leaves, the cork oaks appear unhealthy and have led many unacknowledged observers to predict low vitality for the trees.

The cork oak flowering and fructification begin at a tree age around 15–20 years. The flowering season extends from April until the end of May. The female flowering buds are located in the insertion axils of the new leaves while it is in the shoots of the previous year that reddish buds develop as male catkins (Fig. 4.5b). Pollination occurs in spring, but it is only one and half months later that the ovules complete their differentiation and fertilisation occurs, with only one ovule successfully maturing during autumn into a monospermic seed (Boavida et al., 1999). Summer or autumn flowering occurs only in limited extent in some years and in a scarce number of individuals (Machado, 1938; Boavida et al., 1999; Díaz-Fernández et al., 2004). The pollination occurs with pollen from the same tree as well as from neighbouring trees and the off-springs from one tree may differ considerably from one another.

The cork oak acorns vary largely in form and size, with a length from less than 2 cm to over 5 cm (Fig. 4.5c). Because of the long flowering period, the acorns do not ripe at the same time and precocity or delay in flowering events plays a crucial role on acorn maturation. Annual and biennial cycles of acorn ripening are observed (Díaz-Fernández et al., 2004). The annual acorns grow mainly in late summer and autumn and the complete acorn maturation is attained in November (Merouani et al., 2003). Biennial acorns do not experiment any significant growth during their first vegetative period, and their growth only recommences towards the beginning of the spring of their second year. Biennial frequency may be related to the length of the vegetative period, i.e. shortening caused by intense and long summer drought.

The amount of acorns produced by one tree has large variation among years, with 2–3 years of high fruit production out of 10 years. Cold spells during flowering may cause catkin damage and may be one of the factors influencing the high inter-annual variability of acorn production (Garcia-Mozo et al., 2001).

4.3. Ecology

The cork oak is ecologically plastic and grows in warm humid and sub-humid conditions from sea level to 2000 m, but with optimum growth occurring until 600 m of altitude. It is considered a semi-tolerant species, well adapted to mild climates, namely to the Mediterranean climates with Atlantic influence, with mild winters and hot and dry summers. The species shows a high plasticity and it is able to adapt its phenology and physiological activity to changing environmental conditions such as severe summer drought and high temperatures.

The cork oak grows well with mean annual precipitations of 600–800 mm, but still survives in years with very low precipitation under 400 mm. It is usually considered that the minimum annual precipitation for a balanced tree development should be 500 mm. The cork oak admits higher precipitations up to 1700 mm, but it is very sensitive to water logging. As regards the seasonal distribution of rainfall, the cork oak is adapted to the Mediterranean type of climate with precipitation concentrating in late autumn and winter (October–March) and very few, if at all, summer rains. Figure 4.7 shows precipitation and temperature diagrams for some regions within the cork oak distribution area.

Figure 4.5. Cork oak leaves (a), male flowers (b) and acorns (c) (see Colour Plate Section).

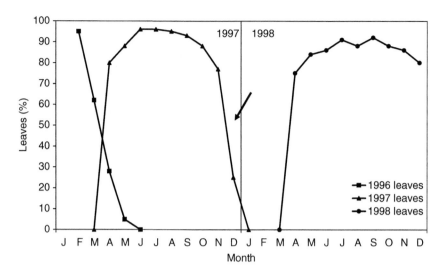

Figure 4.6. Time evolution of cork oak leaves in young cork oaks during three growing seasons (1996–1998). In 1997, very heavy rains occurred in October–December (arrow) (drawn with data from Fialho et al., 2001).

The optimum mean annual temperature is in the range 13–16°C, but cork oak survives until 19°C. The temperature optima for the photosynthetic functions was estimated as 33–34°C in cork oak seedlings and the response to temperature increase showed a large ability to acclimate to drought and elevated temperatures (Ghouil et al., 2003). This is an important aspect since summer temperatures above 40°C are frequent in many cork oak regions. As regards low temperatures, the mean temperature of the coldest month should not be below 4–5°C and the absolute minimum survival temperature is -12°C. The leaves of cork oak have a freezing tolerance that allows withstanding the frequent but mild freezing events in the Mediterranean area (Cavender-Bares et al., 2005). In general, the temperatures of -5 and 40°C may be considered as the limits for cork oak growth.

In relation to soils, the species is very tolerant with the exception of calcareous and limestone substrates. It may grow on poor and shallow soils, with low nitrogen and organic matter content and it allows a pH range between 4.8 and 7.0. The cork oak occurs preferentially in siliceous and sandy soils, and prefers deep and well aerated and drained soils. It is very sensitive to compaction and water logging.

The cork oak woodlands are renowned reservoirs of biodiversity and home to a variety of threatened species: the endangered imperial eagle (*Aquila adalberti*), nine vulnerable (e.g. black vulture *Aegypius monachus*) and three rare (e.g. black stork *Ciconia nigra*) bird species (Tucker and Evans, 1997). They are also winter habitats for a large number of woodpigeons (*Columba palumbus*), cranes (*Grus grus*) and many passerines.

The cork oaks are frequently associated with holm oaks (*Quercus ilex* and *Q. rotundifolia*) and to a much lesser extent with other deciduous oaks (e.g. *Q. faginea* and *Q. pyrenaica*). The agro-forestry cork oak and holm oak systems (montado and dehesa) constitute a good case study where the combination of human influence and of nature may be analysed, namely in relation to the so-called depletion of biodiversity in managed areas.

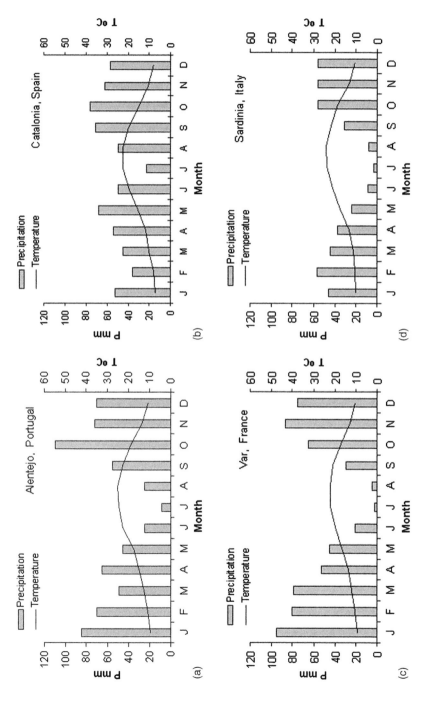

Figure 4.7. Climatic diagrams of monthly mean precipitation and temperature for four regions within the distribution area: Alentejo (Portugal), Catalonia (Spain), Var (France) and Sardinia (Italy).

On the contrary, the traditional management of the land use systems of cork oak of the Iberian peninsula rather supports biodiversity then reduces it, and the systems show a high level of structural diversity both on a within-habitat scale and a between-habitat scale (Plieninger and Wilbrand, 2001; Pereira and Fonseca, 2003).

Fire is a frequent occurrence in Mediterranean forests, but the cork oak is well protected by its cork layer in relation to understorey fires. Because of the thermal insulation properties of cork, the charring only occurs in the 1–2 mm outermost layers even in severe fires. Most trees survive after fire (Pausas, 1997; Barberis et al., 2003). Resprouting occurs from stem buds, with a quick recovery of the canopy, or from basal buds, with regrowth from basal sprouts (Fig. 4.8). The damage caused to the tree by the fire will depend on much cork protects the stem, and trees that were debarked in the year of the fire occurrence or in the previous year will be the most sensitive. Also of high danger for fire damage will be the occurrence of stem wounds with showing wood areas.

4.4. Tree growth

4.4.1. Seasonality of radial growth

The cork oak radial growth has a clear within-the-year seasonality with a period of active growth and a period of dormancy. The growth begins usually in March and extends to October with the maximal increase in June–July, while in August there is usually a decrease of growth rate (Fig. 4.9). The maximum stomatal conductance and transpiration rates were found to occur from March to June (Oliveira et al., 1992). The annual cork oak radial growth may be divided into three phases: (i) an early phase corresponding to the early spring period of March–April; (ii) a main phase from May to August when most of the diameter increment is concentrated, corresponding to 64% of the total annual growth; and (iii) a late autumn phase in September–October (Costa et al., 2002). This growth pattern and intensity is similar in trees of different diameter classes, as shown in Figure 4.10 (Costa et al., 2003).

4.4.2. Influence of climatic factors

The climatic factors influence cork oak growth in two ways: on total annual growth corresponding to a long-term effect, and on the seasonality of increments corresponding to short-term effects.

The inter-annual differences of cork oak growth can be associated to annual weather variation, namely to precipitation of the previous winter and spring, as it was already referred to in Chapter 1 in relation to cork growth. The correlation between the accumulated precipitation between the previous November to July with the growth index of cork oaks is significant and high (Fig. 4.11). On the contrary, the effect of temperature, either of July (to study the possible negative effect of high summer temperature) or of January (for a potential positive effect of warm winters), as of any other month, is not statistically significant. Therefore, the total annual diameter growth of mature cork oaks in their region of occurrence is little influenced by temperature and water seems to be a more limiting factor (Costa et al., 2001).

Figure 4.8. Cork oak sprouting after a severe fire in 2003 in Portugal: (a) stem sprouting after 2 months after fire (see Colour Plate Section) and (b) basal sprouts.

The climate also influences the seasonality of cork oak growth, mainly two-fold: (a) higher temperatures in early spring that promote leaf burst and branch elongation will shift radial growth to later stages in the season; and (b) precipitation has a short-term influence in growth only in relation to summer rainfall that favours the late growth of the cork oak by an immediate use during the dry period, taking advantage of the still warm temperature and long photoperiod (Costa et al., 2002). Therefore, the distribution of the

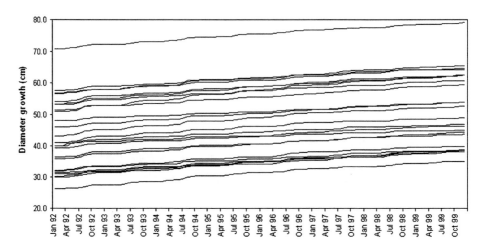

Figure 4.9. Total diameter increments of 50 mature cork oaks during one cork production cycle (1992–1999) (drawn with data from Costa et al., 2002).

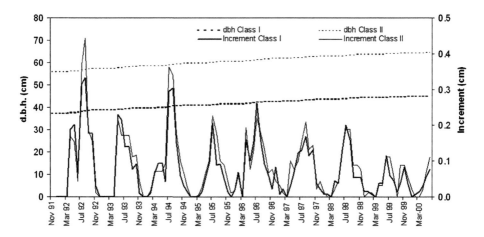

Figure 4.10. Diameter at 1.3 m height (dbh) and monthly diameter increment in 50 mature cork oaks of different diameter classes (class 1, dbh ≤ 55 cm; class II, dbh > 55 cm) during one cork production cycle (1992–1999) (Costa et al., 2003).

diameter increments within the growth season has some month-to-month variations in the different years, as shown in Figure 4.12, corresponding to climate-related differences.

4.4.3. Wood and cork growth

There are very few studies on the quantification of cork oak total radial increments (Fialho et al., 2001; Oliveira et al., 2002) and only one has followed a set of cork oaks along the

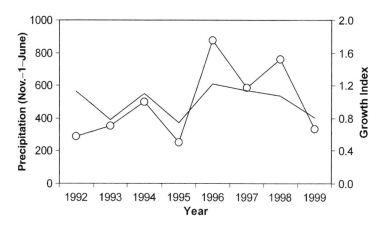

Figure 4.11. Annual diameter growth index (—) and cumulative precipitation in the period November–1–June (open circles) in 50 mature cork oaks during one cork production cycle (1992–1999) (Costa et al., 2001).

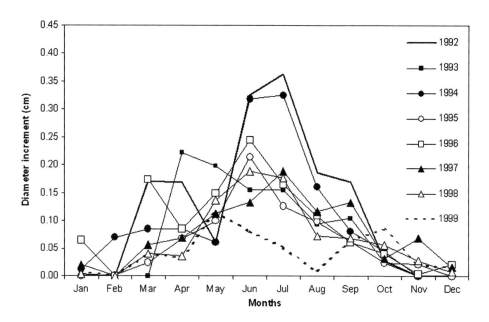

Figure 4.12. Monthly diameter increments of cork oaks during one cork production cycle (1992–1999) (Costa et al., 2002).

9 years of a complete cork production cycle (Costa et al., 2003) and further on to the following cycle (Costa and Pereira, unpublished). Young 11-year-old cork oaks increased their diameter by 9 mm per year in 2-year consecutive measurements (Fialho et al., 2001), and mature trees under cork production between 4 and 8 mm per year after cork stripping (Oliveira et al., 2002). Along the cork production cycle, the diameter increments ranged

from 4 to 14 mm, with a decreasing trend from the year after cork stripping onwards (Costa et al., 2003). The average value was 9 mm per year and this corresponds to a mean annual perimeter increase of 28 mm, leading to a cork oak circumference increase during one cork cycle of 252 mm.

The understanding of the variation of radial growth along the years in the cork oak requires a discussion on the components of the stem increments. The radial stem growth of the cork oak is the result of the activity of the cambium and of the phellogen, and therefore it is the sum of the increments of wood, phloem and of the periderm. When measuring growth (Fig. 4.13), either using band dendrometers (corresponding to circumference data) or point dendrometers (corresponding to radial width data), it is the result of these combined increments that is quantified. In the periderm, the growth is practically only related to the production of cork cells since the phelloderm radial extension is negligible (Chapter 1). As regards the cambium activity, no studies have quantified the production of phloem, but it certainly is much smaller than that of wood. Therefore, one simplification

Figure 4.13. Band and point dendrometers installed on cork oaks in an experimental site of the University of Huelva (Spain). Two point dendrometers were installed: one located over the cork to measure total growth, and the other on the phloem after removal of a small portion of the cork, to measure the wood and phloem growth.

of the cork oak radial growth is to consider that it is made up only of wood plus cork growth.

The distinction between wood growth and cork growth is not an easy task and until now it was not solved by direct measurement. Tentative approaches were made to measure simultaneously the total growth and the wood growth by positioning two point dendrometers, respectively, one over cork and the other in the phloem after removal of the periderm. However, this failed due to the regeneration of a traumatic periderm under the phloem-located dendrometer. An indirect method to estimate annual growths will be to measure the annual growth in the cork (cork rings) and assume the wood growth to be the difference to total growth. The hypothesis of using tree coring for measuring wood rings has not proven feasible due to the poor ring definition in cork oak wood (Gourlay and Pereira, 1998).

Figure 4.14 shows the decomposition of total stem diameter growth into the wood and cork components. The cork increments are the major component of total growth with a mean cork ring width of 3.8 mm year^{-1} in comparison with a mean wood ring width of 1.3 mm year^{-1}. It is also the inter-annual variation with age after cork stripping (phellogen age, see Chapter 1) of the cork layer that results into the pattern of decreasing growth along the cork production cycle. In this case the under-cork perimeter increase from one cork stripping to the following one can be estimated as 70 mm.

Other available measurements of cork oak ring widths made on stem discs indicate 2 mm year^{-1} in young trees (Nunes, 1996) and values ranging from 1 mm to 4 mm year^{-1} in mature trees in cork oak montados (González-Adrados and Gourlay, 1998; Gourlay and Pereira, 1998). A recent measurement using microdensitometry reported a high wood growth rate of 3.9 mm year^{-1} (with 4.2 mm year^{-1} in the first 30 years) (Knapic et al., 2006). A set of cork oaks from dense cork forests in Spain had mean wood ring widths in the approximate range of 1 mm to 3 mm year^{-1} (Sánchez-González et al., 2005).

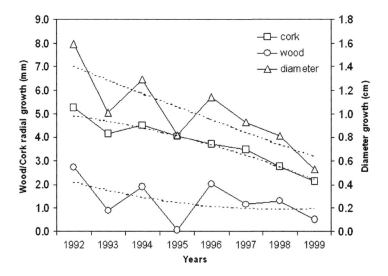

Figure 4.14. Mean annual radial increments of cork and wood and mean tree diameter growth for cork oak during one cork production cycle (Costa et al., 2002).

Tradition has confined cork oak as a slow growth species. However, the actual values shown for the wood growth are quite appreciable, especially if considering the edapho-climatic conditions where the trees grow and the sustained exploitation of cork that is made. The fact that the wood density is very high also results into a great capacity of bio-mass production, thus revealing itself as an interesting species for fixing carbon in such environments.

4.5. Silviculture

4.5.1. Stand regeneration

Most of the existing cork oak trees have originated from natural regeneration of acorn sprouting. The artificial regeneration of cork oak stands is relatively recent and it had a great increment in the 1990s due to the EU policy and incentives for afforestation of set-aside agricultural lands. Either plantation or direct seeding has been used and many thousand hectares have been established with cork oaks in Portugal and Spain during the last 15 years. However, the success of establishment is at the best only moder-ate and mortality during the first years may be very high. The reasons for this are various: (a) the establishment of an efficient root system is essential to guarantee water and survival during the summer months, thus requiring an adequate soil preparation and a good root quality of the nursery seedlings used in planting; (b) the young plants are sen-sitive to water stress and the efficiency in developing physiological mechanisms for drought tolerance seems related with the level of light environment, being smaller in shade-grown seedlings (Pardos et al., 2005; Cardillo and Bernal, 2006); and (c) the acorns are very palatable to small rodents (Herrera, 1995) and the young plants to cattle brows-ing, thus requiring protection, respectively, by repellents and shelters or suppression of grazing.

Site preparation includes weeding, usually made by disc-harrowing, and the improve-ment of soil characteristics for development of the root system. Depending on soil type and slope, the preparation may include: ripping or subsoiling to 60–80 cm depth, espe-cially in soils overlying hard rock, or ploughing and mounding along the contours fol-lowed by deep ploughing and/or disc harrowing. Site preparation may cover the whole area or be concentrated on the planting line. Starter NPK fertilisation is usually applied at 40–100 g per plant.

Plantation with nursery-grown seedlings (Fig. 4.15) and/or seeding can occur either in spring or autumn. Figure 4.16 shows schematically the calendaring of operations. The number of trees per hectare at planting is not as high as in other species that are planted for wood production, and spacings around 4 m × 4 m (625 trees ha^{-1}) are advised combining an adequate initial density with installation costs but wider (8 m × 4 m, 312 trees ha^{-1}) or denser (4 m × 2 m, 1250 trees ha^{-1}) spacings are also practiced.

The mortality during the first years may be very high and the plant is only considered as established when it is over 8–10 years old. The protection of the young seedlings with individual tree-shelters is considered to be beneficial against browsing and for stimulating

Figure 4.15. Cork oak seedlings grown at a nursery to be used in plantations (see Colour Plate Section).

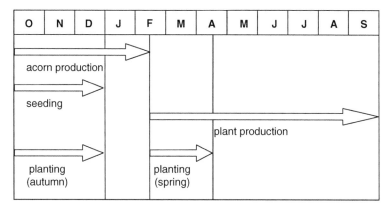

Figure 4.16. Diagram representing the annual distribution of afforestation-related events (adapted from Montero and Canelas, 1999).

initial height growth, but in regions with severe drought and high temperatures, this technique has originated higher mortality rates and the young plants have thinner and less developed wood stems (Quilhó et al., 2003). Whenever spring and summer precipitation is too low, irrigation should be applied 2–3 times during the periods of higher stress. During the first years after planting, weeding is also highly recommended around each seedling or on the whole area.

4.5.2. Juvenile stands

The juvenile phase of a cork oak stand refers to the period until the beginning of cork extraction, which occurs at 25–40 years of age, the lower limit being the usual one in most of the stands in Portugal and southern Spain, and the upper limit in more environmental stressed and dense stands.

The young cork oak trees show abundant ramification and a leading shoot is often not present. Therefore, it is important to carry out a conducting pruning in order to obtain trees with a clear stem height of at least 2.5–3 m. Three pruning operations are usually carried out: a first pruning between 3 and 6 years of age to eliminate all the branches in the first two-thirds of the tree stem, a second pruning between the ages of 12–15 years, and a last conducting pruning after the first debarking. Depending on stand density at planting, the second or final prunings may be complemented by a first thinning to reduce stand density to 400–600 trees ha^{-1}.

Since the cork oaks are very sensitive to competition during their first years of age, weeding should be undertaken, at least during the first 4–5 years and then every 3–4 years until an age of 10–15 years.

4.5.3. Mature stands

The most important silvicultural operations in mature stands are thinning and, naturally, cork debarking, dealt with in detail in the following chapter. Thinning is carried out usually after the first debarking so that the landowner can profit of the cork income from the trees that will be thinned and also to be able to use cork quality as an additional criteria for selection of trees to be thinned.

The thinning is carried out in order to obtain a pre-selected spacing factor. The spacing factor is defined as the quotient between mean distance between trees and mean tree crown diameter. It is usually assumed, even if there is not enough experimental data to confirm this empirical rule, that cork production is affected when inter-tree competition is too high and therefore it is recommended to use a spacing factor that allows each tree enough space to develop its crown without substantial restrictions. The space between trees should be at least half of its mean crown radius, which is equivalent to a spacing factor of around 1.2. Figure 4.17 shows the variation of mean stand density and of tree size measured as circumference over cork in cork oak stands directed towards the production of cork during a rotation cycle of 170 years.

Some forest managers consider application of fertilisers during the period between cork extractions to improve cork production, but the few experimental results available do not show its effect on cork growth and quality.

In past years the cork oaks were pruned to increase fruit yield since the acorns were an economically important product and this practice is still used by some farmers that associate this type of pruning with tree vigour and cork yield, a fact that has never been experimentally proved.

Weed control, depending on the type of underneath cultivation, may also be carried out. It is highly recommended that the mechanical weeding, as any other operation, does not damage the superficial root system of the cork oaks.

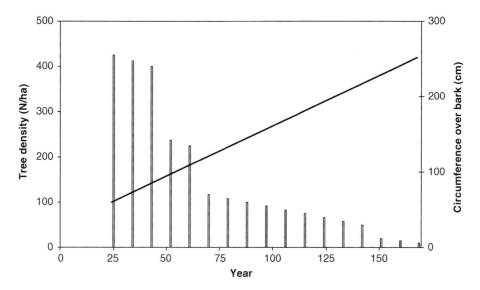

Figure 4.17. Range of variation of cork oak density (number of plants per hectare) with tree average size (circumference over cork) (drawn from data in Montero and Canelas, 1999).

Cork oak trees may attain 250–350 years of age, but as the trees get older the phellogen activity decreases and cork thickness decreases. The age of 150–200 years seems to be the limit for an industrial useful cork production and today's management plans usually consider rotation periods between 100 and 150 years.

Harvesting of cork oaks occurs usually only for sanitary purposes and the present mature stands include a large percentage of old trees. In fact, the cork oak is a protected species and the silvicultural operations are regulated, with tree harvesting being legally forbidden except for a few exceptions.

4.6. Cork oak forests

Cork oaks can be managed according to two different silvicultural systems:

1. cork oak woodlands, with a relatively small number of trees per hectare (a sparse forest with usually 50–150 trees per ha), where forest is associated with pasture for cattle grazing or agricultural crops (Fig. 4.18); this system is called *montado* in Portugal and *dehesa* in Spain;
2. cork oak forests, where the cork oaks are grown in denser stands not allowing the practice of agriculture underneath the trees, namely in mountainous regions.

Both systems are frequently associated with hunting, mushroom-picking and bee-keeping.

Most of the present mature cork oak stands resulted from the management of naturally regenerated stands by generations of landowners or, in some cases, through artificial seeding complemented by natural regeneration. As a result, a large percentage of the stands is characterised by an heterogeneous spatial distribution, and many can be classified as

Figure 4.18. A typical montado landscape in southern Portugal.

uneven-aged. Figure 4.19 shows a crown cover map and tree biometric characteristics for two cork oak montado estates under cork production.

4.7. Conclusions

The cork oaks have a reduced geographical distribution that extends to only approximately 2 million hectares around the western part of the Mediterranean basin. The trees are well adapted to the challenging and diverse climatic conditions of the region, namely by resisting to hot summers without rain and to low fertility and shallow soils, while being plastic in relation to annual precipitation, temperatures and soil pH. The main restrictions to cork oak survival are calcareous soils (alkaline pH), water logging and long frosts.

The main characteristics of cork oak that contribute to the species' long-term ecological success are the following:

- well-developed root system, in depth to tap water in underneath aquifers during drought periods and in surface to ensure enough nutrient supply;
- leaf lifespan and seasonality that allow maintenance of photosynthesis in the beginning of the growth period until full canopy renewal;
- long flowering period and annual and biennial acorn production;
- morphological and phenotypical diversity;
- protective cork layer in the outer bark that imparts considerable fire resistance.

Cork oak trees integrate two types of forest and land-use systems:

- cork oak woodlands as a multifunctional agro-sylvo-pastoral system that constitute the major proportion of the cork oak area in Portugal (montado) and southern Spain

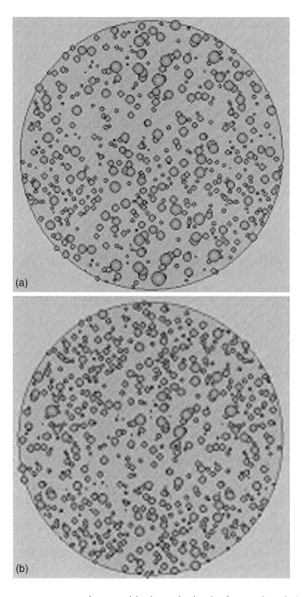

Figure 4.19. Crown cover maps drawn with data obtained after cork oak inventory for two montado cork oak woodlands with mature trees under cork production, in Portugal (Ponte de Sôr). Site A: 95 trees ha^{-1} with a mean diameter of 25.9 cm; Site B: 68 trees ha^{-1} with a mean diameter of 36.2 cm.

(dehesa), where tree density is low, and cork production is associated to cattle and agricultural crops;

• cork oak forests, mostly in the mountainous regions of Portugal and the Spanish Catalonia, and also in France, Italy and north Africa, where tree density is high.

The cork oak woodlands have been intensively managed by the landowners and are important reservoirs of fauna and flora biodiversity.

The characteristics of cork oak stands and of their silviculture may be summarised as follows:

- the existing stands are uneven-aged while the future new stands will be even-aged;
- the establishment of new cork oak stands is by plantation with seedlings and by seeding and the young trees are susceptible to excessive drought and animal damage until about 10 years of age and have to be tended accordingly;
- the forest management is oriented towards cork production and the silvicultural operations include several pruning operations as important stem-forming techniques.

References

Barberis, A., Dettori, S., Filigheddu, M.R., 2003. Management problems in Mediterranean cork oak forests: post fire recovery. Journal of Arid Environments 54, 565–569.

Belahbib, N., Pemonge, M.H., Ouassou, A., Sbay, H., Kremer, A., Petit, R.J., 2001. Frequent cytoplasmatic exchanges between oak species that are not closely related: *Quercus suber* and *Q. ilex* in Morocco. Molecular Ecology 10, 2003–2012.

Boavida, L.C., Silva, J.P., Feijó, J.A., 2001. Sexual reproduction in the cork oak (*Quercus suber* L.). II. Crossing intra- and interspecific barriers. Sexual Plant Reproduction 14, 143–152.

Boavida, L.C., Varela, M.C., Feijó, J.A., 1999. Sexual reproduction in the cork oak (*Quercus suber* L.). I. The progamic phase. Sexual Plant Reproduction 11, 347–353.

Cardillo, E., Bernal, C.J., 2006. Morphological response and growth of cork oak (*Quercus suber* L.) seedlings at different shade levels. Forest Ecology and Management 222, 296–301.

Carrión, J.S., Parra, I., Navarro, C., Munueras, M., 2000. Past distribution and ecology of the cork oak (*Quercus suber*) in the Iberian Peninsula: a pollen-analytical approach. Diversity & Distributions 6, 29–44.

Cavender-Bares, J., Cortes, P., Rambal, S., Joffre, R., Miles, B., Rocheteau, A., 2005. Summer and winter sensitivity of leaves and xylem to minimum freezing temperatures: a comparison of co-occurring Mediterranean oaks that differ in leaf lifespan. New Phytologist 168, 597–612.

Costa, A., Pereira, H., Oliveira, A., 2001. Dendroclimatological approach to diameter growth in cork oak adult trees under cork production. Trees 15, 438–443.

Costa, A., Pereira, H., Oliveira, A., 2002. Influence of climate on the seasonality of radial growth of cork oak during a cork production cycle. Annals of Forest Science 59, 429–437.

Costa, A., Pereira, H., Oliveira, A., 2003. Variability of radial growth in cork oak adult trees under cork production. Forest Ecology and Management 171, 231–241.

Díaz-Fernández, P.M., Climent, J., Gil, L., 2004. Biennial acorn maturation and its relationship with flowering phenology in Iberian populations of *Quercus suber*. Trees 18, 615–621.

Elena-Rosselló, J.A., Lumaret, R., Cabrera, E., Michaud, H., 1992. Evidence for hybridization between sympatric holm-oak and cork-oak in Spain based on diagnostic enzyme markers. Vegetatio 100, 115–118.

Escudero, A. Delarco, J.M., Sanz, I.C., Ayala, J., 1992. Effects of leaf longevity and retranslocation efficiency on the retention time of nutrients in the leaf biomass of different woody species. Oecologia 90, 80–87.

Fialho, C., Lopes, F., Pereira, H., 2001. The effect of cork removal on the radial growth and phenology of young cork oak trees. Forest Ecology and Management 141, 251–258.

Garcia-Mozo, H., Hidalgo, P.J., Galan, C., Gomez-Casero, M.T., Domingues, E., 2001. Catkin frost damage in Mediterranean cork-oak (*Quercus suber* L.). Israel Journal of Plant Sciences 49, 41–47.

Ghouil, H., Montpied, P., Epron, D., Ksontini, M., Hanchi, B., Dreyer, E., 2003. Thermal optima of photosynthetic functions and thermostability of photochemistry in cork oak seedlings. Tree Physiology 23, 1031–1039.

González-Adrados, J.R., Gourlay, I., 1998. Applications of dendrochronology to *Quercus suber* L. In: Pereira, H. (Ed), Cork Oak and Cork. Proceedings of the European Conference on Cork Oak and Cork. Centro de Estudos Florestais, Lisboa, pp. 162–166.

Gourlay, I., Pereira, H., 1998. The effect of bark stripping on wood production in cork oak (*Quercus suber* L.) and problems of growth ring definition. In: Pereira, H. (Ed), Cork Oak and Cork. Proceedings of the European Conference on Cork Oak and Cork. Centro de Estudos Florestais, Lisboa, pp. 99–107.

Herrera, J., 1995. Acorn predation and sedling production in a low density population of cork oak (*Quercus suber* L.). Forest Ecology and Management 76, 197–201.

Knapic, S., Louzada, J.L., Leal, S., Pereira, H., 2006. Radial variation of wood density components and ring width in cork oak trees. Annals of Forest Science, in press.

Lumaret, R., Tryphon-Dionnet, M., Michaud, H., Sanuy, A., Ipotesi, E., Born, C., Mir, C., 2005. Phylogeographical variation of chloroplast DNA in cork oak (*Quercus suber* L.). Annals of Botany, 96, 853–861.

Machado, D., 1938. Poligamia do sobreiro. Publicações da Direcção Geral dos Serviços Florestais e Aquícolas 5, 37–41.

Merouani, H., Apolinario, L.M., Almeida, M.H., Pereira, J.S., 2003. Morphogical and physiological maturation of acorns of cork oak (*Quercus suber* L.). Seed Science and Technolog 31, 111–124.

Molinas, M.L., 1991. The stomata of the cork-oak, *Quercus suber* – An ultrastructural approach. Nordic Journal of Botany 11, 205–212.

Montero, G., Canellas, I., 1999. Manual de reforestación y cultivo de alcornoque. Ministério de Agricultura, Pesca y Alimentación – Instituto Nacional de Investigación y Tecnología Agraria y Alimentaria, Madrid.

Montoya, J.M., 1988. Los alcornocales. Ministério de Agricultura, Pesca y Alimentación – Secretaria Nacional Técnica, Madrid.

Nardini, A., Lo Gullo, M.A., Salleo, S., 1999. Competitive strategies for water availability in two Mediterranean *Quercus* species. Plant Cell and Environment 22, 109–116.

Natividade, J.V., 1950. Subericultura. Ministério da Economia – Direcção-Geral dos Serviços Florestais e Aquícolas, Lisboa.

Nunes, E., 1996. Estudo da influência da precipitação e temperatura no crescimento juvenil de *Quercus suber* L. através dos anéis anuais de crescimento. Ms. D. Thesis, Instituto Superior de Agronomia, Lisboa.

Oliveira, G., Correia, O., Martins Louçao, M., Catarino, F.M., 1992. Water relations of cork oak (*Quercus suber* L.) under natural conditions. Vegetatio 100, 199–208.

Oliveira, G., Correia, O., Martins Louçao, M., Catarino, F.M., 1994. Phenological and growth patterns of the mediterranean oak *Quercus suber* L. Trees 9, 41–46.

Oliveira, G., Martins Louçao, M., Correia, O., 2002. The relative importance of cork harvesting and climate for stem radial growth of *Quercus suber* L. Annals of Forest Science 59, 439–443.

Pardos, M., Jimeneza, M.D., Aranda, I., Puertolas, J., Pardos, J.A., 2005. Water relations of cork oak (*Quercus suber* L.) seedlings in response to shading and moderate drought. Annals of Forest Science 62, 377–384.

Pausas, J.G., 1997. Resprouting of *Quercus suber* in NE Spain after fire. Journal of Vegetation Science 8, 703–706.

Pereira Coutinho, A.X., 1939. Flora de Portugal. Bertrand Irmãos Ltd, Lisboa.

Pereira, P., Fonseca, M., 2003. Nature vs. nurture: the making of the montado ecosystem. Conservation Ecology 7, 7 (on line).

Plieninger, T., Wilbrand, C., 2001. Land use, biodiversity conserevation, and rural development I the dehesas of Cuatro Lugares, Spain. Agroforestry Systems 51, 23–34.

Quilhó, T., Lopes, F., Pereira, H., 2003. The effect of tree shelter on the stem anatomy of cork oak (*Quercus suber*) plants . IAWA Journal 24, 385–395.

Robert, B., Caritat, A., Bertoni, G., Vilar, L., Molinas, M., 1996. Nutrient content and seasonal fluctuations in the leaf component of cork oak (*Quercus suber* L.) litterfall. Vegetatio 122, 29–35.

Sánchez González, M., Tomé, M., Montero, G., 2005. Modelling height and diameter growth of dominant cork oak trees in Spain. Annals of Forest Science 62, 633–643.

Toumi, L., Lumaret, R., 1998. Allozyme variation in cork oak (*Quercus suber* L.): the role of phylogeography and genetic introgression by other Mediterranean oak species and human activities. Theoretical and Applied Genetics 97, 647–656.

Tucker, G., Evans, M., 1997. Habitats for Birds in Europe. Their Conservation Status for a Wider Environment. Bird Life Conservation Series No. 6, Bird Life International, Cambridge.

Vicioso, C., 1950. Revision del genero Quercus en Espana. IFIE, Madrid.

Chapter 5

The extraction of cork

The exploitation of the cork oak as a producer of cork requires its periodical removal from the stem and branches in an extent that is considered compatible with the maintenance of the tree in good physiological conditions. The extraction of cork, or cork stripping, is done manually by cutting large rectangular planks and pulling them out from the tree. The operation takes advantage of the fragility of the phellogen and of the newly formed cork cell layers in order to be able to tear the cork layer out of the tree without damage to the inner bark and cambium. Therefore it requires the cork oak to be physiologically active and it is strictly seasonal in late spring and early summer. The cork stripping also requires technical expertise to avoid wounding of the tree.

The industrial requirements for the raw material derive from its use, which nowadays is directed in the first option to the production of stoppers. Therefore a sufficient thickness is necessary, i.e. the cork planks should be over 27 mm thick, and defects such as excessive discontinuities, i.e. deep fractures, cannot be present. This is the reason why virgin cork and second cork, the corks obtained, respectively, from the first periderm and from the first traumatic periderm, are not used for stoppers.

The operation of cork stripping and of the subsequent field and mill yard storage are detailed in this chapter and graphically exemplified in a sequence of pictures. The effect of cork removal on the tree physiology and radial growth is also analysed. The characterisation of the different types of cork raw materials that are obtained from the exploitation of cork oak stands is presented in conjunction with their implications in the industrial processing.

5.1. Cork stripping

5.1.1. Process of extraction

The cork layer that covers the cork oak stem is removed by pulling it out during the period of periderm activity when the phellogen mother cells and the recent formed phellem cells are turgid and the cell walls are thin and fragile. In this state it is easy to separate the cork layer at the level of the phellogenic active zone by applying a moderate tensile force in the radial direction. This is made after cutting through the cork layer in order to be able to grab it and pull it out. The timing of this operation is essential to guarantee that the underlying phloem and cambium are not harmed. Therefore cork extraction is made only during the period of strong phellogen activity, which occurs by mid-May to the beginning of August. Climatic variations dictate the scheduling, i.e. cold spells, periods of drought or warm weather may cause either a delay or an anticipation of operations.

The observation of the onset of new spring growth, visible in the formation of shoots and leaves, is usually an indication of the tree physiological activity and a prerequisite to plan the beginning of the cork stripping operation. One usual practice is also to test in a few cork oaks how easily cork may be pulled out, or "giving away" in the cork producers' jargon. In a tree that is not giving away cork properly, the cork layer cannot be removed with the usual practices; if the applied tensile pressure is too high, rupture may occur at the cambium level causing irreversible damage. However, due to the variability of tree physiological status within a stand, it may occur that during the cork extraction operation, some of the trees will not yield the cork, i.e. weakened or diseased trees, or even in one tree it may be impossible to strip the cork from some parts, i.e. due to partial crown attack. The expertise of the cork strippers is a decisive factor to carry out the operation technically well and to make the correct decisions.

The cork stripping is done manually. Usually the work team is constituted by two cork strippers who strip together the same tree. The cork extraction is shown in Figure 5.1. The tool used is a stripping axe with a curved cutting blade and a relatively long wooden arm that has a chiselled end to be used as a lever for the separation of the cork planks. The cork stripping starts with the cutting through the cork layer, first horizontally around the tree perimeter at a height level of about breast height or slightly above, then followed by two or three vertical cuts depending on tree perimeter. The cut is made by balancing the axe and dosing the strength in order to cut through the cork layer down to the inner bark but not to penetrate into it. The successive cutting strokes are done following a straight line around or along the stem. The separation of the cork layer is made by introducing the axe's arm in the cut and levering it out. The cork plank is then pulled out and usually separates easily. The cork strip at the bottom end of the stem in contact with the soil is removed leaving a clean stripped surface. The operation is continued in a similar way upwards along the tree stem and in the main branches until the limit of cork extraction. Ladders, formerly wooden but now mostly aluminium, are used to reach the upper parts of the tree. The cork stripping with experienced workers is carried out rather fast and the team of two cork strippers achieve an average yield of about 900 kg of cork in a working day (Costa, 1990).

In large trees with low stem lengths and many spreading thick branches, several groups of cork strippers are involved in the cork stripping and they have to work up in the tree by standing on the branches (Fig. 5.2). These trees yield a large quantity of cork amounting to several hundred kilograms. The largest and oldest cork oak tree in production in Portugal, the Whistler tree, planted in 1783, was stripped in 2000 and the cork obtained totalled 650 kg, while in the previous cork extraction in 1991 it yielded 1200 kg of cork.

Figure 5.1. Different aspects and sequence of procedures in the stripping of cork: (a) the cork stripper uses an axe to cut through the cork along horizontal and vertical lines; (b) a detail showing how the axe penetrates in the cork plank; (c) the axes' arm is used as a lever to help separate the cork plank (see Colour Plate Section); (d) extraction of cork in the upper part of the stem requires the use of ladders; and (e) general view of a cork stripping operation (see Colour Plate Section).

The cork stripping is carried out by a large group of workers, i.e. about 100 of persons. They are organised under a manager, and in addition to the cork strippers, these groups include also workers who collect the cork planks and transport them to a central point and those who pile them up. The role of women in this operation is reserved to the gathering of planks from the ground for tractor loading and distributing drinking water to the cork strippers (Fig. 5.3).

The cork stripping is a picturesque event that is part of the cultural and social history of these regions. It is often depicted in popular handicrafts such as pottery or hand-painted tiles (Fig. 5.4).

There have been some experiments to mechanise the cork stripping. Rather sophisticated techniques such as the use of a high pressure water jet or a laser beam were already tested but proved unfeasible out of practical or economic considerations. Some powered hand-carried sawing machines have also been proposed for the cutting operation and used in field demonstration and comparative testing (Antolín et al., 2003). The results seem promising regarding the precision of the cutting and the operational costs although limited to the extraction of vertical and short stems. Mechanical cork stripping has, however, no practical expression at the moment.

Immediately after the removal of the cork layer, the cork oak stem has a golden brown colour and a smooth appearance. The colour darkens during the subsequent weeks turning into a dark reddish brown while the outer tissues of the phloem dry out quickly and give a rough touch to the stem. With weathering during the following years, the stems acquire a dark greyish brown colour and develop numerous fissures that go more or less deep depending on the growth intensity of the underlying cork layer.

Figure 5.2. Cork stripping in large-sized cork oaks.

(a)

(b)

Figure 5.3. The role of women within the cork stripping working group: (a) gathering scattered cork planks for tractor loading; (b) distributing drinking water to the workers.

Figure 5.4. The cork stripping is depicted in popular handicrafts such as these decorative painted tiles on the front wall of a country house in Portugal.

5.1.2. Cork stripping intensity

The intensity of the cork stripping refers to the area of the tree where the cork layer is removed and it is defined in relation to the tree size. The most usual way to express the extent of cork removal is by using the height until which the stripping of cork is made, the so-called debarking height. The debarking coefficient is defined as the quotient between the debarking height and the perimeter over cork at breast height (at 1.3 m of height) ($pbh_{overcork}$):

$$\text{Debarking coefficient} = \frac{\text{debarking height}}{pbh_{\,overcork}}$$

The debarking coefficient is regulated by legislation or by good practices procedures. In Portugal, the maximum debarking coefficient is legally enforced and depends on the stage of development of the tree: 2.0 for the 1st cork stripping, 2.5 for the 2nd cork stripping and 3.0 for subsequent cork strippings. The cork cannot be removed before the tree has an over-cork perimeter at breast height of 70 cm (22 cm of diameter) and this is also the limit of cork removal in branches. In Spain, the limit is lower at a perimeter at breast height of 60 cm (19 cm of diameter) When the cork oak stripping includes branches, the debarking height includes in addition to the stem height also the branch length measured along the longest debarked branch. The age at which a cork oak may enter production depends on its radial growth, but it is considered that the required diameter is attained at 20–25 years of age in most of the growing conditions or at 30–40 years in less favourable environments.

The definition of the limits to the cork stripping intensity was not based on results of research but rather on an empirical common sense on which proportion of stem and branches surface the tree can endure to loose its periderm without irreversible loss of vitality.

The intensity of cork stripping is also related to the frequency of the cork extractions defined as the number of years between two cork extractions, the so-called cork production cycle. There are no studies that quantify the effect of the duration of the cork cycle in the tree growth. The growth of cork within the cycle is well studied (see Chapter 1, Fig. 1.18), and it seems that it is the need to obtain a certain thickness in the cork plank for the subsequent industrial utilisation that has dictated the duration of the cycle. The duration therefore may vary from region to region depending on the corresponding mean annual cork growth. In most regions of Portugal and southern Spain, a total cork plank thickness of about 3–3.5 cm is attained in 9 years, while in the Spanish region of Catalonia or southern France, more than 15 years are necessary to attain a similar thickness.

In Portugal and Spain, there is a legal limit for the minimum duration of the period between cork extractions, i.e. 9 years. This is also the legal age of the cork that can be seen on a cork section by counting the number of rings, with the years of extraction counting as two half years, as represented on Figure 5.5.

The cork extraction of a stand may be done either simultaneously in all the trees – even-aged cork – or in only a part of the trees, therefore cork age being not the same for the trees in the same stand – uneven-aged cork. In the uneven-aged cork stands, two cork stripping rotations are defined: the tree and the stand cork stripping, depending on how many different cork ages are present in the stand. For forest controlling purposes, it is usual to mark the year of cork removal on some cork oaks by painting the last year's digit in white on the stem after removal of the cork. The present management favours even-aged cork management areas and in large stands 2 or 3 cork ages.

In the past there was also the practice of doing a fractionated stem cork stripping: about half of the stem height was debarked in 1 year, and the remaining area was stripped only after 4–5 years. Therefore one tree had two cork cycles with 3–5 years interval between them. This procedure is nowadays forbidden, or strongly unadvised, since the repeated

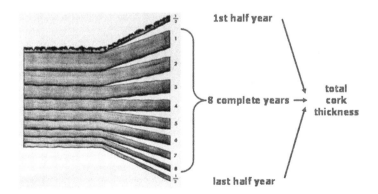

Figure 5.5. Schematic representation of the counting of years and half-years to determine the age of a cork (adapted from Natividade, 1950).

cutting on the same region of the stem originates wound responses that form a belt around the stem with a very low quality cork (Fig. 5.6.a). These cork oaks are still a striking image in the montado landscape.

5.1.3. The effect on tree growth

One direct and obvious impact of cork extraction on the tree results from the technical expertise of the cork stripping operation itself. It is recognised that the cork stripper has to have experience and be careful in order to cut through the cork without harming the

(a)

(b)

Figure 5.6. Tree reaction to wounding: (a) formation of a continuous wound callus in a zone with repeated stripping cutting wounding; (b) in the previously wounded areas, the risk of damage is higher, as shown here with the tearing out of the phloem.

tree. In fact the cutting stroke has to be powerful and instantaneous, or the cork will absorb most of its energy, yet not too strong to penetrate into the phloem or, worse, to attain the cambium.

The wounding of the tree is therefore the first eventual effect of cork stripping. Wounds constitute roads for infection and biological attack, and the tree responds by the formation of wound reaction calluses that protrude in the stem, sometimes very notoriously, as it is the case of the border line between extractions in the fractionated stripping (Fig. 5.6a). The repeated cutting lines in the same place in successive cork strippings increase this occurrence. In these reaction zones, the production of cork is smaller and the subsequent cork stripping is technically more difficult, thereby increasing the risks of further wounding (Fig. 5.6b). The effect of damages done to the tree during cork extraction was measured in the following cork extraction (Costa et al., 2004). The overall radial growth of the trees decreased by approximately 13% in relation to undamaged trees and the production of cork decreased by 14%, the effect being more pronounced in the 2 years following the cork extraction.

A study on cork stripping and pruning damages carried on in four transects in the Spanish Extremadura region showed very high incidence of damages, ranging from 31 to 47% of the trees, while significant correlations were found between damage and holes due to attack of the *Cerambyx* beetles and subsequent fungal infection by *Biscogniauxia mediterranum* (Martín et al., 2005).

The need to use experienced labour for the cork stripping, the disinfection of the cutting tools between the stripping of individual trees and strict operational control are therefore recurrent statements in cork oak management, albeit not always enforced.

The cork stripping itself has a direct physiological effect on the tree with an increased water loss through stem transpiration in the area where cork was removed while stoma close quickly in the hours following cork stripping (e.g. after 24 h stomatal activity is nil), leading to interruption of the nutritional functions, only returning to normal after 24–30 days (Santos, 1940; Oliveira, 1995). During this period the traumatic phellogen is formed and some layers of cork cells are produced thereby protecting the active phloem from further water losses. Because of the very high consumption of reserves necessary for this process, the activity of the vascular cambium decreases and the wood growth stops during this period (Natividade, 1938).

As regards the effect of cork removal or its intensity on the radial growth of the tree, there are very few results from research. The fact that it is difficult to observe the annual rings in the cork oak wood does not allow to follow the effect of cork removal on the subsequent growth rings using analysis of stem discs, although it seems that once the tree enters into cork production the anatomy of wood becomes disturbed with formation of smaller cells, thicker walls, higher number of fibres and less parenchyma, and with the wood elements irregularly distributed and loosing the typical gradual earlywood/latewood transition (Lupi, 1938; Gourlay and Pereira, 1998). One study suggests a reduction in wood growth in the 2 years that follow the removal of cork in mature trees, indirectly estimating it in about 15% in a 9-year production cycle (Leal et al., 2006). However, in very young cork oaks growing under favourable conditions, the removal of cork did not cause a reduction in the tree's total radial growth (Fialho et al., 2001).

5.2. Post-harvest operations

5.2.1. Field storage

The cork planks that are stripped are left on the ground by the cork strippers. They will be collected afterwards with a tractor and led to a field yard where the cork piles are erected. In hilly conditions with difficult accessibility, the transport of the cork planks is made with a belt tractor or with mules (in Spain).

The construction of a cork pile is a rigorous task that starts with the selection of location, which has to be even, dry and with a good road accessibility, and with a careful marking of its rectangular emplacement in the soil using rope guidelines. The cork planks are curved following the form of the stem, with the corkback to the convex side. They are piled on each other with the cork back facing up, and disposed carefully in order to make a straight and solid outside wall for the pile. Regular shaped planks are chosen for this purpose, while the irregular and smaller planks are put in the interior of the pile. The dimensions of the piles may vary but usually they are 8–10 m wide, 2–2.2 m high and extend to 30–50 m. Figure 5.7 shows photographs of field collection and pile construction. In Morocco the cork piles are usually smaller with about 2–3 m of width.

The pile is erected as the cork planks are coming from the stand. Therefore it includes cork planks with different thickness and quality due to the variability that exists between trees in the same stand and no separation is made at this stage. Separate piles are made only of cork pieces (a cork plank has a minimum area of 400 cm^2) and of virgin and second cork.

The cork pile masters are also experienced workers who construct the pile taking into account that it will be scrutinised by potential buyers. In fact the commercialisation of cork and the agreement on the price to be paid are made after visual appreciation of the pile. In some cases, samplings from the interior of the pile are made previous to an offer, but the visible cork planks always play an important role. The price is established by weight, the usual unit being 15 kg, but a further agreement has to be reached in relation to the water content of cork and to the corresponding weight discount, as detailed further on.

The time of field storage is variable and may range from a few weeks to nearly 1 year. It depends mostly on the agreements between producer and client, and often it is the logistics at the industrial mill that determine when the cork is collected from the field to the yard. In recent cases, the storage in the field and the pile construction are avoided, and the cork planks are transported directly to the industrial yard after the cork stripping.

5.2.2. Water content variation

During the stripping of the cork oaks, it can be observed that the inner part of the cork planks is humid, as it is also the tree stem underneath. This is to be expected since the phellogen is active and the cells in this region are turgid.

The water content of cork planks after the stripping is on average around 25%, but a large variation occurs between individual samples ranging from less than 10% to more than 50% (Velasco et al., 1990; Gonzalez-Adrados and Haro, 1994; Sousa, 1997). The water is not distributed homogeneously within the cork plank: the outer part of the cork

Figure 5.7. Collection of cork planks from the cork oak stand and construction of the cork plank pile: (a) a tractor transports the cork planks to the emplacement of the pile, which is in a central location with easy access for truck loading; (b) beginning of the construction of one pile showing the selection of the cork planks for an orderly arrangement (see Colour Plate Section); (c) one corner of a pile showing the regular disposition of the cork planks, making up vertical columns of parallel compacted planks with the cork back up; notice the mechanical strengthening at mid-height by using a cross-orientation of the planks; (d) general aspect of the field storage with one complete pile in the back, the start of a second pile in front and in-between the storage of cork pieces.

plank is dry and in equilibrium with the ambient air relative humidity, while the inner part is moist. After the separation from the stem, it is clearly visible that the innermost cork layer of about 1 mm is translucent as typical of a water-filled tissue. Determinations made in a cork plank with an average water content of 32% showed the following distribution going from the external part to the interior: corkback 9%, outer cork rings 8%, inner cork rings 18%, innermost cork layer (belly) 121% (Fernandes, 2004).

It is not surprising that the water content in cork at the moment of extraction (green cork) is on average low in comparison for instance with wood at tree harvesting (with more than 50% moisture content). This is the result of the very low hygroscopicity of cork (see Chapter 8) and the already long period of drying that the outer layers of the periderm went through during the current cork cycle.

When left at ambient conditions, the cork planks start to lose water very rapidly. It is estimated that in about 9 days the water content of cork planks piled in the field is less than 14% and may be considered as commercially dry cork. The fully equilibrated cork to ambient air conditions is between 6 and 10%. The storage of the cork planks in the field piles does not have any effect on the width of the cork growth rings nor on their porosity.

When making the commercial agreements regarding the transaction of the cork pile, usually a price is set for the unit weight of the cork at the moment of transport to the yard (when it is weighed). When the cork is transported after the cork stripping or shortly afterwards, there is also a discount for the water content of the cork, in the range of 20%, and this value may be a matter of vivid discussion. In fact, the amount of water in different conditions may be estimated and the calculation of how much this represents of the cork weight at that moment is easily made. Considering a moisture content of 25% at the moment of stripping, 14% after 9 days and 10% after 1 month, 100 kg dry cork (0% moisture) will weight 125, 114 and 110 kg, respectively, after stripping, 9 days and 1 month. This means that the "discount" to the weight due to water would be (a) in relation to absolute dry cork 20, 12.3 and 9.1% of the cork weighed, respectively, after stripping, 9 days and 1 month and (b) in relation to air dry cork (the usual practice) with 10% moisture, the discount would be 10, 2.3 and 0% of the cork weighed, respectively, after stripping, 9 days and 1 month.

5.2.3. Yard storage and preparation for processing

The industrial mills have always in their premises a large area devoted to the air storage of the cork raw materials. The storage area of the cork planks, or stabilisation area as it is also called, has a small slope and it is paved to guarantee that rainwater will not accumulate near the ground. This care is taken in order to avoid microbial development or contamination that could introduce or intensify mouldy flavours in the cork planks. This is an aspect that the modern industrial mills are very attentive to, since taints and off-flavours in the wine caused by cork stoppers are a key quality issue (see Chapter 14).

The piles of cork planks in the yards may be organised regularly following the procedure used with the field piles or they may be more random when they are constructed directly by unloading of the truck (Fig. 5.8). The raw-material origin is preserved at the

Figure 5.8. Pile of cork planks at the mill yard in a modern industrial unit.

mill, and the storage piles are separated for the different provenances, as a first step of the traceability of the cork within the industrial process.

The duration of the storage is very much in relation to the requirements of raw-material flow into the industrial process. In past times, long durations of 2 years, later reduced to 1 year, were practiced and claimed to be necessary to stabilise chemically and structurally the cork planks. As with many empirical rules, also this one has no scientific basis. The structural relief of tensions is made during the subsequent operations of boiling in water (see Chapter 10), while the drying of the inner layers of cork already occurred in the field together with the air oxidation of metabolic products, not forgetting that these are limited to the few cell layers near the phellogen. At present, the official good practices advise a storage period of 6 months, but very often the raw material is processed with shorter storages due to economic and logistic factors related to the seasonality of cork production and the very high costs of the raw material. A report on the influence of the duration of storage of cork planks from the moment of cork oak stripping to processing into stoppers from 0 to 6 months showed no differences on cork ring width, cork porosity and stoppers' mechanical properties (Fernandes, 2004).

The preparation for the industrial processing is carried out at the storage yard and it consists in assembling the cork planks in pallets for the operation of boiling in water (see Chapter 12). During this operation, the planks that have an obvious extreme low quality in the whole area are separated as refuse raw cork planks and the cork planks where mouldy inclusions and yellow stain (see Chapter 7) are observed are also segregated. Some of the planks from the lower part of the cork oak stem contain the collar of cork that was inserted in the soil at the junction of the tree roots, here called the footer. This part is cut away from the cork plank with a height of approximately 20 cm because of the risk of soil-derived microbial attack. This operation also starts to be carried out in the field before the pile construction.

5.3. Forest-to-industry chain

5.3.1. Types of raw cork

The cork produced in the first periderm of the cork oak, the virgin cork, has the typical appearance shown in Figure 1.12 that can be seen on the young cork oaks before the first cork extraction or on the unstripped branches of the mature cork oaks. The first extraction of cork therefore yields virgin cork. Because of to the deep fractures, virgin cork is used for triturating for production of agglomerates.

After the first cork extraction, the new traumatic periderm develops until the second cork layer (second cork) is removed in a subsequent cork extraction. Second cork also shows in most cases profound cracks (Fig. 1.12) that make it still only usable for triturating. The following periderm, as well as all those that may be formed later on, contains the reproduction cork. This is the raw material that will be used by the industry for the production of stoppers.

In the normal exploitation of a cork oak stand, virgin cork is obtained from the young trees that are coming into production and also from the trees in production where the height of cork striping can be increased as a result of the tree radial growth. In this case, a strip of virgin cork is taken in the upper part of the stem until the desired height limit.

In the case of trees still in a phase of important height growth, the same cork plank may contain a portion of second cork together with the reproduction cork. This could be seen in Figure 1.13 that showed a picture of a young cork oak after cork stripping where the limits between cork extractions can be noticed on the exposed innerbark: virgin cork was removed from the upper part, second cork from the next below region and reproduction cork from the remaining lower part of the stem.

Virgin cork may also be taken from branches that are pruned or from young trees that were cut in thinning operations. The pruning of cork oaks is done in winter and therefore the cork layer adheres firmly to the inner bark. The removal of cork is done either manually with a small axe or with a debarking machine (Fig. 5.9). In either case, the cork is obtained in pieces with variable dimensions from a few centimetres to longer (about 20 cm) strips, but it contains a large proportion of the innerbark. This type of cork is called corkwood (or winter cork). For the industry it is a less valuable raw material since it has to undergo severe cleaning after trituration to separate the woody particles from the cork granules.

Occasionally other types of cork raw material are available. They may be constituted by under-aged thin reproduction cork that result from the felling of cork oaks at the end of their life or otherwise dead, sick or weakened trees. In this case the proportion of cork is also small due to the relative importance of the cork back.

Table 5.1 summarises the types of cork that are available from the exploitation of the cork oak trees, their main characteristics and industrial main uses.

5.3.2. Industrial requirements

The cork industry produces different products with variable incorporation of cork and technological transformation, but the economic feasibility of the whole sector is determined by the production of stoppers of natural cork to be used in the bottling of wines. Nowadays it is the suitability of the cork raw material for this production that establishes its commercial value and the objectives for the forest manager.

Figure 5.9. Removal of the virgin cork from cork oak branches manually using an axe.

The thickness of the cork plank is the most important variable when analysing the raw material suitability for processing. Therefore it is not surprising that the observation of the field piles of the cork planks by the potential industrial client takes the thickness distribution in great care. There is a large variation in the thickness of cork planks obtained from different trees in the same stand as well as between stands in the same region, as shown in Figure 5.10 for four piles. The proportion of cork planks with a thickness that allows the production of stoppers, corresponding to a minimum of 27 mm, is the determining factor. In the case of the piles shown in Figure 5.10, this proportion was 80 and 63% for the two piles from the same stand, and 71 and 48% for the other two stands (Cumbre, 1999). The cork planks that are thinner are directed for the production of cork discs to be used for technical stoppers, i.e. for champagne bottles (see Chapter 12).

The thickness of the cork plank depends on environmental conditions and tree genetics, but also on the duration of the cork cycle: the longer the period between cork extractions, the thicker the cork layer that is produced. This is a management variable that is available to the foresters and should be considered for optimising production value taking into account the industrial requirements.

Table 5.1. Characterisation and use of the different types of cork that are obtained from the exploitation of cork oak trees.

Type of cork	Origin and characteristics	Principal use
Virgin cork	First cork extraction of young trees; it has deep fractures and a distorted structure	Trituration for agglomerates; some complete cork tubes are used for decorative purposes
	Increase of cork stripping height from stem and branches, during the juvenile and first age of mature trees; it has deep fractures and a distorted structure	Trituration for agglomerates
	From branches of pruned trees or of felled trees; it is obtained manually with an axe in chunks or as strips with a debarking machine; it contains a large proportion of inner bark and wood; also called winter virgin cork	Trituration for expanded agglomerates
Second cork	Second cork extraction; it contains deep fractures	Trituration for agglomerates
Reproduction cork	From the third cork extraction onwards	Planks for production of stoppers and discs; refuse, pieces and very thin planks for trituration for agglomerates
	Under-aged cork from tree fellings; the cork may be obtained either by stripping of the standing tree or by axe removal after felling; in this case it contains portions of inner bark	Trituration for agglomerates

Attention is also given to the quality of the cork planks regarding the homogeneity of the cork tissue, as it will be discussed in detail in Chapter 7. Excessive porosity from lenticular channels and other discontinuities such as fractures or insect galleries largely reduce the yield and quality of the production of stoppers and may even hinder its use. In this case such planks are considered as refuse and are directed for granulation for production of cork agglomerates.

The raw cork planks have variable dimensions, depending on the tree size and operational conditions at the tree cork stripping. They are approximately rectangular with dimensions usually falling in the range of 1–1.8 m of height and 0.4–0.8 m of width. Measurements done on cork planks for production of stoppers showed average dimensions of 1.19 m height and 0.47 m width, with an average area of 0.37 m^2 (Costa and Pereira, 2004). Similar measurements made in cork planks directed for production of

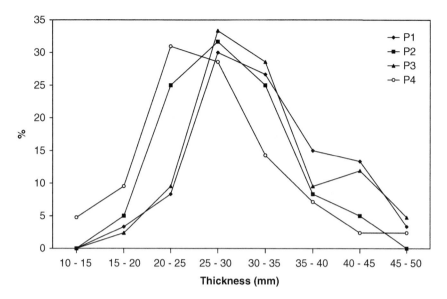

Figure 5.10. Distribution of thickness of 9-year-old cork planks in field piles in the same region. Piles 1 and 2 are from the same cork oak stand, piles 3 and 4 from two other stands in the same region (calculated and drawn from Cumbre, 1999).

discs showed average dimensions of 0.75 m in height and 0.35 m in width with an average area of 0.19 m^2 (Fernandes, 2005). Cork pieces that are under 400 cm^2 are not considered as planks and are piled separately, their utilisation being also the production of granulates.

The classification of cork planks regarding thickness and quality is made after boiling in water, as described in Chapter 11. As raw material from field production the main requirement towards the cork planks is their adequacy for the production of stoppers.

5.4. Conclusions

The supply of cork raw material to the industry is based on the periodic removal of cork from the cork oak stem and branches with an intensity considered compatible to the maintenance of the tree vitality. For the industry, the most important material is the reproduction cork, obtained from the 3rd extraction on, because it is suited for the production of stoppers for the wine bottling.

The cork stripping is an operation characterised by the following:

• strict seasonality, with a 2–3 months window in late spring and early summer, when the phellogen is active;
• an immediate effect on the tree physiology, with stomata closure and development of a traumatic periderm, and in the following 2 years a small reduction in tree radial growth;

- technical skill requirements in order not to wound the functional phloem and cambium of the tree;
- legal enforcements regarding its extent (measured by a debarking coefficient and perimeter) and frequency (measured by the cork duration cycle).

As a consequence of the seasonal production of cork, the continuous year-round supply of raw material to the industrial processing requires the storage of the raw cork planks either in the field or in the mill yard.

References

Antolín, P.G., Díaz, A., González, J.A., Guerra, M., Iglesias, J.M., Maestre, A., Peralta, A., Pianu, B., Del Pozo, J.L., Robledano, L., Rodríguez, M.A., Sánchez, L., Santiago, R., Sanz, J., Vasco, A., 2003. La máquina IPLA para el descorche. IPROCOR, Junta de Extremadura, Merida.

Costa, A., 1990. Metodologias para o ordenamento do montado de sobro. Graduation Thesis, Instituto Superior de Agronomia, Lisboa.

Costa, A., Pereira, H., Oliveira, A., 2004. The effect of cork-stripping damages on diameter growth of *Quercus suber* L. Forestry 77, 1–8.

Costa, A., Pereira, H., 2004. Caracterização e análise de rendimento da operação de traçamento na preparação de pranchas de cortiça para a produção de rolhas. Silva Lusitana 12, 51–66.

Cumbre, M.F., 1999. Avaliação da qualidade tecnológica de pranchas de cortiça por amostragem em pilha. Graduation Thesis, Instituto Superior de Agronomia, Lisboa.

Fernandes, P., 2004. Influência do período de estabilização da cortiça e da cozedura na largura dos anéis de crescimento, no coeficiente de porosidade da cortiça e em algumas características tecnológicas das rolhas de cortiça natural. Graduation Thesis, Instituto Superior de Agronomia, Lisboa.

Fernandes, R., 2005. Estudo da influência do calibre e da qualidade das pranchas de cortiça delgada no rendimento do processo fabril de produção de discos de cortiça natural. Graduation Thesis, Instituto Superior de Agronomia, Lisboa.

Fialho, C., Lopes, F., Pereira, H., 2001. The effect of cork removal on the radial growth and phenology of young cork oak trees. Forest Ecology and Management 141, 251–258.

Gonzalez-Adrados, J.R., Haro, C., 1994. Variación de la humedad de equilibrio del corcho en plancha con la humedad relativa. Modelos de regression no lineal para las isotermas de adsorción. Investigación Agraria, Sistemas y Recursos Forestales 3, 199–209.

Gourlay, I., Pereira, H., 1998. The effect of bark stripping on wood production in cork oak (*Quercus suber* L.) and problems of growth ring definition. In: Pereira, H. (Ed), Cork Oak and Cork. Proceedings of the European Conference on Cork Oak and Cork. Centro de Estudos Florestais, Lisboa, pp. 99–107.

Leal, S., Nunes, E., Pereira, H., 2006. Variation of wood ring width and vessel characteristics in *Quercus suber* trees with climate and cork harvesting. *European Journal* of *Forest Research.,* submitted.

Lupi, V.B., 1938. Descortiçamento dos sobreiros. Contribuição para o estudo da sua influência no desenvolvimento lenhoso. Dissertation, Instituto Superior de Agronomia, Lisboa.

Martín, J., Cabezas, J., Buyolo, T., Patón, D., 2005. The relationship between *Cerambyx* spp. damage and subsequent *Biscogniauxia mediterranum* infection on *Quercus suber* forests. Forest Ecology and Management 216, 166–174.

Natividade, J.V., 1938. Técnica cultural dos sobreirais II. Descortiçamento. Junta Nacional da Cortiça, Lisboa.

Natividade, J.V., 1950. Subericultura. Ministério da Economia, Direcção-Geral dos Serviços Florestais e Aquícolas, Lisboa.

Oliveira, G., 1995. Autoecologia do sobreiro (*Quercus suber* L.) em montados portugueses. Ph. D. Thesis, Faculdade de Ciências da Universidade de Lisboa, Lisboa.

Santos, J.B., 1940. Consequências fisiológicas do descortiçamento. Boletim da Junta Nacional da Cortiça 20, 5–9.

Sousa, M.A., 1997. Sobreiro – Caracterização do crescimento e avaliação da produção de cortiça. Ms. D. Thesis, Instituto Superior de Agronomia, Lisboa.

Velasco, F., Gonzalez-Adrados, J.R., Pertierra, A., 1990 Mapa subericola de Espana. I. Província de Cáceres. Instituto Nacional de Investigacion Agraria, Madrid.

Chapter 6

The sustainable management of cork production

The whole cork chain from the forest to the consumer relies on the regular and sustainable production of cork and therefore on an adequate management of the cork oak forests. This is more so since the total area of cork oak is relatively restricted both in terms of geographical distribution and extent. In relation to other forest systems, the cork oak forests differ very clearly because the product is not the wood but rather a component, the outer bark, that can be extracted during the tree's lifetime and whose growth derives from a traumatic response of the tree, as it has been detailed in Chapter 1. The forest operation of cork extraction is demanding and rather unique within the usual forest exploitation context, as seen in the previous Chapter 5.

The growth of cork is one of the most important variables to the forest owner and should be, therefore, one of the main criteria for management. It is also a determining characteristic of the raw material for the industrial processing, given the fact that in the first option cork is oriented towards the production of wine stoppers and these require a minimum thickness from the cork planks. Cork production yields depend on the growth of cork and on tree growth, as well as on management variables such as the intensity of cork extraction, i.e. the stripped area, and the interval between stripping. The inventory of cork oak forests, therefore, requires additional information in relation to other tree species. Prediction models for cork production have been used for quite sometime while modelling of cork oak growth and cork production in a concept of stand management has been developed only recently.

The sustainability of the cork oak forests and of the cork chain is a matter of general concern, given their role in environmental protection against soil erosion and desertification and in preserving biodiversity, as well as in the social and economic framework of the impacted populations and regions. Cork has also shaped social history and left an important cultural legacy that goes far beyond its areas of growth. It is within this context that cork oak sustainable management planning and certification are gaining recent importance.

6.1. Cork growth and productivity

6.1.1. Cork growth

The growth of cork is the tree physiological output with the highest practical interest for the producer and the industrial consumer of the raw material, since it determines the thickness of the cork plank that is available for processing.

The cork plank includes two types of tissues that are the result of the process of traumatic phellogen formation and of its activity, as it has been described in detail in Chapter 1: (a) the cork layer made up by the successive cork increments added each year by the meristematic activity of the phellogen and (b) the remains of the phloemic tissue that stayed to the outside of the traumatic phellogen, and constitute the so-called cork back. The total thickness of the cork plank (its calliper, as called in the sector's terminology) is, therefore, the sum of the cork back and the cork layer.

The cork back has a thickness of a few millimetres and mean tree values of 2–4 mm have been reported (Taco et al., 2003). This width is related to the depth of regeneration of the phellogen inside the non-functional phloem (see Fig. 1.11) and, therefore, it is usually larger in the second cork than in the subsequent reproduction corks.

In relation to the annual deposition of cork layers, it is well established that the width of the cork rings decreases with phellogen age, more rapidly in the first 2–4 years, and then slower until reaching a rather constant annual growth rate after approximately 5 years (Fig. 1.19). This age-related pattern results into a characteristic assembly of annual cork rings of decreasing width from the cork back to the inner part of the plank, which is very similar in trend in trees from different regions, as exemplified in Figure 6.1 for two sites. Climatic variations or other external factors (i.e. disease or tree damage) may cause occasional deviations to this pattern, as discussed in Chapter 1.

The comparison of total cork growth between different trees has to be made for the same number of growth years and corresponding to the same phellogen age. Since the duration of the cork production cycle is most often 9 years, it is usual to use the cumulative growth of 8 complete years of growth after cork stripping or the average annual growth for this period. Figure 6.2 shows the cumulative growth curves of the corks that were represented in Figure 6.1 for their annual increments. It is interesting to notice that the curves are very similar for all the corks, showing a steady increase along time, more accentuated in the first years and rather linear subsequently. There are some differences in growth intensity among the trees in the same location but the differences are not very large and the coefficients of variation of the mean are usually below 10%. This is certainly lower than the variation usually found for the radial growth of the wood stem of trees where this coefficient is well over 50%.

In a recent European research project, a comparative characterisation of cork was made for different locations in Portugal, Spain, France and Italy (Corkassess, 2001), including an analysis of annual cork growth. In Portugal, 30 different locations were sampled from North to South: the annual growth was on average 3.5 mm, ranging between 2.1 and 4.6 mm. In Spain, the sampling was made in 20 sites in Andalucia and 11 sites in Catalonia: the overall average of annual growth was 3.1 mm, ranging from 2.0 to 4.8 mm. In France, the sampling was made in two sites in the South and two sites in Corsica: the annual growth rate was on average 3.4 mm, ranging between 2.8 and 4.4 mm. In Italy, the sampling was made in two sites in Sardinia: the average was 2.6 mm (2.3 and 2.8 mm).

The data collected on the variation of cork growth show that there are differences between sites that have statistical significance. However, the range is moderate when compared, for instance, with the variability found in wood growth for most species. This is more striking given the differences between some of the locations in relation to edapho-climatic conditions, and stand and management characteristics (e.g. the montado/dehesas of the Portuguese Alentejo and Spanish Andalucia and the hilly cork oak forests of the

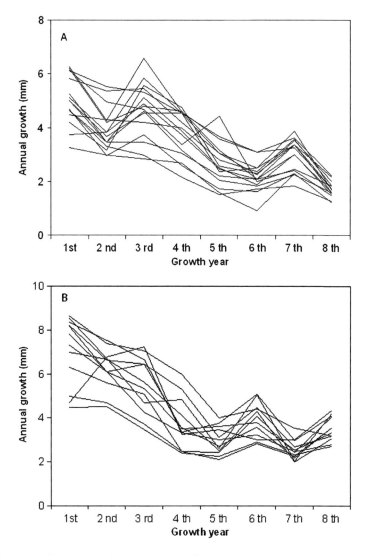

Figure 6.1. Annual increments of cork during the 8 complete years of growth after cork stripping in trees in the same location: (a) site A: Benavente, Portugal; (b) site B: Grândola, Portugal.

Spanish Catalonia and southern France). The fact that cork growth is the result of the activity of a traumatic phellogen as a response to wounding (the cork stripping) in rather a short-term basis (the duration of the cork production cycle) imparts more uniformity.

For the industrial user of the cork planks, it is the thickness of the cork layer that is the determining factor to discriminate the corks that may be used for the production of wine stoppers (the most valuable products) from the other that are too thin. The most usual wine stopper diameter is 24 mm, which would advise a production with about 26 mm clean cork. With a growth during 8 years (9-year cork cycle), this thickness would require a

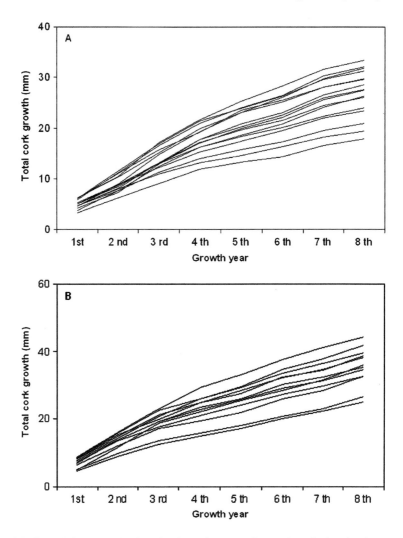

Figure 6.2. Cumulative growth of cork along 8 years after cork stripping in the same trees represented in Figure 6.1: (a) site A: Benavente, Portugal; (b) site B: Grândola, Portugal.

mean cork growth of approximately 3.2 mm year^{-1}. As shown in the examples, this cork growth is attained in the major cork producing regions of Portugal and Spain.

6.1.2. Cork productivity

The production of cork from a tree, on a mass quantitative basis, depends on the surface that is stripped, therefore on the tree dimensions, and on the unit mass of cork produced by a unit surface area. This cork productivity may be defined as a productivity index, in kg m^{-2}, as the mass of the cork plank calculated per area of the belly surface (representing the area of the tree's stripped surface).

The variation of cork productivity for the sampling throughout Europe referred above showed the following values (Corkassess, 2001): in Portugal an overall average of 8.3 kg/m^2, with site means ranging from 7.0 to 11.2 kg m^{-2}; in Spain an overall average of 7.1 kg m^{-2}, with site means ranging from 5.9 to 9.0 kg m^{-2}; in France an overall average of 8.7 kg m^{-2}, with site means ranging from 7.6 to 11.0 kg m^{-2}; and in Italy with site means of 7.5 and 7.6 kg m^{-2}.

However, for comparisons between productivities in different trees and sites it is also important to consider the same duration of the cork cycle. In fact, a calculation made by dividing the productivity by the number of years of the cycle is not directly comparable between cork planks with different number of years since the density of the cork plank is not uniform because the cork rings decrease along the cycle and because there is a constant component of the cork plank, the cork back. In the cork tissue, the radial density gradient along the cork plank is not important and the differences in density between inner cork rings and outer cork rings are not significant. However, the difference is significant in relation to the cork back which has about three times the density of cork: 150–225 kg m^{-3} in the cork layers and 495–608 kg m^{-3} in the cork back (Barbato, 2004).

By grouping corks per duration of the cork cycle, a better insight into the mass productivities calculated on an annual basis (kg m^{-2} year^{-1}) can be obtained. The results are shown in Table 6.1. They indicate a decrease of the annual productivity index with the

Table 6.1. Variation of the annual cork productivity (in mm year^{-1} of complete growth) with the duration of the cork cycle in different countries.

Cork cycle duration (years)		Mean annual growth (mm year^{-1})	Country
Total	Complete growth		
9	8	0.97	Spain
		0.94	France
10	9	0.96	Portugal
		0.88	Spain
11	10	0.87	Portugal
12	11	0.80	Portugal
13	12	0.63	France
		0.62	Italy
14	13	0.63	France
		0.58	Italy
15	14	0.50	Spain
		0.60	Italy
16	15	0.52	Italy
27	26	0.44	France

increase of the duration of the cork cycle, from about 0.94 kg m^{-2} year^{-1} for cycles with 8–9 years to 0.54 kg m^{-2} year^{-1} for cycles with 14–15 years.

The productivity index related to the weight of cork that may be obtained on extraction by a unit area of stripped surface has a very practical interest to the forest owner, since the transaction of cork is made by weight. Since estimates on the total stripped surface can be made for a stand if an inventory of trees is available (diameter and cork extraction height), the use of the productivity index for the stand would allow the forest owner to estimate the weight of his total cork production. The basis of calculation has to have a fixed cork cycle, and it is proposed to use the more usual 9-year-cycle as the calculation basis. Therefore, the cork productivity index would be the cork produced in a 9-year-cycle per unit area of stripped surface. A stand with a higher productivity would in principle bring higher revenue if the quality of production is not very unfavourable. It is also clear that it is more interesting from an economical point of view to have shorter cork production cycles once that a suitable cork plank thickness is attained.

Therefore, the forest manager has interest in knowing and crossing two types of information regarding the stand: its productivity index, which allows estimating production quantity, and the cork growth index, which allows estimating quality in terms of plank thickness and a value component.

6.2. Cork oak and cork growth modelling

6.2.1. Prediction of cork production

The prediction of cork production from cork oak trees and stands has been the object of several modelling exercises since the early 1950s (e.g. Natividade, 1950; Guerreiro, 1951) to the present (e.g. Fonseca and Parresol, 2001; Ribeiro and Tomé, 2002; Vasquez and Pereira, 2005) for cork production in different regions. The cork weight prediction models are important management tools both for the forest manager as well as for the industrial supply. The variables that directly determine the yield of cork are: (a) on the individual tree, the total surface that is stripped (m^2) and the unit surface yield of cork (kg m^{-2}); (b) on a stand basis, the sum of the individual trees, or the tree density and diameter distribution, or the basal area.

The cork weight that is predicted as the response variable in the models may correspond to cork with different contents of water: (a) fresh cork weight measured immediately after extraction (as considered in most models); (b) air-dry cork weight, after a determined period of air-drying; and (c) oven-dry cork weight. The modelling using oven-dry cork is more correct and it is now the practice in the more recent works (Vasquez and Pereira, 2005), since it avoids the variability of moisture contents at extraction and during air-drying. The calculations to any desired moisture content may be made subsequently.

When considering the prediction of cork production at the individual tree level, the independent variables in all models are based on the measurement of variables related to tree size and to the intensity of the stripping, and the main differences between models refer to their measuring complexity. The simplest is to use only the perimeter over cork at breast height (1.3 m height) and the most complex is to use a measured stripped surface on the stem and branches of the cork oaks. However, the usual approach is to make an approximation to the exact stripped surface by using the perimeter at breast height multiplied by the maximum stripping height, as a compromise between simplicity and approximation to the real stripped surface (Ferreira et al., 1986; Ferreira and Oliveira, 1991). The accuracy of the approximation varies with the type of cork stripping: when the stripping is made only in the stem the accuracy is higher, but when it includes stem and branches there is more variation namely related to the number of branches that are stripped. For practical reasons, the perimeter that is usually measured is the perimeter over cork instead of the perimeter under cork.

Table 6.2 shows some of the equations that are available for prediction of cork production. Most are linear equations of the type Cork weight $= a + b$ (perimeter $\times$ stripping height), that were developed based on measurements in individual trees from specific regions in Portugal and Spain. In several cases stratification by stripping type (stem only, or stem + branches), by stripping intensity, or by site quality is made and individual equations were developed. In addition to the tree measurements, data collected directly from a cork sample such as cork thickness, surface density and density have also been incorporated into the more recently developed models with an improvement of the fitting and validation criteria (Vasquez and Pereira, 2005).

The feasibility of using cork production models depends strongly on the complexity and cost of the tree measurements that constitute their input information. Therefore, the objectives underlying the model development and application should be the guidelines, namely through the distinction between research models and management models, as proposed by Vasquez and Pereira (2005).

The estimates of cork production at a stand level have been less studied than those at the tree level and Table 6.3 summarises the corresponding equations. The dependent variable is in most cases the basal area of the stand, defined as the total cross-sectional area

Table 6.2. Examples of equations proposed for prediction of cork production.

Equation		Reference
$W_g = 72.5C_{oc} + 3.2H + 0.7\,D_c - 54.9$		Natividade (1950)
$W_g = 70.5C_{oc} + 2.7C_{oc}^2 + 43.7$		
$W_g = 76.9C_{oc} - 47.2$	(trees with NB = 2 and SC < 2.5)	
$W_g = 36.76C_{oc} + 7.4H + 0.09CC - 32.2$	(only stem stripping)	Guerreiro (1951)
$W_g = 46.06C_{oc} + 12.2H + 0.46CC - 65.6$	(stem + branch stripping)	
$W_g = 50.89C_{oc} + 9.7H + 0.07CC - 46.47$	CD < 3	Alves (1958)
$W_g = 49.52C_{oc} + 9.9H + 47.91$	CD < 3	

Notes: W = cork weight, in kg (W_g, green weight; W_{ad}, air-dry weight; W_{od}, oven-dry weight).
C = circumference at breast height, in m (C_{oc}, over cork circumference; C_{uc}, under cork circumference).
H = stripping height of stem, in m.
CC = cork plank calliper, in cm.
NB = number of stripped branches.
SC = stripping coefficient.
D_c = crown diameter, in m.

Table 6.3. Examples of equations proposed for prediction of cork production at the stand level.

Equation		Reference
$W_g = 0.93$ CCP $- 4.81$	CCP < 2000 m² ha⁻¹	Madureira (1981)
$W_g = 36.76C_{oc} + 7.4H + 0.09$CC $- 32.2$	(only stem stripping)	Montero and Grau (1989)
$W_g = 46.06C_{oc} + 12.2H + 0.46$CC $- 65.6$	(stem + branch stripping)	
$W_g = 50.89C_{oc} + 9.7H + 0.07$CC $- 46.47$	CD < 3	Alves (1958)
$W_g = 49.52C_{oc} + 9.9H - 47.91$	CD < 3	

Notes: W = cork weight, in kg (W_g, green weight).
ABAH = basal area, in m² ha⁻¹.
C = perimeter at breast height, in m (C_{oc}, over cork circumference).
H = stripping height of stem, in m.
CCP = crown cover projection, in m² ha⁻¹.
CC = cork plank calliper, in cm.

at breast height level of the stand per unit area (m² ha⁻¹) but the crown projection area or the product of perimeter over cork by stripping height have also been used.

6.2.2. Growth and yield modelling

The first management oriented growth and yield model for the cork oak, the SUBER model, was developed only recently (Tomé et al., 1998, 1999). The model was constructed using existing data, published results and empirical knowledge on the cork oak system, with the objective of designing a methodology suitable to simulate the development of cork oak stands for management purposes. Due to the particular characteristics of this forest, a growth and yield model for cork oak stands, able to predict also cork yield and quality, is more complex than a model for estimating wood production. The SUBER model is an individual tree model, which means that stand growth is predicted by the joint simulation of each individual tree growth.

In the SUBER model, cork oak growth is classified into three main phases: (a) a regeneration stage, since the plant emerges or is planted until it attains a diameter at 1.3 m of height of 5 cm or a total height of 3 m; (b) a juvenile stage from the end of the regeneration stage until the first cork extraction of the virgin cork; and (c) the mature stage from the first cork stripping onwards.

The tree growth in the mature stage includes the growth of under bark diameter, which is based on wood growth curves that were derived from stem analysis (Fig. 6.3) as well as height–diameter relationships (Sanchez-Gonzalez et al., 2005). The diameter growth is slow but trees can attain very large dimensions as a consequence of the large rotation period. The cork growth modelling uses a two-step approach: first, the accumulated thickness of cork in complete years as a function of cork age and of the cork growth index; then the calculation of this value as total cork thickness by inclusion of the two "half-years" and the cork back (see Fig. 5.5) and its subsequent transformation into weight predictions.

The stand development depends on site characteristics and on forest management. Table 6.4 gives an example of the predicted development of a cork oak stand in a site

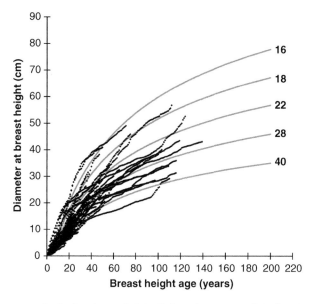

Figure 6.3. Diameter under bark at breast height (1.3 m above ground) in function of age at breast height for different site quality index (age required to attain the legal size for the extraction of virgin cork, here varying between 16 and 40 years).

Table 6.4. Yield table for an even-aged stand of mean site productivity managed to maintain a spacing factor of 1.2 and using the legal cork extraction limits.

Age (years)	Stand characteristics after thinning				Cork weight (kg ha^{-1})	
	N (ha^{-1})	G (m^2 ha^{-1})	d_g (cm)	d_{crown} (m)	Virgin cork	Mature cork
27	250	7.1	119.0	5.0	833	1153
45	114	12.0	36.7	7.8	639	3834
63	76	14.6	49.5	9.5	24	4656
81	59	16.3	59.4	10.8	43	4853
99	50	17.8	67.3	11.7	60	4994
117	44	19.0	74.1	12.5	63	5171
135	40	19.9	79.6	13.1	41	5294
153	37	20.5	84.0	13.6	39	5394

Source: Pereira and Tomé (2004).
Notes: N, number of trees per hectare; G, basal area; d_g, quadratic mean diameter; d_{crown}, crown diameter.

of mean productivity, managed as an even-aged cork stand, maintaining a spacing factor close to 1.2 and following the maximum legal debarking coefficients (Pereira and Tomé, 2004). The number of trees per hectare is quite low in comparison with other species that are planted for wood production, leading to low values of stand basal area. The production of virgin cork is highest in the younger trees until about 50 years of age, corresponding to the cork removed in the first debarking and to the increment in the stripping height, which is significant in these ages of higher stem radial growth. The amount of virgin cork decreases subsequently and it is most probable that increments of debarking height are not made in the mature corks over 60–80 years of age; this means that in practice virgin cork will not be taken after this age. The amount of reproduction cork increases with age rapidly until about 60 years of age and more slowly afterwards. In the stand conditions of this example, it may be considered that production of reproduction cork at each cork extraction period remains rather constant at about 5 tons per hectare.

Figure 6.4 shows examples of cork yield and stand characteristics in four different stands with the same percent crown cover of approximately 58%, using an even-aged silviculture at stand age of 27 and 153 years, and a continuous cover silviculture with two different structures of diameter distribution. The production in even-aged stands is very much dependent on the tree age, as already discussed based on the data of Table 6.4; also the regeneration of the stand has to be considered at the end of the rotation and this following phase will be characterised by a significant period without cork production. In the case of uneven-aged stands as hypothesised in Figure 6.4, a more uniform production of reproduction cork is attained, although under that obtained in the final phase of even-aged stands; however, the total value calculated for the whole rotation will be similar, with the benefit of a better situation regarding the regeneration of the stand. This is an important concern related to the sustainability of these systems, as discussed further on, and advises to consider the concept of applying continuous crown silviculture to the cork oak stands.

In any case, the production of a stand is very much dependent on the large diameter trees that are present. Therefore, preservation of the mature trees in good health and with a low mortality is a requisite for maintaining high levels of production. This stresses the importance of carefully performing the cork extraction due to the potential risks of endangering the tree, as discussed in Chapter 5.

6.2.3. Field sampling for inventory and production estimates

Forest inventory in cork oak stands requires additional information in relation to the biometric data of diameter and height that are usually gathered for other species. The data regarding the exploitation variables, namely the stripping height and information on the stripped branches, are required as well as data on the cork status regarding the number of years of growth (date of last cork extraction) and the measurement of cork thickness.

Field sampling has also been in use to estimate cork production in terms of cork plank thickness and quality. The idea is to have an adequate number of cork samples extracted from the trees that will be classified in terms of total thickness and of quality. The criteria underlying this classification, which is related to the cork value as an industrial raw material, are detailed further on in Chapter 7. The suitability of the cork plank for the production of wine stoppers nowadays establishes its commercial quality and is the production objective for the forest manager. Therefore, a good estimate of the quality characteristics of his production allows him to better evaluate it in the price negotiation phase.

An underlying difficulty is the variability shown by cork of different trees that requires an adequate intensity of sampling. The first samplings started to be made in the 1990s and were carried out using a zigzag path covering the stand and sampling every

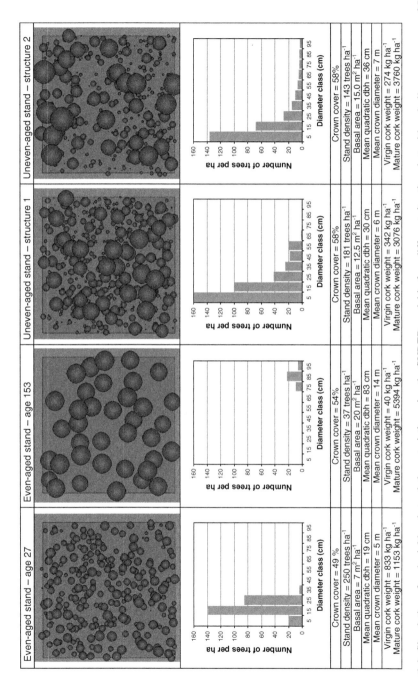

Figure 6.4. Simulation of cork yield and stand characteristics using the SUBER model in four different situations but with a similar crown cover of 58%: an even-aged stand at the beginning of cork exploitation (age 27); an even-aged stand at the final stage of cork exploitation (age 153); uneven-aged stand with maximum tree diameter restricted to 67.5 cm; and even-aged stand at the final stage of cork exploitation (age 153); uneven-aged stand with maximum tree diameter restricted to 97.5 cm (Pereira and Tomé, 2004).

tree at a fixed distance. However, a cluster sampling with a fixed number of trees proved less biased and with smaller standard errors, while being operationally more convenient (Paulo et al., 2005).

6.3. The sustainability of cork oak forests

6.3.1. The multifunctionality of cork oak systems

The cork oak forests have a long tradition of multifunctionality, namely in the sparse systems in the Portuguese montado and the Spanish dehesa. They are typical agro-forestry systems where the forest exploitation of the tree is associated with pasture and agriculture as undercover, usually with a rotation scheme that includes fallows. The relative weight of each component – forest, agriculture and animal production – in the overall economic return of the system have changed along time. Creating a habitat for hunting has once been a major function of the past oak lands, nowadays revived in the leisure hunting of modern times. Agriculture was responsible for the opening up of vast areas of forest and cereal cropping in the cork oak stands was carried out extensively during the last century. Cattle, either feeding on natural vegetation and on acorns or on improved pasture, has been and still is one of the important productions supported by the cork oak stands. Other uses of cork oak stands are based on their rich biodiversity: beekeeping, mushroom picking and aromatic plants.

Considerable attention is also given to the environment-related functions of the cork oak lands in terms of protection against soil erosion by wind and water and desertification, which is a major concern in the Mediterranean basin, as well as of the support given by the trees and shrubs to a large animal biodiversity. Recently, the WWF environmental organisation has launched a campaign in support of the cork oak forests, considered a unique reservoir of biodiversity and environmental benefits, with a social and cultural role that might be endangered by a market decrease of wine stoppers (WWF, 2006).

This calls attention to the present role of the tree exploitation as a cork producer. In fact, cork is now the major economic asset of the cork oak lands which makes them sensitive to market changes, namely to alternative bottle closures. In the past, the cork oaks have been used largely as a timber species, and the cork oak wood was highly prized because of its high density, hardness, resistance to impact and friction and durability and applied for demanding uses such as shipbuilding. In fact, cork oak wood already crossed the world in the Portuguese *caravelas* of the 16th century during the period of the Discoveries.

Recently, the European research project Suberwood investigated the quality of cork oak wood and the concept of a management integrating the wood and cork productions (www.Suberwood.com). At present, the trees are felled at the end of the rotation or when dead or diseased, and in these conditions the wood is used only as fuel (Fig. 6.5). However, sound cork oaks have a dense wood with a strong aesthetical value due to its anatomical features that is suited for high value wood products (Knapic et al., 2006; Leal et al., 2006). It can, therefore, be envisaged to manage the cork oak stands taking into

Figure 6.5. The cork oaks that are harvested are sawn to be used as a fuel wood.

account that the stem may be directed towards production of quality timber, leading to a silviculture that combines both options.

6.3.2. The sustainable management of cork oak forests

The main concern on the sustainability of the cork oak forests regards the montado and dehesa systems due to lack or insufficient regeneration (Pulido and Díaz, 2003). In these sparse systems, there is very little natural regeneration and the existing stands have a predominantly mature and over-aged structure. Although in spring time an abundant number of plantlets can usually be observed around the mature cork oak trees from the sprouting of the acorns that scattered in the area, most will disappear by the next year. The young seedlings need shade and protection to survive and this is usually not the case in such unsheltered sunbathed and hot conditions. Quite a different situation is found in the dense cork oak woodlots where natural regeneration is vigorous under the protection of the dense tree cover and the under storey shrub vegetation. In these stands, the number of young plants is high even if growth is slow due to competition. However, the area of cork oak forests in these conditions is minor compared to the extension of the montado and dehesa system.

For the continuity of the cork oak stands it is therefore necessary to consider their artificial regeneration by either seeding or plantation of nursery grown seedlings (Montero and Canellas, 1999). A large effort of afforestation with cork oaks has been carried out in Portugal and Spain, during the last 20 years, fostered with incentives given by the European Common Agricultural Policy. In these afforestation projects, a high density of over 600 plants/ha is applied after soil preparation but in several cases the survival still is not satisfactory, with the first 5–8 years being determining for the tree successful establishment.

Another factor that jeopardises the natural renovation of the cork oak stands directly relates to its multifunctional character, namely in what refers to animal browsing. Cattle will finish destroying the young plants that have resisted the harsh summer conditions, and therefore animal production will have to be banned from the areas to regenerate until the young trees are over 10 years.

A sustainable forest management is a European and a world commitment. The role of forests in the planet has been recognised and put forward in successive international res-olutions, specifically including their multiple roles (societal, economical, environmental and cultural) which contribute to the sustainable development of societies, namely in the rural areas, to the production of renewable goods and to environmental protection. The goal of a sustainable forest management is supported by a framework of criteria and asso-ciated indicators that constitute the guidelines for action. Criteria consist in aspects rele-vant to forest management and through which the performance of management is assessed. The indicators are measures or parameters of quantitative, qualitative or descrip-tive nature that with a periodical measurement or control will make evident the change of a criterion. The criteria and indicators that are presented in Table 6.5 follow the guidelines put forward in the pan-European agreements for sustainable forest management.

The sustainable management of cork oak forests is for enhanced reasons a matter of importance due to the environmentally and social sensitive regions where they spread. This concept starts to breed and develop at the public and stakeholders level. The certifi-cation of sustainable forest management for cork oak stands has just started, and a few cases have already completed the process.

Table 6.5. Summary of the criteria and of their characterising indicators that maintenance and enhancement give the framework for a sustainable forest management.

Criteria	Indicators
1. Resources and carbon cycles	Forest area/growing stock/age structure and/or diameter distribution/carbon stock
2. Ecosystem health and vitality	Deposition of air pollutants/soil condition/ defoliation/forest damage
3. Productive functions	Increment and fellings/roundwood/non-wood goods/services/forests under management plans
4. Biological diversity	Tree species composition/regeneration/ naturalness/introduced tree species/deadwood/ genetic resources/landscape pattern/ threatened species/protected forests
5. Protective functions	Soil, water and other ecosystem functions/ infrastructure and managed natural resources
6. Other socio-economic functions	Forest holdings/contribution to GDP/net revenue/expenditures for services/forest work force/occupational safety and health/wood consumption/trade in wood/energy from wood/ accessibility for recreation/cultural and spiritual values

6.4. Conclusions

The sustainable production of cork is the basis for the whole cork chain from the forest to the consumer. The management of the cork oak forests differs in several aspects from other forests as

- the main product is cork, and not wood, and its exploitation is carried out during the trees' life;
- the age of productive trees is frequently high and the maintenance of mature trees with good vitality along the production is important;
- most of the present cork oak systems are agro-forestry sparse forests but their past multifunctionality has shrunk to cork production and animal browsing;
- natural regeneration is low and insufficient to guarantee the long-term sustainability of the forest.

The role of cork oaks in environmental protection against soil erosion and desertification and in preserving biodiversity, as well as in the social and economic framework of the countries involved has triggered them into the limelight of environmentalist concerns and considerable efforts are put into new cork oak afforestation in southern Europe using seeding and planting, as well as in promoting a sustainability oriented management.

The management of the cork oak stands for cork production has to take into account not only the requisites for sustainable trees and ecosystem but also the industrial objectives and requirements for the product, in terms of quantity and of quality. Modelling of tree growth and cork production is therefore an important tool and knowledge on production-related variables such as the cork growth index and the productivity, and on management variables such as intensity of cork stripping and cork cycle duration allows better estimates of production.

References

Alves, A.M., 1958. Tabelas de previsão de peso de cortiça para o sobreiro nos xistos do carbónico. Boletim da Junta Nacional da Cortiça 237, 233–243.

Barbato, F., 2004. Variação radial da densidade e da porosidade em pranchas de cortiça. Graduation thesis, Instituto Superior de Agronomia, Lisboa.

Corkassess, 2001. Field assessment and modelling of cork production and quality. Final report. Contract FAIR.CT97.1438. Brussels: European Commission, Research Directorate General, Life Sciences, Agriculture, Agro-industry, Fisheries and Forestry.

Ferreira, M.C., Oliveira, A.M.C., 1991. Modelling cork oak production in Portugal. Agroforestry Systems 16, 41–54.

Ferreira, M.C., Tomé, M., Oliveira, A.M.C., Alves, A.M., 1986. Selecção de modelos de previsão de peso de cortiça. In: 1° Encontro sobre os montados de sobro e azinho. Évora: Sociedade Portuguesa de Ciências Florestais, pp. 65–83.

Fonseca, T., Parresol, B.R., 2001. A new model for cork weight estimation in Northern Portugal with methodology for construction of confidence intervals. Forest Ecology and Management 152, 131–139.

Guerreiro, M.G., 1951. Previsão do peso da cortiça explorável de um montado. Publicações da Direcção Geral dos Serviços Florestais e Aquicola XVIII(I), 57–83.

Knapic, S., Louzada, J.L., Leal, S., Pereira, H., 2006. Radial variation of wood density components and ring width in cork oak trees. Annals of Forest Science, in press.

Leal, S., Sousa, V.B., Pereira, H., 2006. Within and between-tree variation in the biometry of wood rays and fibres in cork oak (*Quercus suber* L.). Wood Science and Technology 40, 585–597.

Madureira, M.A.B.B.V., 1981. Produção de cortiça em montados. Companhia as Lezírias E.P. Ensaio prelimi-
nar. Graduation thesis, Instituto Superior de Agronomia, Lisboa.

Montero, G., Canellas, I., 1999. Manual de reforestación y cultivo de alcornoque. Madrid: Ministério de
Agricultura, Pesca y Alimentación – Instituto Nacional de Investigación y Tecnología Agraria y Alimentaria.

Montero, G., Grau, J.M., 1989. Producción de un alcornocal en Santa Coloma de Farnés, Gerona. Sciencia
Gerundensis 1, 131–139.

Natividade, J.V., 1950. Subericultura. Lisboa: Ministério da Economia, Direcção-Geral dos Serviços Florestais
e Aquícolas.

Paulo, M.J., Tomé, M., Otten, A., Stein, A., 2005. Comparison of three sampling methods in the characterization of cork oak stands for management purposes. Canadian Journal of Forest Research 35, 2295–2303.

Pereira, H., Tomé, M., 2004. Cork oak. In: Burley, J. (Ed), Encyclopedia of Forest Sciences. Oxford: Elsevier
Ltd., pp. 613–620.

Pulido, F.J., Díaz, M., 2003. Dinámica de la regeneración natural del arbolado de encina y alcornoque. In:
Pulido, J., Campos, P., Montero, G. (Eds), La gestión forestal de las dehesas. Hitoria, ecologgá, selvicultura
y economía. Mérida: Instituto del Corcho, la Madera y el Carbón, pp. 39–62.

Ribeiro, F., Tomé, M., 2002. Cork weight prediction at tree level. Forest Ecology and Management 171,
231–241.

Sanchez-Gonzalez, M., Tomé, M., Montero, G., 2005. Modelling height and diameter growth of dominant cork
oak trees in Spain. Annals of Forest Science 62, 633–643.

Taco, C., Lopes, F., Pereira, H., 2003. La variation dans l'arbre de l'épaisseur du liège et du dos des planches
pour des chênes-liège en pleine production. Anais do Instituto Superior de Agronomia 49, 209–221.

Tomé, M., Coelho, M.B., Lopes, F., Pereira, H., 1998. Modelo de produção para o montado de sobro em
Portugal. In: Pereira, H. (Ed), Cork Oak and Cork/Sobreiro e cortiça. Centro de Estudos Florestais, Lisboa,
pp. 22–46.

Tomé, M., Coelho, M.B., Pereira, H., Lopes, F., 1999. A management oriented growth and yield model for cork
oak and cork oak stands in Portugal. In: Amaro, A., Tomé, M. (Eds.), Empirical and Process-Based Models
for Forest Tree and Stand Growth Simulation, Edições Salamandra, Lisboa, pp. 189–271.

Vasquez, J., Pereira, H., 2005. Distance dependent and distance independent models to estimate tree cork weight
in Portugal. Forest Ecology and Management 213, 117–132.

WWF., 2006. Cork skrewed? Environmental and economic impacts of the cork stoppers market. Report. Rome:
WWF/MEDPO.

Colour plates

Formation of Cork

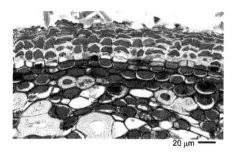

Figure 1.6. (c). A young periderm with some phellem layers; the result of anticlinal division of the phellogen initial can be observed.

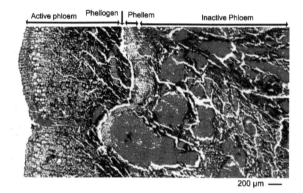

Figure 1.11. Formation of a traumatic periderm in the phloem of the cork oak showing the circumferential path of the phellogen and the first layers of cork cells.

The Cork Oak

Figure 4.4. (a). Typical cork oak architecture in the managed agro-forestry systems.

Figure 4.3. (b). Dense forest stand.

Figure 1.8. Photograph of a young cork oak stem showing the areas still covered with the epidermis (smooth appearance) and longitudinally running fractures that expose the cork tissue underneath.

Figure 4.5. (c) Acorns.

Sustainability

Figure 4.8. (a) Stem sprouting after 2 months after fire.

Figure 4.15. Cork oak seedlings grown at a nursery to be used in plantations.

Cork Extraction

Figure 5.1. (c). The axes' arm is used as a lever to help separate the cork plank.

Figure 5.1. (e). General view of a cork stripping operation.

Figure 5.7. (b). Beginning of the construction of one pile showing the selection of the cork planks for an orderly arrangement.

Cork and Industry

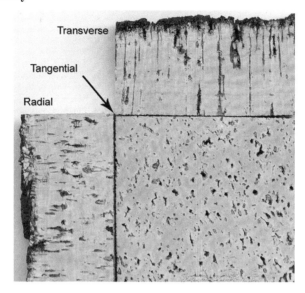

Figure 7.1. Lenticular channels of cork observed in (a) transverse section; (b) radial section; and (c) tangential section.

Figure 14.10. General view of one hall of an industrial unit producing cork stoppers where the degree of automation is very high.

Part III

Cork properties

Chapter 7
Macroscopic appearance and quality

Cork has a macroscopic aspect that singularises it from other lignocellulosic materials such as wood, although the plant origin can usually be recognised in the raw cork planks because of the presence of successive annual layers (cork rings) in a way similar to the rings found in the stem wood. The pure cork mass itself appears as homogeneous and without any differentiation of cell types. At the macroscopic level the cellular structure is not observed due to the small dimensions (under 50 μm) of the cells (Chapter 2).

The most conspicuous characteristic of cork is the natural occurrence of lenticels that are present as lenticular channels crossing the cork layers from the outside to the inner tissue of the phellogen and are filled with a darker coloured, non-suberified material. These features are in fact the distinctive macroscopic mark of cork.

In addition to their physiological role in the tree, the lenticular channels also have a practical importance in the cork sector since they constitute the major factor to establish the quality grading of the raw cork and cork products. The use of optical vision systems coupled with image analysis have allowed to characterise in detail this so-called porosity of cork. As a natural material, cork may include occasional defects of internal or external biological origin or resulting from mechanical failures. These are also factors that determine the adequacy of the cork planks to specific uses, e.g. to the production of wine stoppers. The quality classification of the raw material, as well as of the products, namely of stoppers and discs, as detailed in Chapter 12, is one important aspect related with the economy of the sector, since there is a large price difference between different cork quality grades.

In this chapter the macroscopic features that characterise the cork planks are illustrated, and the porosity revealed by image analysis of the cork surfaces is detailed because of its importance in the quality grading of cork. The separation of cork planks by thickness and by quality, as used in the industry, is presented and the underlying rationale is discussed.

7.1. The appearance of cork

7.1.1. Lenticular channels

The lenticels that are formed in certain areas of the phellogen (the lenticular phellogen, see Chapter 1) are natural and biologically necessary features of the suberised periderm that allow gas exchange between the environment and the underneath living tissues. The activity of the lenticular phellogen continues year after year, in parallel with that of the surrounding phellogen. The lenticels therefore take up a radially aligned tubular form, stretching from the outside to the phellogen.

Figure 7.1 shows the aspect of lenticular channels as seen in the three main sections of the cork plank. They are commonly referred to as pores and their number and dimensions characterise the so-called porosity of cork.

When observed in transverse sections, the lenticular channels appear as more or less thin lines crossing the cork plank radially. Often only portions of one channel are observed resulting from the fact that the observation plane does not coincide with the lenticular channel development axis in all its length. The radial sections of cork are similar to the transverse sections and the lenticular channels are sectioned in the same manner. When there are important tangential tensions due to tree growth, the direction of the lenticular channels is distorted, as seen clearly in virgin cork (Fig. 7.2). In tangential surfaces, the lenticular channels are sectioned perpendicularly to their axis and appear with an approximate circular form.

Figure 7.3 shows schematically the three-dimensional development of the lenticular channels through cork and their sectioning in three orthogonal planes.

The tangential section of the cork clearly differs from the transverse and radial sections. These seem equal on a first impression. However, most lenticular channels are vertically elongated and their dimensions in the tangential and axial directions are not the same. As a result the pores are larger in the radial section than in the transverse section. The dimensional characteristics of the lenticular channels are detailed later on using image analysis techniques.

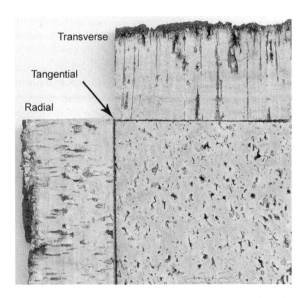

Figure 7.1. Lenticular channels of cork observed in (a) transverse section; (b) radial section; and (c) tangential section (see Colour Plate Section).

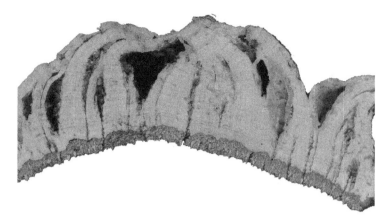

Figure 7.2. Cross-section of virgin cork showing the radial distortion of the lenticular channels due to the tangential-growth stress.

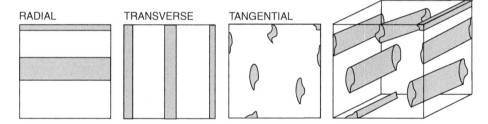

Figure 7.3. Schematic representation of lenticular channels within a cork sample and their sections in three orthogonal planes (tangential, radial and transverse).

The lenticular channels are conspicuous because the cellular tissue that fills the lenticels clearly differs from the surrounding phellemic tissue at anatomical and chemical levels, showing a dark brown colour that contrasts with the lighter shaded cork cells. The filling material somewhat resembles brown earth, and corks that have very large lenticular channels are sometimes called "earthy" corks.

The anatomical characteristics of the lenticular filling tissue are shown in Figure 7.4. The cells have a loose structural arrangement, either separating as individual cells or, more often, as a small multi-cellular particle aggregate. The lenticular cells are rigid and have comparatively thicker walls. Fractures and intercellular voids occur to a large extent. Therefore the lenticular channels make an easy pathway for the air passage, as well as a preferential impregnation route, namely for liquid flow or microbial penetration. The lenticular channels often have lignified and thick-walled cells at their borders as well as collapsed cells. The region bordering the lenticular channels therefore has a higher density than the surrounding material.

Chemically the lenticular cells are not suberified and the cell wall is lignocellulosic, resembling the wood cell wall. Table 7.1 shows the summative chemical composition of

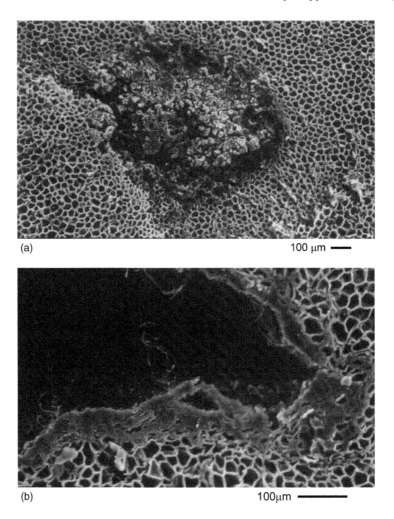

Figure 7.4. SEM observations of lenticular channels and of the lenticular filling tissue. (a) Multi-cellular particle aggregate of the lenticular filling tissue; (b) Thick-walled and collapsed cells bordering one lenticular channel.

the lenticular filling tissue and the cork tissue, as well as the composition of their cell wall structural components, on an extractive-free basis. The lenticular cells contain a very high amount of extractives, mainly polar extractives, that make up about one-third of the material. Polysaccharides amount to 21% of the cell wall material (on an extractive-free basis) and lignin to 47% (compared with 15 and 29%, respectively in cork).

The chemical distinction between cork and lenticular tissues is also clearly seen by comparing their FTIR-spectra (Fig. 7.5 for the filling tissue and Fig. 3.24 for the pure cork tissue). While the cork spectrum is dominated by the absorptions distinctive of suberin corresponding to the aliphatic C–H bonds with two peaks at 2927 and 2854 cm^{-1} and to C=O ester bonds at 1742 cm^{-1}, and the bands at 1087 and 1035 cm^{-1} arising from

Table 7.1. Chemical composition of the lenticular filling tissue and of the pure cork tissue on an initial oven-dry material and on an extractive-free basis.

Component	Lenticular filling tissue		Cork	
	% total	% extr. free	% total	% extr. free
Ash	3.7		2.2	
Extractives				
Total	31.8		12.8	
Dichloromethane	11.2		6.2	
Ethanol	11.9		2.1	
Water	8.7		4.4	
Suberin	5.6	8.2	39.1	44.8
Lignin				
Total	32.2	47.2	25.0	28.7
Klason lignin	30.1	44.1	24.3	27.9
Soluble lignin	2.1	3.1	0.7	0.8
Polysaccharides	14.6	21.4	12.8	14.7

Monosaccharide composition	% polysaccharides	
Glucose	57.7	57.9
Xylose	22.7	18.3
Arabinose	7.3	10.2
Mannose	3.4	3.9
Galactose	7.4	8.4
Rhamnose	1.5	1.3

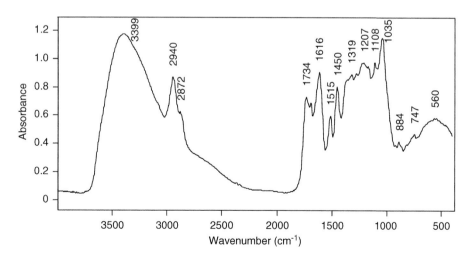

Figure 7.5. Fourier transform infrared spectra of lenticular filling tissue (compare with Fig. 3.24 or pure cork tissue).

polysaccharides are of smaller intensity, in the case of the lenticular material the FTIR spectrum is characterised by a dominating band corresponding to the O–H absorption with a maximum at 3401 cm^{-1} and by the peaks corresponding to polysaccharides at 1110 and 1035 cm^{-1}, with a small intensity C=O peak at 1737 cm^{-1} from pentosans. The spectrum of lenticular material only shows small peaks at 2925 and 2854 cm^{-1} corresponding to absorption due to C–H bonds.

7.1.2. Colour

The colour of cork can be described as light brown or honey brown. It is not uniform though, and the latecork regions are darker than the earlycork regions due to a higher material density. Therefore, a stripped succession of darker and lighter regions appears in the transverse and radial sections of cork, while the tangential section sometimes shows a spotted pattern when the observation plane cuts through earlycork and latecork regions.

Some differences occur between corks regarding their colour shade, with cork planks with thin cork rings usually darker than those with wide cork rings. Figure 7.6 shows the reflectance spectra taken between 360 and 740 nm for the tangential sections of two cork planks. The colour may be characterised by the corresponding CIELAB parameters: L^* (lightness) that varies between 0 (pure black) and 100 (pure white), a^* between $+a^*$ (red) and $-a^*$ (green) and b^* between $+b^*$ (yellow) and $-b^*$ (blue); in absolute figures a^* and b^* vary between 0 and 60. Cork shows the following average values: L^* is 55.7, a^* is 12.5 and b^* is 24.9.

The lenticular filling tissue has colour characteristics that are different from the pure cork tissue. Macroscopically it is seen with dark brown, sometimes reddish shading,

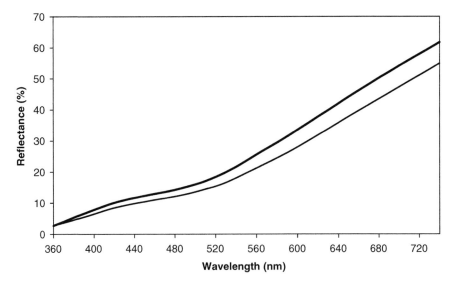

Figure 7.6. Reflectance spectra of tangential section of cork.

which gives it its earthlike look. It is this darker colour of the lenticular channels that allows them to be singularised from the cork background when using image scanning and analysis, as discussed later.

Cork planks with a higher content of lenticular channels will be overall darker than those with less. This can be observed in Figure 7.6 since the two corks correspond to differences in the amount of lenticular channels.

7.1.3. The cork back

The back of the cork planks is made up of the phloemic lignified tissues that have stayed to the outside of the differentiated traumatic phellogen (Chapter 1). This tissue contains sclerified nodules and phloemic rays, as well as fibre aggregates. Subject to the environment, it dries out, fractures and weathers into a dark brown colour due to the oxidation of the phenolic and tannin compounds.

The cork back is a layer with about 2–4 mm width representing on average 9.6% of the total thickness of the cork planks. It is however variable between trees, i.e. from about 5 to 13% (Taco et al., 2003).

The macroscopic pattern of the cork back may be classified as (a) reticulate, when there are horizontal and vertical fractures, or (b) fibrous, when the vertical fracture lines predominate (Fig. 7.7). In either case the fractures may vary in depth, from superficial lines to profound furrows that go down to the cork tissue, and in texture from regular to irregular.

Chemically the cork back is lignocellulosic with a chemical composition close to wooden materials. The FTIR spectrum of the cork back is dominated by the band at 1047 cm^{-1} from C–O stretching in polysaccharides and by the C=C absorption at 1625 cm^{-1}. The O–H absorption band with a maximum at 3428 cm^{-1} is also very large.

7.2. Defects of cork

The cork may include occasionally features of biological or external origin that correspond to defects in the structure and that impact the raw-material quality to a degree that depends on their type and extent. These defects may be grouped as discontinuities (fractures), inclusions of non-phellemic tissue and stains, and the presence of high water content regions.

7.2.1. Discontinuities

Some discontinuities may occur in the cork tissue which devaluate the cork planks and disqualify them for production of cork stoppers in the worse cases.

One discontinuity is seen as an empty pocket in the cork plank (Fig. 7.8a) that corresponds to a radial fracture, called "lung" in the cork technical jargon. It resembles a large lenticular channel but differs because it has no filling tissue and usually develops only in a part of the thickness of the cork plank. This discontinuity should result from fragility in the

(a)

(b)

Figure 7.7. Different patterns of fissures in the back of cork planks: (a) reticulate pattern; (b) fibrous pattern.

radial cell bonding of the cork tissue, probably due to some periodic disfunction of a few cells in the phellogen. With the cork growth and the corresponding tangential stress the tissue is fractured and the hole expands tangentially, more in the earlycork region where the cells may collapse than in the latecork. This gives to the lung discontinuity its specific indented contour.This fracture may attain appreciable dimensions, i.e. average area of 32 mm^2 with a radial length of 19 mm and tangential width of 2 mm, although much larger voids are found, especially in cork planks with very fast growth and wide growth rings (Gonzalez-Adrados et al., 2000). The cork planks that contain an appreciable extent of lung are, for obvious reasons, devaluated and cannot be used for the production of cork stoppers.

Another discontinuity develops tangentially, corresponding to an exfoliation of the cork plank and separation between growth rings (Fig. 7.8b). It is the result of a loss of activity of the phellogen during a period due to a stress situation: for instance, an extremely dry period, the occurrence of fire, or a biological attack, i.e. the crown defoliator *Lymantria dispar*. This defect, named "exfoliation" or also "dry year" occurs only rarely in the cork-production regions.

More frequent is the development of deep fractures in the back of the cork planks that run vertically and penetrate into the underneath cork tissue. The fracture lines correspond to a tissue failure due to the growth tangential stress and therefore are more frequent in trees with a large radial growth. This is typically the case of second cork but it occurs also when the cork thickness is large.

The cork oak may also be attacked by insects that develop galleries in the cork tissue, namely ants and the coleopteron *Coroebus undatus* F. The areas of the cork planks with insect attacks are unsuitable for the production of cork stoppers.

Several species of ants inhabit the cork oak bark, the most usual being the *Crematogaster scutellaris* OI. These ants excavate galleries that run through the cork plank in random directions (Fig. 7.8c). The galleries are empty apart from occasional impurities such as sand particles, insect excrements and fungal hyphae (Liese et al., 1983). In cross-section they appear approximately circular with an average dimension of about 13 mm² while the tangential section may show sections of galleries with 48 mm² (Gonzalez-Adrados et al., 2000).

Coroebus undatus F is a coleopteron that lays its eggs in the cracks of the cork back. The larvae develop to a considerable size of approximately 3 cm and feed on the phellogen layer where they excavate long galleries that criss-cross the phellogen surface (Fig. 7.8d). The galleries are approximately elliptical in cross-section and are stuffed with excrement material and remains of chewed phloemic material. With the growth of cork, these galleries become embedded in the cork tissue, developing vertically and tangentially in the cork plank in the growth ring that corresponds to the year of attack. In the tangential section the galleries may correspond to large areas, i.e. 123 mm² with an average 26 mm² in cross-section (Gonzalez-Adrados et al., 2000).

The occurrence of larvae within the cork may also induce attacks from woodpeckers that bore out small cylindrical holes through the cork plank to catch their meal!

7.2.2. Inclusions and stains

The inclusion of small portions of lignified cells within the cork tissue is a frequent occurrence, named technically as "nail" due to their hardness. These inclusions are usually of small dimensions and in this case have no practical meaning. In some cases, however, they may be of larger dimensions, e.g. 18 and 13 mm² in the transverse and tangential section of cork, respectively (Gonzalez-Adrados et al., 2000). These cells have in general thick walls, often almost without an open lumen, and a lignocellulosic chemical composition. The density of these inclusions is therefore well above the density of the cork tissue and a cork plank with many nail inclusions is comparatively heavy. The lignocellulosic inclusions occur because sclerified cells and portions of the phloemic rays in the inactive phloem may be pushed out into the cork; occasionally a few cells of the phellogen may stop their activity, and a new phellogen portion is regenerated underneath, thereby including the overlying lignified cells in the cork mass (see Fig. 2.18).

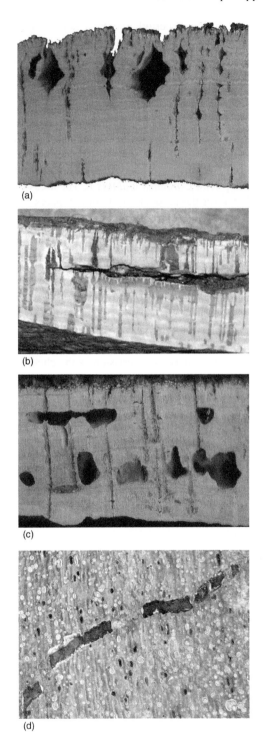

(a)

(b)

(c)

(d)

Figure 7.8. Defect of cork corresponding to discontinuities in the cork tissue: (a) "lung"; (b) exfoliation; (c) ant galleries; (d) galleries of *Coroebus undatus*.

The cork planks occasionally may show some stains that result from microbial attacks: yellow stain and marble stain. In both cases, the cork planks are not used for the production of stoppers to avoid eventual contamination with off-flavours and taints.

The yellow stain is caused by the *Armillaria mellea* (Vahl. Ex Fr.), a saprophytic basidiomycetes, that grows on soil and lignocellulosic materials. It is the cork cells that avoid the pathogenic invasion of the tree. In these regions the cork acquires a greyish colour and the surrounding tissues show a white-yellowish discoloration with a strong and easily detectable mouldy odour. The chemical composition of cork is altered with a decrease of lignin and polysaccharides, and an increase of polar extractives. A comparison of the chemical composition of cork with yellow stain with unstained cork showed the following values, respectively: extractives 17 and 13%, klason lignin 20 and 23%, polysaccharides 20 and 26%, suberin 38 and 37% (Carriço, 1997). The yellow stain occurs only seldom in cork but it is more frequent in the bottom part of the tree stem, near the soil. It is however an important defect and care is taken to prevent the stained corks from entering the process for the production of wine stoppers. The footer of cork planks is nowadays removed either during the cork extraction or in the mill yard in the post-harvest processing, and all stained corks are taken out from production.

Marble cork corresponds to a bluish staining of cork resulting from a fungal attack (*Melophia opiospora* Sacc.) that may occur in the external layers of cork and near lenticular channels. The hyphae penetrate the cork cells and develop in the radial and tangential directions, but the cork cell wall remains intact apart from the site of perforation and do not affect cork properties (Liese and Parameswaran, 1974; Liese et al., 1983). Such a cork is not used for the production of wine stoppers but may be valued for decorative purposes due to the marble-like random distribution of the stain and its dark contour lines.

7.2.3. Wetcork

One of the most important defects of cork if undetected before processing, is related to the occurrence of some regions in the cork plank with a very high proportion of water (400–500%). The reasons for the occurrence of this wetcork are yet unknown both at the tree and the cell level. The frequency of occurrence of wetcork is neither related to specific sites or climatic conditions, nor to specific trees, and it happens that trees produce wetcork in one production cycle and not in the following one. Wetcork is also not correlated with cork growth rate or with the magnitude of lenticular porosity.

The wetcork areas are distinguished macroscopically in the transverse or radial sections of the freshly extracted cork planks by a translucent darker colour typical of water-filled plant tissues, which after drying remain as lighter-coloured spots (Fig. 7.9). The wetcork forms discontinuous spots located in the inner side of the cork plank near the phellogen and extending to about one-third of the plank thickness. The wetcork regions have a contour line with an approximate wedged form, directed towards the exterior (the cork back).

Because wetcork contains much more water than the surrounding cork, it takes considerably more time to dry to the usual processing moisture content (about 14% or below) than normal cork. With drying, the cork volume shrinks by about 30% and the cells wrinkle heavily and collapse in some cases (Fig. 7.10). The dried spots of wetcork therefore

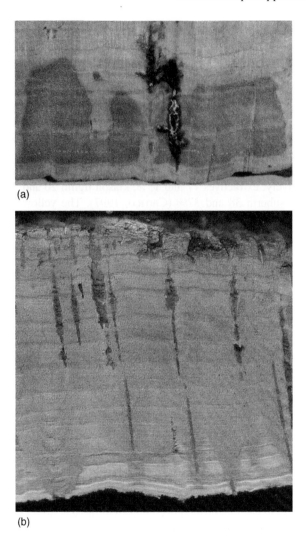

Figure 7.9. Wetcork areas as observed in the transverse section of cork immediately (a) after cork extraction and (b) after drying.

appear as conspicuous concave areas. Should a cork stopper be produced from an unde-tected wetcork area, it would show such a concave area on its cylindrical external surface that could question its performance as a sealant in the bottleneck.

Once the wetcork is dried, it has an equilibrium moisture content similar to normal cork and the same permeability to water (Rosa et al., 1991). Even the large volume shrink-age of the cells during drying is similar to what occurs to normal cork when it is made to absorb large water contents and is subsequently dried.

Until now no differences in cell structure and chemical composition could be found to explain the occurrence of wetcork (Parameswaran et al., 1981). A recent study that

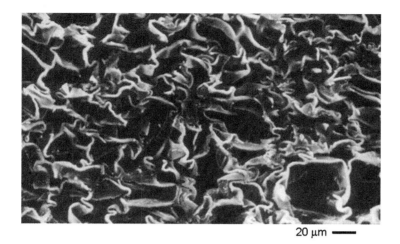

20 μm ━━

Figure 7.10. Scanning electron micrograph of a wetcork region after drying showing the cell shrinkage and collapse.

chemically compared the tissue in the wetcork region with the surrounding normal tissue, at an individual cork-plank level (Palma, 2002), found no differences in relation to the suberin content and to its monomeric composition as well as to the non-polar extractives. The only difference was a higher content of polar ethanol and water extractives in wetcork by about 30%, but these compounds should not have a role as a barrier to water transport.

The industry has solved the problem of wetcork by a careful screening of its occurrence at the post-harvest processing phase and all the wetcork containing planks are either let to air dry for about 2 years or are kiln-dried in a 3–4 day programme (see Chapter 12).

7.3. Image analysis of cork surfaces

The image of a cork surface may be captured with photographic or video digital equipment and subsequently processed by image analysis techniques. The objects that can be distinguished from the background by different colour intensity or composition may be extracted from the image by an adequate thresholding of the grey level. This is a powerful tool to characterise the lenticular channels and other defects in cork sections and to quantify their occurrence.

In image analysis the grey-level scale goes from 0 to 254, corresponding respectively to pure black and white. The cork tissue has grey-level values between 100 and 120, while the material filling the lenticular channels has grey levels in the range of 170–200, allowing to extract them as objects by adjusting a threshold at about 150–170. In a colour RGB system, the grey-level thresholds are approximately the following: Red 126–164; Green 114–158 and Blue 118–161.

The image-extracted pores can be measured and quantitatively characterised as individual objects, while a defined observation area may be characterised by the average characteristics of its pores or by their concentration. Table 7.2 summarises the variables of the

Table 7.2. Variables that characterise each pore of cork as measured by image analysis.

Variables (unit of measurement)	Definition
Dimension type	
Area (mm^2)	Area of the pore
Length (mm)	Maximum projection on the axis of elongation (i.e. radial direction in the transverse section, axial direction in the tangential section)
Width (mm)	Maximum projection on the axis perpendicular to the axis of elongation
Diameter (mm)	Orthogonal distance between two parallel lines that completely include the pore
Equivalent diameter (mm)	Diameter of the circle that has an area equal to the area of the pore
Shape type	
Shape factor	4μ area/perimeter^2, measures the roundness of the pore
Esfericity	Elongation of the pore by using central moments (value 1 for a perfect circle)
Aspect ratio	Maximum ratio between width and length of a bounding rectangle of the pore
Concentration type	
Next neighbour distance (mm)	Distance to the closest pore

pores in the cork surfaces that can be measured with image analysis systems and that have been applied to cork, classified into dimension, shape and concentration-type variables.

The dimension-type variables include: area, width, length, diameter (defined as the orthogonal distance between two parallel lines that completely include the pore) and the equivalent diameter (defined as the diameter of the circle that has an area equal to the area of the pore). The shape-type variables include: shape factor (sf), measuring the roundness of the pores, esfericity (Φ) that describes the elongation of the pore (value 1 for a perfect circular particle), aspect ratio (ar) as the maximum ratio between width and length of a bounding rectangle of the pore, and orientation, measuring the angle of the longest chord linking the gravity centre to the periphery. The concentration-type variables include the next neighbour distance (nnd, mm), defined as the distance to the closest pore.

When characterising a sample, average values of the above referred to variables are used as well as maximal values. The number of pores per unit area, as well as the proportion of the area occupied by pores are also important sample variables. The last one is called the coefficient of porosity, expressed as percent of area of pores in the total area of the sample, and is one of the most important variables to characterise the porosity in cork surfaces. Usually the data obtained are filtered to exclude the very small pores (e.g. pores below 0.5–0.8 mm^2), because small porosity is functionally and aesthetically irrelevant and only brings higher variance and variability to the sample (Gonzalez-Adrados and Pereira, 1996; Pereira et al., 1996; Lopes and Pereira, 2000).

Several studies have investigated the use of image analysis techniques for quantification of porosity in cork planks (Molinas and Campos, 1993; Gonzalez-Adrados and Pereira, 1996; Pereira et al., 1996; Gonzalez-Adrados et al., 2000), in cork discs (Lopes and Pereira, 2000) and in cork stoppers (Costa and Pereira, 2005, 2006).

The porosity of cork has been observed in the three principal sections of cork: tangential, transverse and radial sections. The appearance of the lenticular channels in the tangential section is very different from that shown in the other two sections, as a result of their formation and development within the cork plank, as already discussed.

7.3.1. Tangential section

Figure 7.11 shows a tangential section of cork and the corresponding digitalised image after thresholding. The pores appear as more or less circular objects but their size varies from minute spots to pores with a few square millimetres. Cork planks with different quality grades have large differences in relation to their porosity coefficient, number of pores, their average area or the maximum pore area (see Table 2.3, Pereira et al., 1996). The form of the pores is on average ellipsoid, elongated with an average aspect ratio of 0.5, meaning that their length is on average double the width. The shape factor shows that most pores are rather regular with values around 0.7. The longer axis of the pores is directed close to the axial direction of the cork plank, as exemplified in Figure 7.12 for the distribution of the orientation of the pores in one sample.

7.3.2. Transverse and radial sections

Figure 7.13 shows a transverse section of cork and the corresponding digitalised image after thresholding. In the case of transverse and of radial sections of cork planks, the

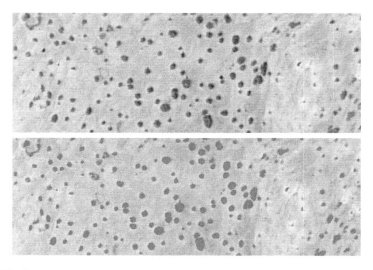

Figure 7.11. Cork seen in the tangential section: photography (top) and digitalised image after thresholding (bottom).

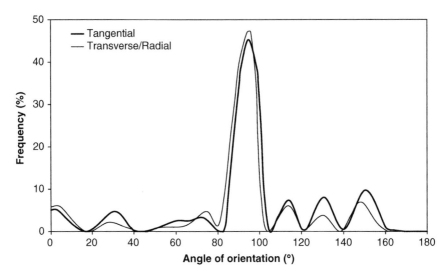

Figure 7.12. Frequency distribution of the orientation of the longer axis of pores: 90° corresponds to the vertical tree axis in the tangential section and to the tree radial direction in the transverse/radial sections.

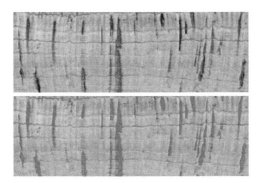

Figure 7.13. Cork seen in the transverse section: photography (top) and digitalised image after thresholding (bottom).

image includes a dark outer layer corresponding to the cork back. Before measurement of the pores, the cork back has to be singularised manually and subtracted from the sample.

The pores appear as channels running in the radial direction (Fig. 7.12) but only a few cross the total width of the sample. As already seen for the tangential section, there is a large range of variation of the pores' characteristics (refer to Table 2.3), although comparatively the number of pores per unit area is lower and the average pore area larger than for the tangential section. The aspect ratio and the shape factor of pores in the transverse and radial sections are similar with those obtained for the tangential section.

The features shown by the lenticular channels in transverse and radial sections are very similar and on a first approach almost undistinguishable. However the form of the channels shows that on average they have a larger width in the vertical direction than in the tangential direction, which leads to an overall higher porosity in terms of coefficient of porosity and average pore area in the radial section in comparison with the transverse section. The ratio of the coefficient of porosity in the radial and transverse sections is in the range of 1.1–1.5.

7.3.3. Porosity variation

The use of image analysis of a cork surface to estimate the porosity features is linked to some aspects of variability of the lenticular channels' distribution (within the tree in axial, tangential and radial directions) that have to be taken into account in sampling and in establishing the observation protocols.

One aspect regards the size of the observation area that is necessary to estimate the characteristics of a larger unit, for instance, of a cork plank or a tree, because of the variability in the distribution of the lenticular channels within the tree. A study that calculated the mean error associated to the determination of the porosity of a cork plank when using samples of different size showed that in order to obtain an error below 15%, the samples observed should have a minimum area of approximately 225 cm^2 (e.g. 15 cm $\times$ 15 cm) for the tangential section or a minimum length of 15 cm for the transverse and radial sections (Pereira et al., 1996).

When considering the variation of the porosity at the level of one tree, it was found that there is no systematic variation around the stem circumference when measured in 15 cm $\times$ 15 cm samples (tangential variation), while there is a trend to find a higher porosity at the bottom of the tree and smaller pores in the upper part of the stem (axial variation). However between-tree variation is high and a unique pattern of axial development of porosity is not found.

The porosity also varies radially within a sample, from the inner part (the belly side) to the outer part (the cork back). In this case, higher porosity is always found near the cork back, decreasing afterwards until reaching the lowest porosity near the belly (Fig. 7.14). From about half the thickness of the cork plank to the belly side the porosity remains rather stable (Fig. 7.15).

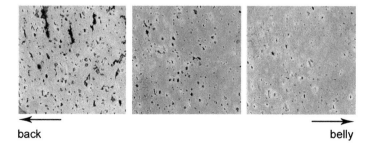

back   belly

Figure 7.14. Radial variation of the porosity of cork shown by successive images of tangential sections of one cork sample taken at approximately 20, 50 and 80% of the thickness from belly to cork back.

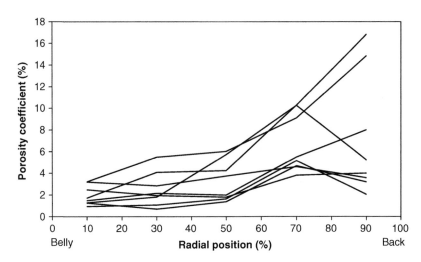

Figure 7.15. Radial variation of the porosity of cork along different cork samples from belly to cork back.

There is a large variability of cork porosity between trees in the same location and between sites. Measurements carried out in the transverse section of cork planks from a total of 680 trees (20 trees in each of 34 locations) showed a large variation between individual trees from a minimum value of 0.5% to a maximum of 14.6%. As the mean per site, the porosity ranged between 3.2 and 8.8% but the within-site variation was high with coefficients of variation of the mean around 40% (Fig. 7.16). This between-tree variation in the number and area of the lenticular channels has been repeatedly reported for all regions within the cork oak area.

The porosity is not significantly related to cork growth (Fig. 7.17) and there are corks with low coefficient of porosity in all thickness classes of cork planks, the same being true also for samples with a large porosity. This is sometimes against the human subjective perception because for the same coefficient of porosity large pores in thick cork planks have a higher visual impact.

7.4. Quality classification of cork planks

7.4.1. Thickness classes

The cork-plank thickness is the first and determining factor to establish the end use of the raw material and consequently it has a first-hand influence on the economy of the industrial processing. Therefore one of the main variables taken into account in the commerce of cork planks is their thickness. Classification of cork planks by thickness, or calliper, as given by standards (Table 7.3) is the usual practice in the industry.

The classes were established taking into account the potential suitability for further processing. The most adequate thickness classes for the production of stoppers are 27–32 mm and 32–40 mm, because they allow the punching out of the stoppers and give the best mass

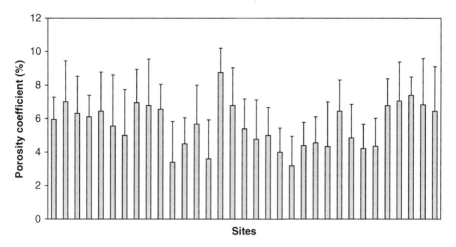

Figure 7.16. Variation of the coefficient of porosity of cork measured in the transverse section between different locations in Portugal. Mean of 20 trees per site and standard deviation as bar.

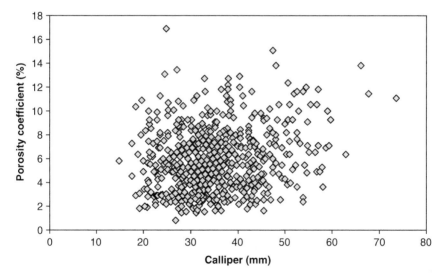

Figure 7.17. Coefficient of porosity measured in the transverse section and thickness of cork planks.

and quality yield (see Chapter 12). Therefore the goal of cork production is the production of cork planks with this thickness range. The thickness of the cork planks depends on the duration of the cork production cycle and on the annual cork growth (see Chapter 1). In the major cork producing regions, a cork production cycle of 9 years is adequate for the main raw material industrial requirements. Three examples of the calliper distribution of cork planks are shown in Figure 7.18, representing respectively a field sampling from 34 different locations in Portugal covering the production area, the raw-material supply in an industrial

Table 7.3. Commercial thickness classes for cork planks as used in the cork industry.

Commercial class	Thickness (mm)
Extra thin	9–22
Thin	22–27
Half standard	27–32
Standard	32–40
Large	40–54
Extra large	>54

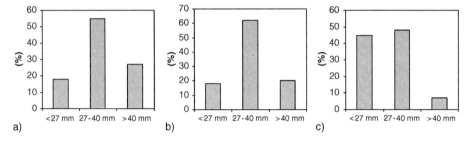

Figure 7.18. Distribution of cork planks by three thickness classes (<27 mm, 27–40 mm, >40 mm) for (a) 34 different locations in Portugal covering the full production area; (b) the raw-material supply to one cork industrial unit corresponding to a total of 1500 tons (Lopes, 2001); and (c) the production of one large cork estate (Cumbre, 1999).

unit, and the production of one large cork estate. It is clear that the majority of the samples fall in the 27–40 mm thickness class, although there is also a substantial proportion of thinner cork planks.

The thickness of the cork planks is the overall measure taken from belly to back. Therefore the outside layer of the back must be taken into account when evaluating the suitability of a cork plank as an industrial raw material. The lignous tissues included in the back of the cork planks may differ in thickness but on average represent less than 10% of the total thickness of the plank (Taco et al., 2003).

The cork plank calliper classes differ considerably in price: although very dependent on market opportunities, in general the unit price (per kg) of planks in the 27–32 mm and 32–40 mm range is the same and serves as reference for the other classes. The price of planks in the 22–27 mm class is nearly half of this reference value, less than one-third for the thinner planks under 22 mm, and approximately two-thirds for the thick planks in the 40–54 mm class.

7.4.2. *Quality classification*

The cork planks are commercially classified by quality into different classes. This evaluation of quality is made by the visual appraisal of the cork plank on the transverse and radial sections as well as on the belly surface. The first criterion of quality is related to the porosity given by the lenticular channels, in number but mainly in area, i.e. a good quality cork plank will have small lenticular channels. The second criterion is related with the presence of defects and to the impact that they may have on the processing. Underlying the classification of one cork plank is always the integrated concept of its potential yield in good quality products of natural cork, stoppers or discs.

There are six quality classes for cork planks, from 1st to 6th, and a refuse class. Since the evaluation is made manually by an operator, there is an important degree of subjectivity and the classification of a specific cork plank may vary from one operator to another. It is consensual and easy to recognise very good and very bad corks, but this is not true in relation to medium quality corks. Therefore there are important differences between the classifications made by different experts. An experiment with a reference catalogue with 480 cork samples (Corkassess, 2001) that were classified by five independent experts resulted into only a 5.3% coincidence in the classification of the individual samples (but with 100% coincidence regarding 1st quality and refuse!).

In the present practice, this classification of cork planks into six qualities (and refuse) is not applied and instead more aggregated quality classes are used: a current grading is to use a 1st–3rd assortment, representing good quality corks, a 4th–5th assortment for medium quality and a 6th class for poor quality. Another option, and probably the one that is nowadays mostly used in the industry is to separate the cork planks into only two quality classes: a 1st–5th assortment representing the cork planks that are adequate for the production of cork stoppers, and a 6th class that includes the planks with a lower quality and a poorer performance for the production of stoppers (see Chapter 12).

The relation between the quality grading and the porosity quantified by image analysis has been studied, and as was expected, the porosity and the dimension of the pores increased from the best to the worst quality classes. This is true for the observation in the tangential section (Table 7.4) and for the transverse or radial sections (Fig. 7.19). However there is a large variation within each class and the differences between mean values of contiguous classes have no statistical significance.

Table 7.4. Average porosity parameters for cork planks with different commercial quality grades, observed in the tangential section.

	Quality class				
	1st	1st–3rd	3rd	4th–5th	6th
Porosity coefficient (%)	3.3 (1.2)a	4.5 (0.9)ab	6.0 (0.8)bc	6.7 (2.0)c	12.4 (3.6)d
No. of pores/100 cm^2	568 (186)a	584 (157)a	785 (234)b	860 (153)b	815 (194)b
No. of pores >0.8 mm^2/100 cm^2	47 (18)a	84 (30)ab	111 (36)b	121 (52)b	176 (64)c
Average pore area (mm^2)	0.58 (0.15)a	0.83 (0.28)a	0.81 (0.21)a	0.79 (0.24)a	1.57 (0.53)b
Maximum pore area (mm^2)	19.7 (18.2)a	23.1 (12.2)a	21.1 (18.5)a	26.3 (15.0)a	82.0 (31.7)b

Note: Average of 40 samples per class and standard deviation in parentheses. Means in one line followed by the same letter are not significantly different (Pereira et al., 1996).

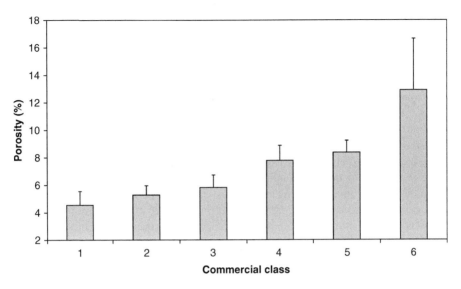

Figure 7.19. Mean values and confidence intervals of the means for the coefficient of porosity for the different commercial quality classes of cork planks (Corkassess, 2001).

The grouping into only three classes allows for a better differentiation and to a more consistent grading. The coefficient of porosity may be used to classify the cork planks, for instance, using the following rules to establish the class limits: the good quality cork planks must have a coefficient of porosity below 6%, and the medium quality planks a porosity below 10%. The dimensional or shape characteristics of the objects seen in the image of the cork surfaces may also be used to differentiate between the lenticular channels and other defects, for instance insect galleries (Gonzalez-Adrados and Pereira, 1996; Gonzalez-Adrados et al., 2000).

7.4.3. *Integrated value-index of cork*

Since the value of a cork plank is given by a combination of thickness- and porosity-related quality, the selection of a combined variable that gives an indication of the cork value has been proposed on the basis of the relative commercial value of the different combinations of calliper and quality classes (Table 7.5). This may be applied to estimate the value of production for a particular cork stand, taking into account the distribution of the cork planks by thickness and quality.

The following quality index was proposed:

$$Q = \sum_{k=1}^{n_k} Q_k p_k = \frac{\sum_{i=1}^{n_t} Q_i}{n_t}$$

where Q_k is the index price for cork quality of class k, p_k the proportion of cork-sampling units in cork quality class k and n_k the number of cork quality classes,

Q_i the index price for the sampling unit extracted from tree i and n_t the total number of trees sampled.

The index prices are indexed to the price of the most valuable cork-quality class, and Table 7.5 shows an example established by an expert's panel. It is clear that the figures may be adjusted to changing cork market or to specific trade and processing conditions.

7.5. Conclusions

The visual aspect of cork is characterised by the presence of annual rings, a light-brown colour and the presence of lenticular channels. The lenticular channels are the most notorious visual characteristic of cork and its specific mark. They constitute the porosity of cork and have the following characteristics:

• they cross the cork planks from the inner side to the outside;
• in tangential sections they appear as small rounded pores and in transverse and radial sections as thin strips;
• they contain a filling tissue with a darker colour that allows their easy identification visually and by image analysis techniques;
• the lenticular filling tissue is cellular but differs anatomically and chemically from the cork tissue;
• they are the basis for the quality classification of cork planks.

The cork planks are classified in relation to their adequacy for industrial processing and the potential value yield of the products, with the wine cork stoppers as the main target. The cork planks are characterized by the following:

• thickness classes are standardised, with the 27–32 mm and 32–40 mm as the most valuable;
• quality grading is based on the lenticular porosity and on the presence of occasional defects of biological or external origin;
• classification can use quantified measurements of porosity by image analysis techniques and selection of adequate thresholds of the coefficient of porosity for class separation;
• an integrated quality index combining thickness and quality can be used to evaluate the value of a raw-material assortment.

Table 7.5. Index prices for the different industrial cork thickness and quality classes.

Calliper class (mm)	Quality class			Refuse
	1st–3rd	4th–5th	6th	
14–18	22	10	8	8
18–22	31	13	8	8
22–27	50	30	13	8
27–32	100	60	28	8
32–40	100	60	28	8
>40	66	33	17	8

References

Carriço, S., 1997. Estudo da composição química, da estrutura celular e dos componentes voláteis da cortiça de *Quercus suber* L. Ph. D. Thesis, Universidade de Aveiro.

Corkassess, 2001. Field assessment and modelling of cork production and quality. Final report. Contract FAIR.CT97.1438. European Commission, Research Directorate General, Life Sciences, Agriculture, Agro-industry, Fisheries and Forestry, Brussels.

Costa, A., Pereira, H., 2005. Quality characterization of wine cork stoppers using computer vision. International Journal of Vine and Wine Sciences 39, 209–218.

Costa, A., Pereira, H., 2006. Decision rules for computer-vision quality classification of wine natural cork stoppers. American Journal of Enology and Viticulture 57, 210–219.

Cumbre, F., 1999. Avaliação da qualidade tecnológica de pranchas de cortiça por amoostraggem em pilha. Graduation thesis, Instituto Superior de Agronomia, Lisboa.

Gonzalez-Adrados, J.R., Lopes, F., Pereira, H., 2000. Quality grading of cork planks with classification models based on defect characterisation. Holz als Roh- und Werkstoff 58, 39–45.

Gonzalez-Adrados, J.R., Pereira, H., 1996. Classification of defects in cork planks using image analysis. Wood Science and Technology 30, 207–215.

Liese, W., Gunzerodt, H., Parameswaran, N., 1983. Alterações biológicas da qualidade da cortiça que afectam a sua utilização. Cortiça 541, 177–297.

Liese, W., Parameswaran, N., 1974. Natureza do marmoreado da cortiça. Cortiça 425, 46–47.

Lopes, P., 2001. Avaliação da qualidade industrial da cortiça no processo fabril de produção de rolhas. Graduation thesis, Instituto Superior de Agronomia, Lisboa.

Lopes, F., Pereira, H., 2000. Definition of quality classes for champagne cork stoppers in the high quality range. Wood Science and Technology 34, 3–10.

Molinas, M., Campos, M., 1993. Aplicación del análisis digital de imágenes al estudio de la calidad del corcho. In: I Congreso Forestal Español. Sociedad Española de Ciencias Forestales, Lourizan, Pontevedra, pp. 347–352.

Palma, S., 2002 Caracterização do verde na cortiça e da sua ocorrência no abastecimento da matéria pima à indústria. Graduation Thesis, Instituto Superior de Agronomia, Lisboa.

Parameswaran, N., Liese, W., Gunzerodt, H., 1981. Characterization of wetcork in *Quercus suber* L. Holzforschung 35, 195–199.

Pereira, H., Lopes, F., Graça, J., 1996. The evaluation of the quality of cork planks by image analysis. Holzforschung 50, 111–115.

Rosa, M.E., Matos, A.P., Fortes, M.E., Pereira, H., 1991. Algumas características da cortiça verde. In: Actas do 5 Encontro Nacional da Sociedade Portuguesa de Materiais. Sociedade Portuguesa de Materiais, Lisboa, pp. 2, 737–746.

Taco, C., Lopes, F., Pereira, H., 2003. La variation dans l'arbre de l'épaisseur du liège et du dos des planches de liège pour des chênes-liège en pleine production. Anais Instituto Superior de Agronomia 49, 209–221.

Chapter 8

Density and moisture relations

Cork is a light material that floats on water with large buoyancy and that absorbs water only very slowly. This is the result of its cellular structure formed by hollow and closed cells with a small solid fraction that is concentrated in thin cell walls, without any inter-cellular communication channels, as described in detail in Chapter 2. Floating utensils for fishing and boating are among the first uses of cork that took advantage of these properties. The cork hydrophobic character and its very low permeability to liquids are also among the properties that led to its still main use as sealant for bottles and other liquid containers.

This chapter describes the physical properties of cork related to density and the factors involved in its variation, and to its behaviour with water vapour and liquid water, such as the equilibrium moisture content of the material in different environments, the absorption of liquid water at various temperatures and the drying process. Such properties are involved in the industrial processing of cork, from the field drying of the cork planks to the boiling of cork in water and subsequent moisture variation along the production line. Several of the other properties of cork, namely mechanical properties, are influenced by the water content of the material. The wettability of cork surfaces by liquids is also discussed.

8.1. Density

8.1.1. Definitions and cork density

The density of the air-dried cork tissue is low, on average in the range of 150–160 kg m^{-3} but with values that may go from below 120 to over 200 kg m^{-3}.

The mass of a solid per unit volume is its density (ρ) expressed in kg m^{-3} units, and calculated as $\rho = m/V$, with m as the mass (in kg) of volume V (in m^3). In materials that absorb water and therefore will have different mass and volume values for different water contents, the density is specified at defined moisture contents, or humidity environments. This is the case of lignocellulosic materials, such as wood, and also of cork. Therefore density is referred to specified standard conditions such as (a) oven dried (fully dried), considered to correspond to 0% moisture in the solid, or (b) equilibrated to a specific moisture content in the solid, such as 12%, or (c) equilibrated under certain air conditions, such as natural air conditions (ambient equilibrium), frequently called air-dried, or under controlled conditions of air relative humidity (RH) and temperature, for instance RH 50% and 20°C. In these last cases, the water content in the material should be known.

In cellular solids, the density depends on the density of the solid that makes up the cell walls (ρ_s) and of the solid fraction, that is the fraction that is occupied by the cell walls in the total volume (f), as $\rho/\rho_s = f$. Cellular materials have in common a solid fraction below 30%, and in cork the solid fraction was calculated using average cell dimensions to be 8–9% in the earlycork region and 15–22% in the latecork region (see Chapter 2).

The density of the solid cell wall material of cork was estimated as 1150 kg m^{-3} (Gibson et al., 1981) from its chemical composition and the specific gravity for the cell wall components: cellulose 1.5, lignin 1.4, suberin 0.90, glycerol 1.20, aliphatic extractives 0.85–1.00, tannin 1.00. This is a value that is typical for polymeric materials. A highly compressed cork, made from high pressure (*ca.* 55 MPa) densification of cork dust, showed a density of 1200 kg m^{-3} but still contained some areas with voids between the cell walls, so that the cell wall density should be higher, i.e. about 1250 kg m^{-3} (Flores et al., 1992). No large variation in cell wall density is likely to occur within the natural range found for the chemical composition of cork (see Chapter 3). Therefore the range of densities found for the cork material will result from its specific structural features, leading ultimately to differences in the fraction of the volume occupied by the cell walls.

8.1.2. *Factors of variation of cork density*

The density of cork will vary due to the following structural features: (a) the size of the cells, especially in the earlycork region where the largest variation is found; (b) the proportion of earlycork and latecork in the annual ring, which is in relation to the ring width; (c) the corrugation of cells; (d) the extent of the porosity derived from lenticular channels and (e) the presence of woody inclusions and discontinuities.

The geometry and the dimensions of the cells determine the solid fraction of the structure. Considering the cells as hexagonal prisms, and taking into account that the cell wall thickness is much smaller than the other cell dimensions, the following equation was derived for the average density (ρ_m) of cork (Gibson et al., 1981; Rosa and Fortes, 1988):

$$\frac{\rho_m}{\rho_s} = \frac{e_m}{l_m}\left(\frac{2\sqrt{3}}{3} + \frac{l_m}{h_m}\right)$$

where e_m is the average cell wall thickness, l_m the average base edge and h_m the average prism height. Considering average dimensions of $e_m h_m = 35 \times 10^{-6}$ mm^2, $l_m = 15 \times 10^{-3}$ mm and $\rho_s = 1250$ kg m^{-3}, the variation of density can be plotted in function of h_m, the prism height (Fig. 8.1). For a prism height of 15 μm, as typical in latecork, the density will be approximately 420 kg m^{-3}, while for a prism height of 40 μm, as in the large cells of earlycork, the density will be 110 kg m^{-3}.

The annual ring of cork may be considered as composed of two layers, the earlycork and the latecork, and the average density will be the weighed average of their respective

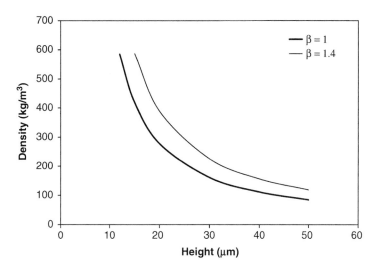

Figure 8.1. Variation of average density of cork with average cell prism height for straight cells ($\beta =1$) and for corrugated cells with $\beta = 1.4$.

densities (ρ_{early} and ρ_{late}) taking into account their proportions. If ρ_{early} and ρ_{late} are averaged at 110 and 420 kg m^{-3}, respectively, and the proportion earlycork:latecork is 75:25 and 95:5, respectively, in a thin and a large cork ring, then the corresponding average density would range between 188 and 126 kg m^{-3}. In fact these average theoretical calculations match rather well with the practice. This explains why thin cork planks with small annual rings are denser than the standard planks.

The corrugation of the lateral prism walls of the earlycork cells is also a factor that impacts on the materials density. There are various degrees of cellular corrugation, from almost straight cell walls to heavily undulated and nearly collapsed cells due to growth stresses (see Chapter 2). The effect of the corrugations may be mathematically considered by the introduction of a corrugation parameter β ($\beta>1$) as the quotient between the length of the corrugated wall and the length of the wall if it was straightened. The density will be directly proportional to β as $\rho = \beta \rho^*$, with ρ^* being the density of cork cells without corrugations. This is also represented in Figure 8.1 where the density for a corrugation factor of 1.4 is plotted in function of the straight cell dimensions.

This effect of the corrugation degree on density explains why treatments that cause cell expansion and straightening result into lower densities (such as the water boiling operation described below and in Chapter 12) or conversely why the increase of cellular corrugation gives denser corks (such as a stopper in the bottle neck or the agglomeration of cork granules, see Chapter 13).

The porosity of cork could be in a first approach thought as a factor to reduce the density of cork. However, the lenticular channels are not voids and instead they contain a filling material, although with numerous fractures, and in most cases they are bordered by thicker cells (see Chapter 7). Therefore, the extent of lenticular porosity is not negatively related to cork density. On the contrary, the general tendency is towards higher density values in corks with more and larger lenticular channels. This is shown in Figure 8.2

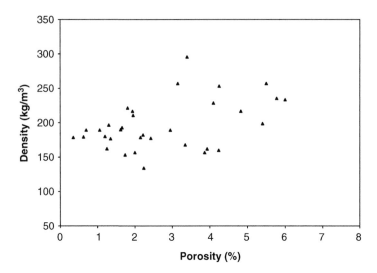

Figure 8.2. Variation of cork density with the coefficient of porosity measured in the tangential section. The samples are approximately 5 mm thin and were cut in the mid-region of cork planks of different calliper.

where thin cork slices were cut from the mid-region of 32 cork planks of variable thickness and their density was determined and the porosity measured by image analysis on the tangential section. Although there is no relevant correlation between these two variables, there is a tendency for higher densities for corks with higher coefficients of porosity.

The corks that present voids in their structure, such as the lung failure or insect galleries, have a lower density which is directly in relation to the void fraction. The lignocellulosic inclusions in cork, the so-called nail, have the opposite effect and they may increase density of cork to values more than 300 kg m^{-3}.

8.1.3. Density of cork planks

The density of the cork planks is the result of its composite nature that includes the cork tissue and the external layer of the cork back. Therefore the densities of both materials have to be taken into account as well as their relative proportion. The cork layer itself is not uniform and shows a radial variation in annual ring width, with the smaller rings in the belly side, therefore with an increasing proportion of the latecork layer from the exterior to the interior. There is also a variation between trees in cork annual growth, resulting in differences in the earlycork proportion.

The variation of density in the radial direction in cork planks was measured for cork planks of two calliper classes and three quality classes (good, medium and low quality) between the belly and the cork back (Table 8.1). The cork back has an average density of 550 kg m^{-3}, well over the density of cork and in accordance with its ligneous nature. The radial variation of density in the cork tissue is small and the pattern is not constant due to the contradicting influence of factors: smaller rings at the belly side leading to a

Table 8.1. Variation of density along the radial direction of cork planks of two calliper classes (27–32 mm and 22–27 mm) and three quality classes (1st, 3rd and 5th) measured from thin slices from the belly to the cork back. Mean of six planks and standard deviation (Barbato, 2004).

Calliper	Radial position	Quality class		
		1st	3rd	5th
27–32 mm	1 (belly)	0.225 (0.021)	0.167 (0.007)	0.194 (0.004)
	2	0.206 (0.011)	0.151 (0.008)	0.170 (0.003)
	3	0.221 (0.020)	0.164 (0.005)	0.186 (0.006)
	4	0.218 (0.029)	0.150 (0.003)	0.185 (0.006)
	5	0.224 (0.035)	0.165 (0.003)	0.176 (0.004)
	6 (back)	0.495 (0.084)	0.524 (0.058)	0.608 (0.036)
22–27 mm	1 (belly)	0.220 (0.016)	0.183 (0.004)	0.184 (0.009)
	2	0.224 (0.019)	0.175 (0.007)	0.173 (0.008)
	3	0.248 (0.026)	0.193 (0.011)	0.187 (0.010)
	4	0.293 (0.029)	0.198 (0.007)	0.183 (0.015)
	5 (back)	0.754 (0.057)	0.461 (0.050)	0.458 (0.103)

higher density in this region, but also higher porosity towards the cork back also increasing the density. It is noteworthy how the lower quality corks have significantly higher density.

The variation of density of cork planks was measured in the already mentioned study of 680 trees that were sampled covering the different locations of cork production in Portugal. The average value of the air-dried cork planks after boiling in water was 251 kg m^{-3} with the individual values ranging from 162 to 474 kg m^{-3}. There was a negative correlation between density and cork plank thickness, with the average density of the thinner cork planks higher than the density of larger cork planks. No correlation was found between the coefficient of porosity of the planks (measured in the transverse section) and their density.

8.2. Moisture content

8.2.1. Water in cork

The water content in cork, as in other materials, is determined by the amount of water per unit of dry mass, calculated as a moisture content (*H*), usually presented in percent as

$$H(\%) = \frac{M_w - M_d}{M_d} \times 100$$

with M_w and M_d as the mass of the wet material and the dry material, respectively. The water content varies from 0%, corresponding to the fully dried material, to substantially higher values such as the 25% that correspond to the average water content of the cork planks at extraction, or the 500% in the areas containing the defect of wet cork.

In the commercial sector, the water content is frequently calculated on a wet material basis, H_w:

$$H_w(\%) = \frac{M_w - M_d}{M_w} \times 100$$

The water content calculated on a wet cork basis varies between 0% (fully dried material) to 100% (only water). Therefore caution should be given when interpreting water content values, and their calculation basis must be clearly understood. H and H_w are mathematically related by the equations:

$$H = \frac{H_w}{100 - H_w} \quad \text{or} \quad H_w = \frac{100H}{1 + H}$$

Their mutual plotting is shown in Figure 8.3. The difference between H and H_w is small for low moisture contents (for instance for $H = 10\%$, $H_w = 9.1\%$), but it is substantial for large water contents (for $H = 100\%$, $H_w = 50\%$).

The water may exist in cork in two different forms: (a) bound water adsorbed to the cell walls by intermolecular forces, namely H-bonding; (b) free water contained in the structural voids. These two types of water are linked to the material with different energy, i.e. it is more difficult (energy demanding) to remove the bound water and even more so as the cell wall has more affinity to water. The hygroscopicity of the different cell wall components is different taking into account the chemical characteristics of their molecular structures (see Chapter 3), in the following order: hemicelluloses > cellulose > lignin > suberin. The presence of non-polar extractives of aliphatic nature also decreases the hygroscopicity of the cell wall material.

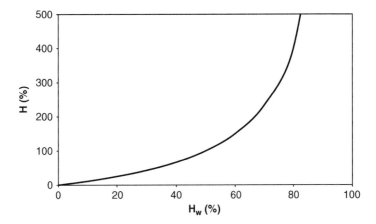

Figure 8.3. Plotting of moisture content expressed on a dry mass basis (H) with moisture content expressed on a wet mass basis (H_w).

The intake of water by cork therefore depends on two aspects:

(a) The chemical composition of the material, and its heterogeneity in the sample, which is related to the bound water in the cell walls; the lenticular filling material is more hygroscopic than the cork cells because it contains more polysaccharides and no suberin; the same applies to the cork back tissues in the cork planks.
(b) The presence of lenticular channels, of voids such as lung or insect galleries, and the volume fraction of the cellular lumina that are the possible reservoirs for the free water; communication paths through the cork material (e.g. lenticular channels) also increase the contact area for the water adsorption.

The maximum moisture content that can exist in a material depends on its density. With increasing density, the lumen volume of the cells and the intercellular voids decrease, and the amount of water that can exist as free water also decreases. The maximum moisture content in cork H_{max} can be calculated by the following equation, where ρ is the relative density and 1.25 the cell wall relative density:

$$H_{max}(\%) = (1.25 - \rho) \times 100/1.25\rho$$

For instance, the maximum moisture content of a cork with a density of 160 kg m^{-3} would be 545%, while for a density of 190 kg m^{-3} it would be 446%.

When the cork planks are stripped off from the tree, they have moisture contents that average 25%, with a large range of values for individual trees from about 5% to more than 50%. There is also a variation of water content within the plank, with higher values in the inner side, in the region of the more recently formed cells, which decrease to about the mid-plank and then remain constant in equilibrium with the air environment. Figure 8.4 shows the radial variation of moisture content in several cork planks after extraction from the tree by measuring successive layers of approximately 5 mm thickness: the phellogen neighbouring region has high moisture contents (50–120% in the cases shown).

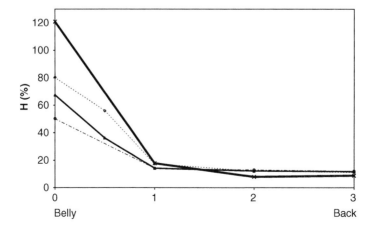

Figure 8.4. Radial variation of moisture content (*H*) in cork planks measured in successive layers of approximately 5 mm thickness along the plank from the belly the cork back.

8.2.2. *Equilibrium moisture content*

When the cork is in equilibrium with its surroundings, it is said to be at its equilibrium moisture content. This equilibrium depends on the relative moisture content of the environment and it will change with changing environments. Its value depends on the chemical composition of the cork, namely on the amount of suberin and of non-polar extractives. The time required to reach the equilibrium depends on the size of the samples and on their structural features.

The equilibrium moisture content of cork after air drying and equilibration in in-door conditions is on average 7%, ranging from about 5 to 10% with very low air RH or humid conditions, respectively. Under controlled conditions of air RH, the following moisture content values were found in cork at ambient temperature: 8% at RH 75%, 10% at RH 85%, 16% at RH 95%. These values are substantially lower than those found for other more hygroscopic cellular materials, such as wood. For instance, cork oak wood has an equilibrium moisture content at approximately 20% for a RH 80%.

8.2.3. *Sorption isotherms*

The moisture status of cork, as well as of other materials, is important because it impacts on its properties, namely on those related to mechanical behaviour. It depends on the temperature and RH conditions of the surrounding environment. The variation of the water content in a material with air RH at a defined temperature is called a sorption isotherm, and it may occur by wetting (absorption of water) or by drying (desorption). The isotherm-adsorbing curve is lower that the desorbing curve, meaning that the moisture status at certain conditions depends if the material is in a wetting or drying process.

Figure 8.5 represents the plot of the cork moisture content in function of the RH for 20°C and 40°C when dry cork adsorbs water and reaches equilibrium at successively

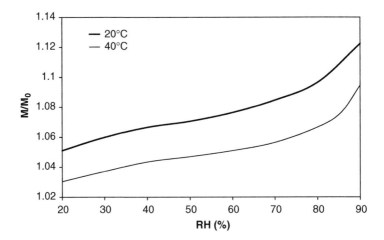

Figure 8.5. Sorption isotherms of cork at 20 and 40°C (adapted from Gonzalez-Adrados and Haro, 1994).

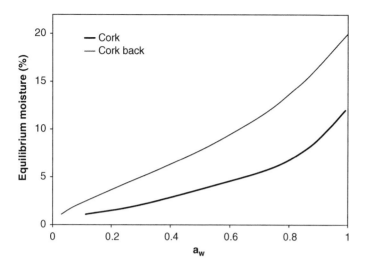

Figure 8.6. Adsorption isotherm at 22°C of the cork back and the cork layers of one cork plank (Pires, 2000).

higher air humidity (Gonzalez-Adrados and Haro, 1994). The water content in cork increases with the air RH with a small and approximately constant rate up to RH 80% after which the absorption rate increases. The temperature also affects the sorption and desorption of moisture in cork and the isotherms decrease with increasing temperature: for instance, in Figure 8.5 there is a decrease of approximately 2–3% in the moisture content of cork between 20 and 40°C. The type of cork (virgin or boiled reproduction cork) may also influence the equilibrium moisture content with differences in the range of 1–2% (Gil and Cortiço, 1998).

The treatment of cork with saturated water vapour at 100°C and atmospheric pressure led to a moisture content of approximately 17% (Rosa and Fortes 1989).

In a cork plank there is a difference of equilibrium moisture content between the cork and the external cork back layer due to its chemical and structural differences. Figure 8.6 shows the isotherm curves at 22°C for the cork back and the cork layers of one cork plank. The rate of moisture content variation is higher for the cork back, and at the same RH the cork back has approximately the double moisture content in comparison with cork.

8.3. Absorption of water

8.3.1. *Diffusion of water*

The adsorption kinetics, i.e. the variation of the water intake along time, depends on the structural characteristics of the material. Because cork is structurally anisotropic, the diffusion of water into it does not proceed at the same rate in the different directions.

Experiments made with thin slices of cork that were cut oriented in the three main directions allowed to calculate the water adsorption when immersed in water since the direction perpendicular to the largest face may be approximated to the diffusion direction

(Rosa and Fortes, 1993). Figure 8.7 shows the kinetics of water absorption measured by the mass variation for the three main directions. The curves have three main regions: a faster initial phase when water is absorbed rapidly with an approximate constant rate, which slows down and is followed by a phase of decreasing absorption rate until the final step when the water absorption is very slow until a constant mass is obtained that corresponds to the maximal intake of water. The absorption along the radial direction is appreciably higher than the absorption along the other non-radial directions. The diffusion in the axial and radial directions is similar in the first phase of absorption, but afterwards the absorption along the axial direction is slightly above the absorption along the radial direction. These differences are the result of the cellular arrangement in cork since more cell

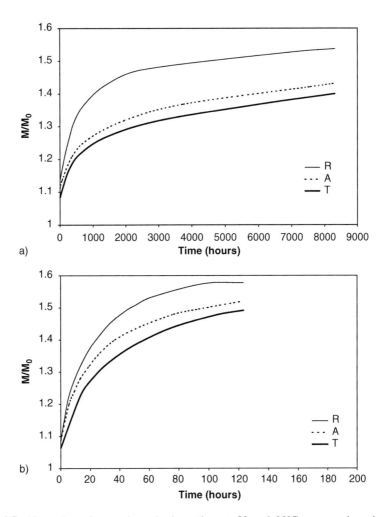

Figure 8.7. Absorption of water in cork along time at 20 and 90°C, measured as the percent mass increase of the samples, in three different penetration directions (radial, axial and tangential): (a) 20°C, (b) 90°C (adapted from Rosa and Fortes, 1993).

wall area is sectioned in a tangential cut, therefore giving a higher diffusion area for the radial penetration.

The absorption rate increases with temperature. As seen in Figure 8.7, the maximal water absorption in achieved after about 8000 h at 20°C and 100 h at 90°C (nearly one year and 4 days, respectively!). In this state the cell walls are saturated and the cell lumina are water filled. This corresponds to water content values more than 500%. In these conditions, the cork density overcomes the density of water and cork no longer floats.

The diffusion of water in cork follows Fick's first law in the first phase of absorption for water contents below 100%:

$$J = -D\frac{dc}{dx}$$

where J is the mass of water that crosses a unit area in a unit time, dc/dx is the gradient of water concentration and D the diffusion coefficient ($m^2\ s^{-1}$). The diffusion coefficients for 20 and 90°C are listed in Table 8.2 (for instance, $2 \times 10^{-11}\ m^2\ s^1$ at 20°C). They are not significantly different for the radial and non-radial directions.

A recent estimation of water diffusivity of cork at 25°C using electrical conductivity measurements of cork samples with a very low water content (3.5%) reports a lower value of approximately $6.8 \times 10^{-13}\ m^2\ s^{-1}$ (Marat-Mendes and Neagu, 2004). The process of water diffusion in cork at this concentration level has two components: an initial phase which is faster and is likely to be related to the physically absorbed water and a subsequent slower phase likely to be related to the chemically bound water (Marat-Mendes and Neagu, 2003).

The diffusion coefficient is a function of temperature as

$$D = D_0\exp(Q/RT)$$

where D ($m^2\ s^{-1}$) is the diffusion coefficient at temperature T (K), Q is the activation energy (kJ mol^{-1}), R the gas constant (8.31 J mol^{-1}) and D_0 a constant that was estimated experimentally for cork as 4.0 $m^2\ s^{-1}$. The activation energy for the diffusion of water was estimated as $Q = 64$ kJ mol^{-1}. This value is similar to others determined for synthetic polymers.

Table 8.2. Diffusion coefficients (D, $m^2\ s^{-1}$) of water in cork in the radial and the non-radial directions at 20 and 90°C (Rosa and Fortes, 1993).

Temperature (°C)	D ($m^2\ s^{-1}$)	
	Radial	Non-radial
20	1.9 (0.2) $\times$ 10^{-11}	2.5 (0.4) $\times$ 10^{-11}
90	2.9 (0.7) $\times$ 10^{-9}	2.6 (0.6) $\times$ 10^{-9}

8.3.2. Dimensional variations

Cork swells when it absorbs water. Because of its structural anisotropy, the dimensional variation is not the same in the three directions. But another structural aspect has to be taken into consideration when analysing the dimensional variation of cork: the corrugation of the cell walls. In fact, the straightening of the cell walls may superpose to the swelling due to water adsorption to the cell walls. One way to circumvent this feature is to test cork samples that have previously been treated to expand by reducing these undulations. Such treatment usually consists in boiling in water at 100°C and it is described in a separate point underneath in this chapter. Since it is also an important operation in the industrial processing of cork, it is further detailed in Chapter 12.

When previously boiled cork is immersed in water, there is a small increase in dimensions. The swelling increases linearly with time and after a certain period it stabilises at a constant value. The swelling is larger, about twofold, in the non-radial directions than in the radial direction, and between the non-radial directions the variation is slightly higher in the axial direction than in the tangential direction. In experiments at 20°C, the maximum swelling was attained after approximately 70 h, with a radial expansion of 0.7% and a non-radial expansion of 1.8%. The volume expansion increases with the absorption of water at a constant rate until a maximum expansion is attained, after which it remains constant with the subsequent increase in water content. This is shown in Figure 8.8 for 20°C water treatment corresponding to a maximum volume swelling of 4.4% that is attained for water content in cork of 58% (Rosa and Fortes, 1993).

Temperature increases the swelling of cork: with immersion in water at 100°C the maximum volume increase is 7.8%, which is attained after 28 h corresponding to a moisture content of 200%. The anisotropy of swelling is maintained and the lowest expansion is found in the radial direction.

In another experiment, previously boiled cork samples were immersed in water at about 22°C during 31 days to a moisture content of 52%, and subsequently dried allowing the

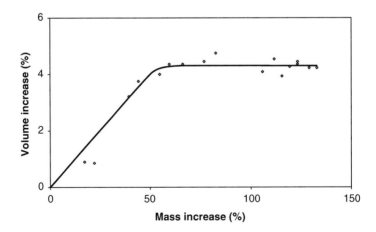

Figure 8.8. Volume expansion of previously boiled cork in function of moisture content during immersion in water at 20°C (Rosa and Fortes, 1993).

plotting of swelling in function of water content. With reference to air-dried cork at 5% moisture content, the dimensional increase for the samples at 15% moisture content were 2.2, 0.9 and 0.4% for the axial, tangential and radial directions, respectively (Reis, 1988).

8.3.3. *Water boiling of cork*

The immersion of the raw cork planks in boiling water is an operation that is undertaken for all planks that will be used in the production of stoppers or discs. The procedure is to put the cork planks in contact with water at 95–100°C in tanks or autoclaves for approximately 1 h, as described in more detail in Chapter 12. The objective is to flatten the curvature of the raw planks, to increase their mechanical workability, to expand the cork tissue and to stabilise it dimensionally, particularly in the radial direction.

The water content immediately after boiling is high with values in the range of 60–75% (Rosa et al., 1990; Costa and Pereira, 1997). A comparison between the water uptake by the cork back and the cork layers in several planks showed an average value of 55 and 44%, respectively (Pires, 2000).

This treatment expands the cork cells and reduces the undulations and corrugations of the prismatic lateral walls. Figure 8.9 shows the effect of the water boiling in the cellular structure where the straightening of the walls is clearly visible. The process is irreversible and the expanded cell form is maintained during the drying to ambient conditions. It is considered as a relief of the internal tensions that caused the cellular corrugation during cork growth (see Chapter 2).

This operation of water boiling causes a dimensional variation of cork that is strongly anisotropic. The expansion is higher in the radial direction, which is the direction with most of the undulations, by about a factor of 2–3 in relation to the expansion in the axial and tangential directions. A study on the kinetics of cork swelling in water at 100°C (Rosa et al., 1990) showed that the maximal expansion is obtained after about 30 min, corresponding to dimensional increases of 15, 6 and 6%, respectively, in the radial, axial and tangential directions, and to a total 30% volume expansion (Table 8.3). The cork density decreases by about 22% as a result of expansion and partial extraction of cork water soluble extractives (Pereira et al., 1979). Further treatment after the maximal swelling has occurred, namely an increase of the duration of the treatment or a second boiling operation does not bring additional expansion.

This swelling that is originated by the straightening of the cell walls is a short-term response that depends on the temperature of water. At 40°C, the linear expansion is less than 1%, and only at 80°C there is a substantial expansion, i.e. of 10% in radial and 4% in non-radial directions (Table 8.4).

The boiling has an effect on the mechanical properties of cork, namely in compression, as described in Chapter 9.

In practical terms the swelling of the cork planks in the radial direction corresponds to an increase of plank calliper, which is a substantial advantage regarding the cork use (see Chapter 7). As with many other properties, there is variability in the radial expansion of different cork planks. The experimental determination of radial expansion with water boiling on 680 planks showed an average value of 11.7%, ranging from almost nil expansion

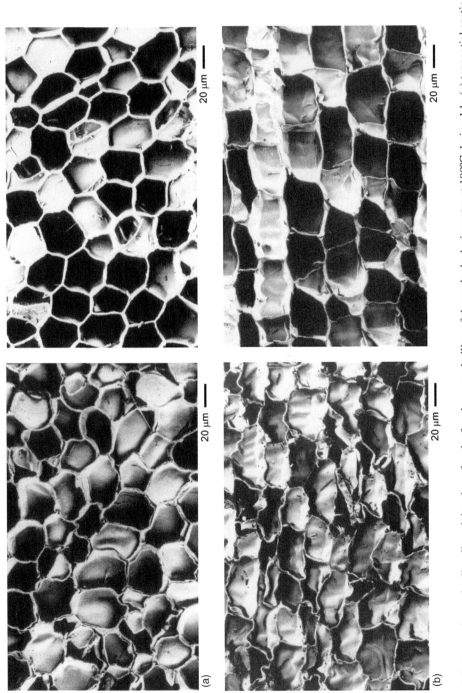

Figure 8.9. Expansion and cell wall straightening of cork after the water boiling of the cork planks in water at 100°C during 1 h: (a) tangential section,

Table 8.3. Effects of the duration of water boiling at 100°C on the dimensional, volume and density changes of cork after air drying (Rosa et al., 1990).

Duration (min)	l/l_0			V/V_0	ρ/ρ_0
	Radial	Axial	Tangential		
5	1.082±0.001	1.031±0.003	1.033±0.005	1.15±0.01	0.87±0.01
15	1.12±0.01	1.050±0.004	1.055±0.005	1.25±0.02	0.80±0.01
30	1.14±0.01	1.062±0.004	1.060±0.004	1.29±0.02	0.78±0.01
45	1.15±0.01	1.063±0.004	1.064±0.004	1.30±0.02	0.77±0.01
60	1.15±0.01	1.061±0.004	1.058±0.004	1.29±0.02	0.78±0.01

Table 8.4. Effect of the water temperature on the dimensional, volume and density changes for a 30-min treatment of cork after air drying (Rosa et al., 1990).

Water temperature (°C)	l/l_0			V/V_0	M/M_0	ρ/ρ_0
	Radial	Axial	Tangential			
40	1.005±0.004	1.003±0.002	1.002±0.002	1.01±0.01	1.019±0.004	1.01±0.01
60	1.05 ±0.01	1.012±0.002	1.01 ±0.01	1.07±0.01	1.025±0.006	0.96±0.02
80	1.11 ±0.01	1.04 ±0.01	1.04 ±0.01	1.20±0.03	1.10 ±0.01	0.92±0.03
100	1.14 ±0.01	1.051±0.002	1.05 ±0.01	1.26±0.02	1.16 ±0.01	0.92±0.02

to 38% calliper increase. The radial expansion was not correlated with the cork plank thickness or its porosity.

The radial expansion of the cork plank is the result of the summative expansion of all the annual rings. The thinner rings at the inner side showed on average a larger expansion than the larger rings in the beginning of the growth cycle: for instance the 1st ring in the cycle expands on average 13% and the 8th ring 27% (Cumbre et al., 2001). The planks flatten, i.e. they lose the curvature that they kept as a memory of the stem curvature. The porosity decreases with the water boiling, with a reduction of about 50% in the coefficient of porosity as a combined result of a larger cork area and of smaller lenticular channels.

8.3.4. Drying of cork

When drying cork, the water content decreases in a first phase rapidly and at a constant rate; afterwards the drying rate decreases and in the final phase the process is very slow until reaching the equilibrium. The initial phase corresponds to the evaporation of the free water at the surface, while the subsequent phases involve diffusion through the material and the removal of cell wall bound water.

Figure 8.10 shows examples of drying curves for cork planks, and small samples of cork and cork back after immersion in water at 100°C in various conditions. The water loss variation with time depends on the drying conditions and on the size of the material that determines the surface of drying, but the drying curves have a common pattern.

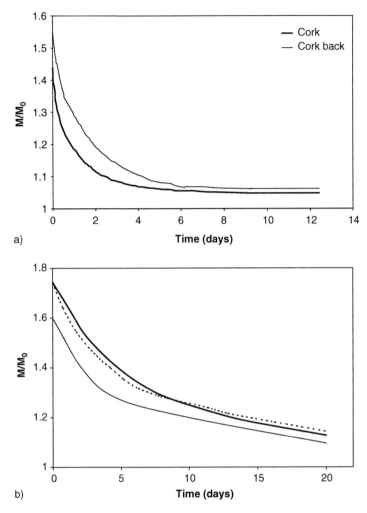

Figure 8.10. Kinetics of mass variation of cork during drying after 1 h boiling in water at 100°C: (a) small samples of cork and cork back, dried at 29°C and RH 38% (Pires, 2000); (b) three cork planks piled in covered conditions and with natural air drying (Costa and Pereira, 1997).

The cork planks that are piled under air natural conditions after the water boiling dry quickly with a decrease of moisture content from about 65 to 40% in the first 3 days. In the case of small samples the drying is more rapid, i.e. from 44 to 9% in 3 days. During drying under these conditions there is no significant dimensional variation in cork cells.

However, when cork has substantial amounts of water, for instance in wet cork with 400–500% moisture content or in cork after long periods of immersion in hot water leading to water contents over 200%, and it is oven dried at 100°C, than the result is a significant volume shrinkage up to 30% (with about 10% decrease in the linear dimensions). This shrinkage is caused by the contraction of cell walls leading to their severe wrinkling (Fig. 8.11).

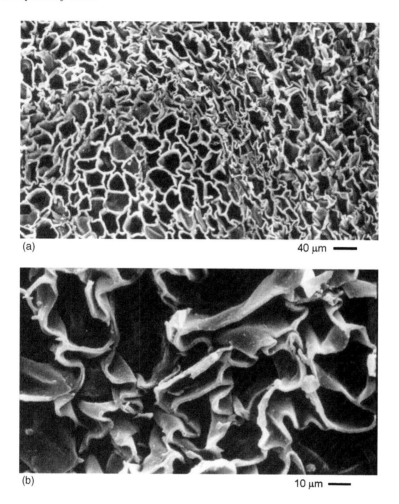

(a)   40 μm ▬

(b)   10 μm ▬

Figure 8.11. Cell structure of wetcork with an initial moisture content of approximately 500% after 100°C oven drying: (a) general view of a tangential section near the belly; (b) heavily wrinkled cells.

8.3.5. Mechanisms of water movement in cork

The absorption of water by the cork seems to proceed by two mechanisms: diffusion of the water molecules along the cell walls and the penetration into the cell lumen.

The diffusion in the cell wall is the first process to occur and it follows approximately Fick's law. The coefficients of diffusion are small and the process is slow. The limit of diffusion can be calculated as

$$X^2 = cDt$$

where X is the length of the diffusion, D the diffusion coefficient, t the time and c a factor related to the definition of the water line and approximately 1. Using $D = 2.9 \times 10^{-9}\,\text{m}^2\,\text{s}^{-1}$

at 90°C, in 1 h, the water penetrates 3.2 mm into the cork; at 20°C, with $D=1.2\times10^{-11}$ $m^2 s^{-1}$, the water penetrates only 0.2 mm in the same time.

The diffusion through the cell walls proceeds until saturation, i.e. the cell material absorbs the maximum water that it can take. This intake of water by the cell wall solid is accompanied by an increase of its volume. Therefore the cells expand with absorption of water until stabilisation corresponding to the maximum swelling and to the cell wall saturation with water. The larger swelling in the axial and tangential directions in relation to the radial direction may be related with the topochemistry of the cell wall and the supramolecular arrangement of the structural components but this is still an unknown matter.

When the cell walls become saturated, water penetrates into the cells by permeation. This is a slow process that proceeds with a much smaller rate than the cell wall diffusion, as it can be seen in Figure 8.7. The process is considered to be a combined evaporation of water from the cell wall followed by its condensation within the cell lumen.

The presence of lenticular channels is important for the water movement into cork. Some studies were made to localise the water in cork using NMR spectroscopy (Gil et al., 2000). After immersion in water during 72 h, and corresponding to an absorbed water content of 170%, it was found that water was restricted to the lenticular channels and to surface bound water. In these conditions water had not penetrated into the cells. The aliphatic extractives of cork, the so-called waxes, do not play a role in the permeability of cork to water, and extractive-free samples showed the same behaviour as non-extracted cork. On the contrary, the removal of suberin even if only in part led to a substantial penetration of water into the cells.

8.4. Conclusions

Cork is a light material due to its hollow cells and thin walls. The variation of cork density between cork planks and within one plank, is as follows:

- the region of latecork has a larger density than the earlycork layer;
- thin annual rings have a larger density than large rings due to a higher proportion of latecork cells;
- the lenticular channels usually have an overall larger density than cork due to the presence of thick-walled lignified cells;
- the cork back is denser than cork.

The hydrophobic behaviour of cork in contact with water vapour and liquid water results from its structure with hollow cells without intercellular communication and a cell wall containing suberin as the major chemical component. On the contrary, the presence of lenticular channels allows them to act as possible reservoirs for water.

The adsorption of water by cork has the following characteristics:

- the equilibrium moisture content is of small magnitude and lower than in lignocellulosic materials;
- the adsorption of water is a slow process, involving in a first phase the diffusion through the cells walls and then a permeation of water into the cellular lumen;

- at 100°C cork swells with an approximately twofold expansion in the radial direction in relation to the non-radial directions;
- the duration of water treatments made in cork processing only allows a water penetration in the external layers of the cork and into the lenticular channels.

The water boiling of cork is a very important operation in cork processing because it brings in the swelling of the material, mostly in the radial direction, a dimensional stabilisation due to stress relief, the decrease of porosity and the flattening of the cork planks.

The drying of cork with water-filled cellular lumina causes shrinkage and cell wall wrinkling. For cell-wall-only saturated cork there is no dimensional variation during drying.

References

Barbato, F., 2004. Variação radial da densidade e da porosidade em pranchas de cortiça. Graduation thesis, Instituto Superior de Agronomia, Lisboa.

Costa, A., Pereira, H., 1997. Humidade e calibre da cortiça após o descortiçamento e durante as fases de cozedura e secagem. In: Pereira, H. (Ed), Cork oak and cork/Sobreiro e cortiça. Centro de Estudos Florestais, Lisboa, pp. 347–353.

Cumbre, F., Lopes, F., Pereira, H., 2001. The effect of water boiling on annual ring width and porosity of cork. Wood and Fiber Science 32, 125–133.

Flores, M., Rosa, M.E., Barlow, C.Y., Fortes, M.A., Ashby, M.F., 1992. Properties and uses of consolidated cork dust. Journal of Materials Science 27, 5629–5634.

Gibson, L.J., Easterling, K.E., Ashby, M.F., 1981. The structure and mechanics of cork. Proceedings of the Royal Society of London A377, 99–117.

Gil, A.M., Lopes, M.H., Pascoal Neto, C., Callaghan, P.T., 2000. An NMR microscopy study of water absorption in cork. Journal of Materials Science 35, 1891–1900.

Gil, L., Cortiço, P., 1998. Cork hygroscopic equilibrium moisture content. Holz als Roh- und Werkstoff 56, 35–358.

Gonzalez-Adrados, J.R., Haro, R., 1994. Variacion de la humedad de equilibrio del corcho en plancha con la humedad relativa. Modelos de regresion no lineal para las isotermas de adsorcion. Investigacion Agraria, Sistemas y Recursos Forestales 3, 199–209.

Marat-Mendes, J.N., Neagu, E.R., 2003. The study of electrical conductivity of cork. Ferroelectrics 294, 123–131.

Marat-Mendes, J.N., Neagu, E.R., 2004. The influence of water on direct current conductivity of cork. Materials Science Forum 455–456, 446–449.

Pereira, H., Ferreira, M., Faria, M.G., 1979. Análise das águas de cozedura do tratamento industrial da cortiça. Cortiça 493, 493–496.

Pires, J.M., 2000. Controlo da maturação de pranchas de cortiça, condição para a obtenção de rolhas de qualidade. Graduation thesis, Instituto Superior de Agronomia, Lisboa.

Reis, A., 1988. Variação dimensional das rolhas com a humidade. Cortiça 596, 135–143.

Rosa, M.E., Fortes, M.A., 1988. Densidade da cortiça. Factores que a influenciam. Cortiça 593, 65–68.

Rosa, M.E., Fortes, M.A., 1989. Effects of water vapour heating on structure and properties of cork. Wood Science and Technology 23, 27–34.

Rosa, M.E., Fortes, M.A., 1993. Water-absorption by cork. Wood and Fiber Science 25, 339–348.

Rosa, M.E., Pereira, H., Fortes, M.A., 1990. Effects of hot water treatment on the structure and properties of cork. Wood and Fiber Science 22, 149–164.

Chapter 9

Mechanical properties

Cork has been singularised as a material because of its properties among which are those related to its mechanical performance. Cork is elastic and allows large deformations under compression without fracture, with substantial dimensional recovery when stress is relieved. Cork is also a light material, rather impermeable to water and other liquids (see Chapter 8). These have been the characteristics that led to its present and past uses as sealant in reservoirs, and as a basis for loads, for instance, in shoe soles and flooring. In this chapter the behaviour of cork under mechanical stress is presented regarding the stress–strain curves, and the factors that induce variation, namely arising from cork treatments such as water boiling.

Compression has been the most studied mechanical property of cork. But other mechanical properties such as tension, torsion and bending also play a role in some applications, i.e. the unbottling of champagne bottles involves the torsion of the cork stopper, and bending is involved in the handling and application of cork boards.

Under compression, cork does not fracture and the result of application of large loads is the compaction of the cellular structure and its densification. Cork fracture happens only when it is stressed under tensile or torsion forces and the deformation overcomes the material's strength.

These properties are presented here. Anisotropy is one characteristic of the material structure and cork's mechanical behaviour has to take this into account, namely by considering the stresses along the different directions. Variation of density and the extent of lenticular channels also influence the mechanical properties and both will be discussed.

9.1. Stress and strain

The mechanical properties of a material are related to its resistance when subjected to an external force and they include the resistance to deformations and distortions (elastic properties) as well as failure-related (strength) properties.

The mechanical behaviour of a material can be described by analysing the corresponding stress–strain curves. Stress (σ) is a measure of the internal forces exerted in a material that result from the application of an external force. It is calculated as the force applied per unit area of the section perpendicular to its direction of application, and measured in Pascal units (Pa). Three types of primary stress exist: compressive stress which pushes or compresses a body, tensile stress which pulls or elongates a body and shear stress which causes two contiguous segments of a body to slide one in relation to the other (Fig. 9.1). Bending stress is a combination of these three primary stresses and causes rotational distortion or flexure. Strain (ε) measures the material's deformation while under stress. It is calculated as the deformation per initial dimensional unit, in percent. The graphical plotting of stress against strain allows drawing the so-called stress–strain curves.

A material is elastic when after being deformed by a stress it can regain the original dimensions once the stress is removed. In this case the stress and strain are directly proportional under Hooke's law: $\sigma = E(\varepsilon)$. On the opposite, a material is said to be plastic when there is no dimensional recovery after removing the stress. When the recovery of dimensions is not complete, i.e. there remains a residual deformation that may evolve with time, the material is said to be viscoelastic. In this case there is no full linearity of the stress–stain elastic curve.

Two main elastic modulii characterise the elastic behaviour:

- modulus of elasticity (E), or Young's modulus, which describes the relationship of stress to deformation strain and
- modulus of rigidity or shear modulus (G), which describes the shearing stress to the angular displacement within a material.

Strength corresponds to the material's ability to resist to applied forces and represents the greatest unit stress a material can withstand without fracture or excessive distortion.

An example for a typical stress–strain curve is shown in Figure 9.2 for wood and for a cellular foam. The elastic region corresponds to the linear portion of the curve where stress and strain are directly proportional. In this region, a perfectly elastic material recovers its original dimensions after it is unloaded. For stresses larger than the elastic point, or yield point, an elastic material will not recover its original dimensions and will maintain

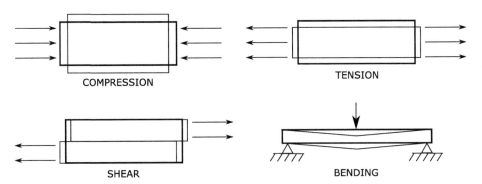

Figure 9.1. Types of loads that may be applied to a body: compression, tension, shear and bending.

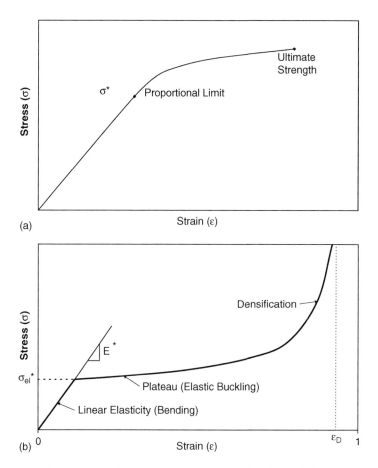

Figure 9.2. Examples of typical stress–strain curves: (a) wood and (b) cellular foam.

a permanent distortion. The final point corresponds to the maximum stress at failure in the case of wood or to densification in the case of the cellular foam.

A structurally anisotropic material will show differences in the relationship between stress and strain for the different directions of load application. Therefore, materials such as wood and cork are characterised in relation to the three main structural directions (radial, tangential and axial).

9.2. Compression of cork

9.2.1. Stress–strain curves

The stress–strain curves for the unidirectional compression of cork when loaded along the radial and non-radial (axial and tangential) directions are shown in Figure 9.3. The compression properties of cork were first characterised in general terms by Gibson et al.

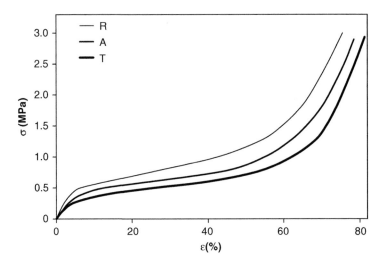

Figure 9.3. Stress–strain curves for boiled cork under compression in the three directions: radial (R), axial (A) and tangential (T).

(1981) and later on detailed in several studies (e.g. Rosa and Fortes, 1988a,b; Fortes and Nogueira, 1989; Rosa et al., 1990; Pereira et al., 1992).

One aspect that has to be taken into consideration is to what type of cork the experiments relate. Most measurements were made on the so-called bottle cork, that means, cork from cork planks that were boiled in water and constitute the material for the production of stoppers. Water boiling is a first step processing operation that is applied to all cork planks except the refuse material (see Chapter 12). It causes a relief of the cells growth stress in the tree and reduces the cell wall corrugation. The physical properties of cork are also altered by the water boiling, i.e. density is decreased and the mechanical properties are influenced as described later on. When referring to cork properties in general, and unless specified, it is meant cork after water boiling, as representative of the cork used in most applications.

The stress–strain curves of cork show three phases that follow the schematic representation given in Figure 9.2b for a cellular foam, and are associated to different deformation processes:

- There is a first phase for small stresses and deformations up to a strain of approximately 5–7%; it shows a linear relationship between stress and strain and corresponds to the elastic deformation of the cells;
- The second region starts after the yield point and it forms a large plateau that is approximately horizontal, or with only a small inclination especially for strains up to about 50%; this region corresponds to the buckling of cells;
- The last phase, starting at a strain of about 70%, shows a sharp increase of stress and a steep slope, corresponding to the densification of the material with the crushing of the cells; in the usual compression tests, maximum strains of 80–85% are attained.

Figure 9.4 exemplifies what occurs in cork at the cellular level during compression. The first phase corresponds to the elastic deformation of cork, which is practically fully reversible. It corresponds to the bending of the cell walls and occurs uniformly within the sample. The second phase corresponds to the buckling of the cells due to their plastic bending. It is a

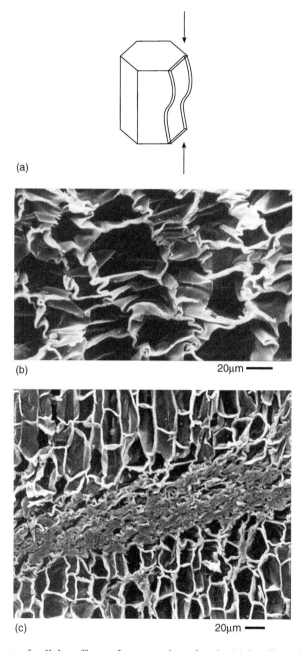

Figure 9.4. Aspects of cellular effects of compression of cork: (a) bending of cell walls in the elastic region; (b) band of buckling cells and (c) cell collapse.

non-uniform process that starts in a localised region (of weaker cells) and further extends throughout in the material. Therefore, in this region of the stress–strain curve, the material is not uniform and contains parts where the cells are collapsed and other parts where the cells have only suffered elastic deformation. The last phase corresponds to the further compression when the buckled cell walls start to touch each other and the empty volume of the cell lumen disappears. The full densification of the material would occur at about a deformation of 85%.

Compression does not cause failure of the cork cells and even in the densification phase the cell walls do dot show fractures.

9.2.2. Anisotropy of compression

The modulus of elasticity of cork is determined as the slope of the linear elastic region in the stress–strain curve. Usually an average modulus is calculated between two points corresponding to small strains (e.g. between 1 and 2%). A range of values between <10 MPa and >20 MPa are reported in the various studies for the compression of cork in the different directions.

The comparison of the stress–strain curves obtained for the compression in the different directions shows that cork strength in the radial direction is higher than in the other directions, and that the axial and tangential directions are more similar. For this reason, sometimes only two compression directions are considered: the radial direction and the non-radial directions. This corresponds to a perfect hexagonal prismatic structure where there is symmetry around the radial direction. However, in cork the two non-radial directions are not totally equivalent and the stress–strain values for the axial compression are higher than those for the tangential compression.

A recent study reported the following average values for the Young modulus of cork: 18.3 MPa for the radial direction, 16.9 MPa for the axial direction and 12.3 MPa for the tangential direction (Anjos et al., 2006).

9.2.3. The Poisson effect

The Poisson effect corresponds to the variation of dimensions in directions perpendicular to the direction of compression. Usually this variation corresponds to an increase of dimensions. It is measured by the Poisson coefficient (υ_{12}) defined as the negative ratio of strains in one perpendicular direction (ε_2) and in the compression direction (ε_1): $\upsilon_{12} = -(\varepsilon_2/\varepsilon_1)$.

For cork the Poisson coefficients are very small, near zero for $\upsilon_{R/NR}$ and $\upsilon_{NR/R}$ and about 0.2 for $\upsilon_{NR/NR}$ (Fortes and Nogueira, 1989). The reason for such low variation of dimensions is related to the material's ability to undulate the cell walls allowing for a large deformation without lateral expansion. When cork is compressed in the radial direction, the staggered and random localisation of the cell bases and the undulation of the lateral cell walls allow the increase in cell wall folding and the alignment of bases in the direction perpendicular to the stress. When the compression is in the non-radial directions, the lateral walls bend and straighten, thereby increasing the dimension in the radial direction ($\upsilon_{R/NR} > 0$), but for large strains the Poisson coefficient becomes negative ($\upsilon_{R/NR} = -0.01$), because the cells are squeezed by the bases of their neighbouring cells and decrease their radial dimension.

9.2.4. The effect of air in the cells

The cells in cork are filled with air that is entrapped in their lumen since there are no intercellular communication channels. It could be thought that these air pockets could contribute to the compressive strength of cork, and this is claimed in non-technical environments. But some calculations on the air pressure that develops in the cells during compression prove that this is erroneous (Fortes and Rosa, 1987).

Considering that the air approximately behaves as an ideal gas, and that it is initially at a pressure of 0.1 MPa (in equilibrium with the atmospheric pressure), if the volume is reduced to half ($\varepsilon = 50\%$) the air pressure doubles to 0.2 MPa. This is a small value considering the compression stress at this deformation level (σ_{50} *ca.* 1.5 MPa). The component of resistance given by the air pressure only has significance for much higher strains: for instance, for $\varepsilon = 80\%$, corresponding to a volume reduction to approximately 20%, the air pressure will be 0.5 MPa.

9.2.5. *The effect of density, porosity and moisture*

The density influences the compressive properties of cork (Fig. 9.5). There is a trend of increasing the Young's modulus (E) with the density, especially for compression in the radial direction, while the effect is less marked for compression in the tangential direction.

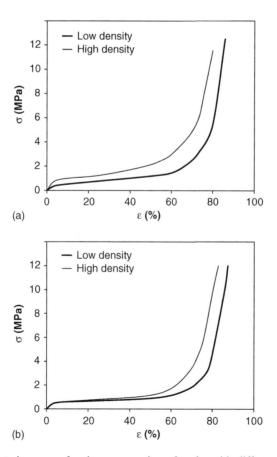

Figure 9.5. Stress–strain curves for the compression of corks with different density: (a) radial direction and (b) tangential direction (Anjos et al., 2006).

As regards the compression behaviour for higher strains, the higher density corks show stress values above those of lower density corks, and the crushing and collapse of the cells corresponding to the steep increase of stress starts for lower strains.

In relation to the influence of the lenticular channels, measured by the coefficient of porosity of the surface perpendicular to the direction of compression, it is found that the stress–strain curves are similar regardless of the porosity of the samples. The porosity does not influence the elastic properties and the plot of the Young's modulus against the coefficient of porosity shows no influence for compression in either direction. However, for higher strains there is a trend for an overall increase of stress with porosity. Figure 9.6 exemplifies this behaviour by showing the variation of the Young's modulus and the stress corresponding to a 25% strain with the porosity.

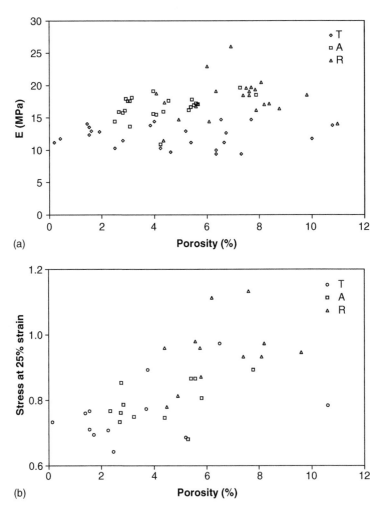

Figure 9.6. Variation of the Young's modulus (a) and the stress for a 25% strain (b) with the coefficient of porosity measured in the section perpendicular to the direction of compression (adapted from Anjos et al., 2006).

The effect of pores cannot be ignored when studying the compressive properties of cork and it may explain some of the unaccounted for differences between samples. However, the effect may be difficult to follow since several superposed effects are involved due to the variability found in the anatomical characteristics of the cork porosity: dimensional variability of the lenticular channels, of the lenticular filling tissue and of the thick-walled lignified cells associated to the pores.

The moisture in cork also influences its compression behaviour. An increase in the moisture content of cork decreases the resistance to compression and reduces its anisotropy (Table 9.1). The recovery of dimensions after the compressive load is removed is also higher when cork has higher moisture content. This effect is caused by the presence of water molecules within the cell walls, most probably bound to the cell wall polysaccharides that act as cell wall plasticisers.

9.2.6. The effect of water boiling

Figure 9.7 shows the stress–strain curves for compression in the radial and non-radial directions of raw cork (unboiled), of cork immediately after 1-h boiling in water at 100°C (with approximately 75% moisture content) and after air equilibrium (about 6% moisture content). The changes that are caused by boiling can be summarised into: (a) there is a general reduction in strength; (b) there is a reduction in the anisotropy, particularly in the elastic region and (c) a sharper yield point appears in the radial compression.

Immediately after boiling, cork is very soft and isotropic, and the Young's modulus is reduced by a factor of 5–9. This explains why the internal stresses of the raw cork may be relieved by water boiling.

The raw unboiled cork shows some differences in the compression curves in relation to what has been described until now, namely in relation to the stress–strain curve for the radial compression. The three regions are not so distinctly visible, namely there is no well-defined yield-point and a plateau region. In boiled cork, the yield point is sharper and the radial compression curve resembles those for the non-radial directions. These changes are related to the attenuation of the corrugations of the lateral cell walls that occur with the boiling. In fact, walls that are already corrugated, such as it is the case in

Table 9.1. Effect of moisture content (H) of cork in the Young modulus (E) and the final strain (ε_f) for a stress of 12.5 MPa in compression in the radial and non-radial directions (Fortes et al., 2004).

	Moisture content			
	6%	9%	12%	33%
E (MPa)				
Radial	15.4	12.7	11.6	5.9
Non-radial	11.6	11.1	10.5	5.7
ε_f (%)				
Radial	81	85	86	87
Non-radial	86	86	87	88

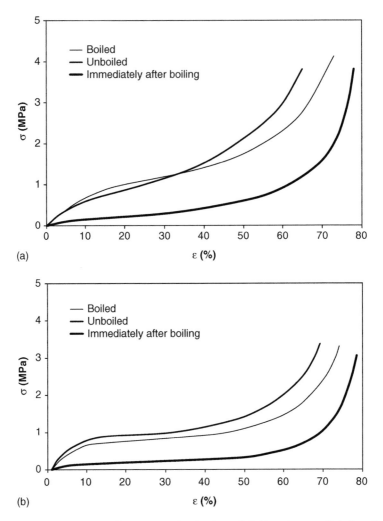

Figure 9.7. Stress–strain curves for compression in the radial (a) and non-radial directions (b) of raw cork (unboiled), of cork immediately after 1-h boiling in water at 100°C (with approximately 75% moisture content) and after air equilibrium (about 6% moisture content) (Rosa et al., 1990).

a large proportion of the cells in raw cork, may yield without buckling, as in the folding of a concertina, and therefore the yield transition and the plateau become less sharp.

By allowing cell straightening and the relief of tensions, the water boiling operation makes the structural arrangement of cork more regular and with a compression behaviour that is closer to the typical compression of a regular cellular solid, as shown in Figure 9.2.

9.2.7. The recovery of dimensions and stress relaxation

The complete recovery of dimensions after the removal of the compressive stress only occurs in fully elastic materials. In cork the dimensional reversibility is not total due to the viscous component of the deformation. Cork only approximates an elastic behaviour and in fact it is a viscoelastic material (Mano, 2002).

Nevertheless, the permanent deformation of cork is very small in the elastic region and only higher for strains above the yield point. Figure 9.8 shows the dimensional variation after radial and non-radial compression to strains of 30 and 80% (Rosa and Fortes, 1988b). A large part of the recovery of dimensions is instantaneous. The recovery proceeds regularly with time and ceases after a few days with a residual strain of approximately −3 and −6%, respectively, for initial compression strains of 30 and 80%. Other studies report somewhat higher residual strains after compression to 80% strain (Anjos, 2000): 8.9, 11.3 and 11.2%, respectively, for compression in the radial,

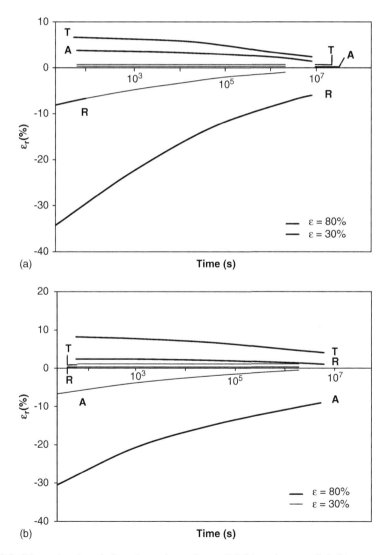

Figure 9.8. Dimensional variation along time after radial (a) and non-radial (b) compression to strains of 30 and 80% (R – radial, A – axial, T – tangential) (adapted from Rosa and Fortes, 1988).

axial and tangential directions, with over 70% recovery in the first 4 min after unloading. The process of dimensional recovery is accelerated by temperature, namely by
water boiling.

When cork is compressed to a certain deformation and this deformation is kept, it is
found that there is a stress relaxation along time, that means, the stress required for a constant strain decreases with time. This is a phenomenon that occurs, for instance, with cork
stoppers when inserted in a bottle: they maintain their strain, given by the bottle-neck
diameter, but the pressure applied to the bottle decreases with time. The stress relaxation
is very quick in the first minutes and slows down afterwards attaining a relaxation limit
after about 10 min that corresponds to 30–40% of the initial stress (Rosa and Fortes,
1988a; Fortes et al., 2004). The stress relaxation in proportion of the initial stress is not
influenced by the moisture content of the cork (Table 9.1).

9.3. Tension

The tensile behaviour of cork has been less studied than the compressive properties, and
some of the effects that will impact the tensile behaviour of cork, for instance the rate of
deformation, temperature and moisture content, have not yet been investigated.

The behaviour of cork in tensile stress is quite different from that shown in compression, as shown in Figure 9.9 with examples of tensile stress curves for the axial,
tangential and radial directions. In comparison with compression, the stress–strain
curves do not show a plateau and the samples fracture at a certain stress corresponding
to relatively small strains (Rosa and Fortes, 1991; Anjos, 2000). Under tensile stress,
the cork cell walls do not bend but are pulled in the direction of the stress and the cell
walls undulations are attenuated. Therefore, the Young's moduli in tensile are higher

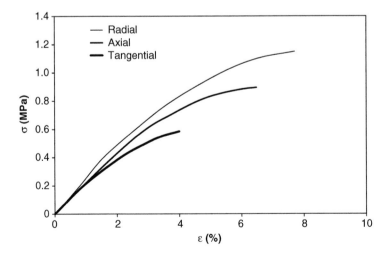

Figure 9.9. Tensile stress–strain curves of boiled cork stressed in the radial, axial and tangential
directions (Anjos, 2000).

than those in compression: 31.7, 23.9 and 31.2 MPa in axial, tangential and radial directions, respectively. While in compression, the amplitude of the undulations of the cell walls increase, in tension it decreases, thereby increasing the stiffness of the cell walls.

When the direction of traction is axial or tangential, the fracture occurs for a stress of approximately 1.0 and 0.6 MPa, respectively, after an elongation of about 5–6%. When traction is in the radial direction, the fracture occurs at stress values of about 1.1 MPa corresponding to elongations of about 8%.

In general, the Young's modulus increases with the density and decreases with the increase of the porosity. Under axial or tangential tension, the fracture initiates near a pore or lenticular channel and the fracture line is, on average, perpendicular to the direction of tension (Fig. 9.10a). The propagation of the fracture is usually along the cell walls through the middle lamella (Fig. 9.10b), although also, in some cases, across the cell walls. For the traction in the radial direction, the fracture occurs across the lateral cell walls and in the more fragile region of the early cork cells (Fig. 9.10c and d).

The presence of lenticular channels is very important for the behaviour of cork under tensile stress, namely for the fracture stress, since pores function as stress concentration points. A study on the influence of porosity characteristics on the fracture stress of cork has shown that the coefficient of porosity, the number of pores and their dimension are negatively correlated with the tensile properties of cork (Anjos, 2000).

9.4. Bending

Bending is a combined stress that includes compressive and tensile stresses: the upper part of the body is under compression in a direction perpendicular to the bending load and the under part is under traction in the same direction. Figure 9.11 shows stress–strain curves for three-point bending in the radial, axial and tangential directions (Anjos, 2000).

The curves are similar to the tensile stress–strain curves with a small initial phase of elastic behaviour until a strain of approximately 5%. For loading in the radial direction, the stress for a given strain is lower when the compression and traction stresses are in the tangential direction than when they are in the axial direction. In this case the Young's modulus is 17.6 and 22.5 MPa, the maximum stress 1.1 and 1.6 MPa for a strain of 12.6 and 18.4%, respectively. A similar relation occurs when comparing loading in the axial direction and compression/traction in the tangential direction with loading in the tangential direction and compression/traction in the axial direction: Young's modulus of 17.9 MPa, maximum stress of 1.3 MPa for a strain of 13.1% in the first case, 21.8 MPa, 1 MPa and 17.6%, respectively, in the second case.

Fracture occurs at the side under tension when the tensile strength is overcome. Therefore, the factors that influence the tensile properties of cork, i.e. tensile strength increases with density and decrease with porosity, also apply for the behaviour of cork under bending. The failure occurs along a lenticular channel, as it also happens in tensile tests of cork.

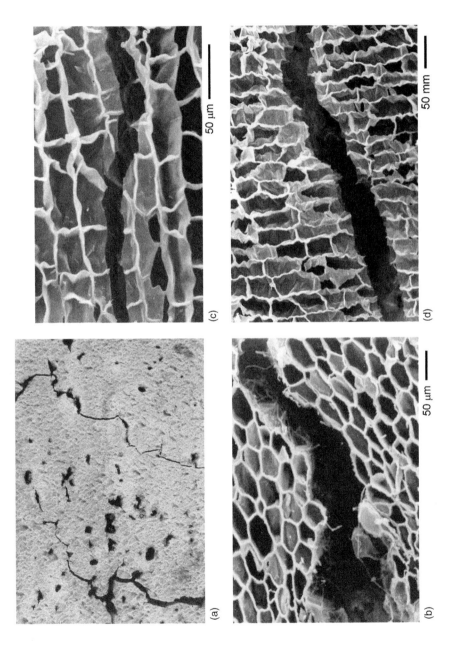

Figure 9.10. Fracture of cork under tensile stress: (a) tension in the axial direction (horizontal direction in the picture) showing the fracture line through the pores and developing approximately in the perpendicular direction; (b) fracture through the middle lamella due to axial traction; (c) fracture under

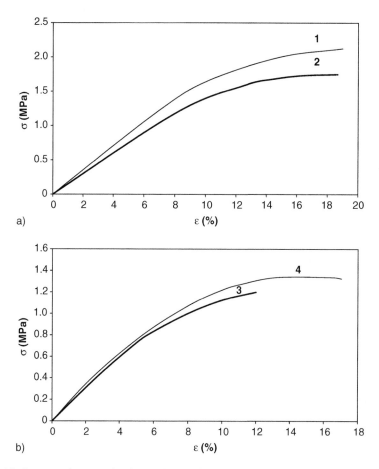

Figure 9.11. Stress–strain curves for three-point bending of cork of boiled cork samples of good quality when the load is applied in the radial direction (a) and in non-radial directions (b): (1) load applied in the radial direction, with the compression/traction in the tangential direction; (2) load applied in the radial direction, with the compression/traction in the axial direction; (3) load applied in the tangential direction, with the compression/traction in the axial direction and (4) load applied in the axial direction, with the compression/traction in tangential direction (adapted from Anjos, 2000).

9.5. Torsion

In torsion, a binary (M_t) is applied that originates an angular deformation (θ), as schematically represented in Figure 9.12 for a cylindrical body. The shear stresses (τ) that originate in the interior of the body and the resulting distortion (γ) are given by

$$\tau = \frac{32M_t r}{\pi D_4} \quad \text{and} \quad \gamma = \frac{r\theta}{L}$$

where D is the diameter, r the distance to the cylinder axis and L the length of the cylinder. The maximum values of shear stress and distortion occur at the surface ($r = D/2$) as

$$\tau_{max} = \frac{16M_t}{\pi D_3} \quad \text{and} \quad \gamma_{max} = \frac{D\theta}{2L}$$

During torsion, the moment is measured with the torsion angle. These stress–strain curves are represented in Figure 9.13 for the torsion of cork stoppers (cylinders) with the axis in the radial and non-radial directions (Fortes et al., 2004). No difference is found between the two directions. In the beginning there is linearity until a maximum shear stress of about 0.7 MPa and a maximum distortion of 0.08, with the shear modulus G of 9.1 MPa. The fracture occurs when the maximum shear stress is attained.

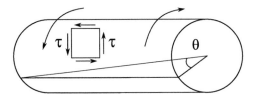

Figure 9.12. Schematic representation of torsion of a cylinder.

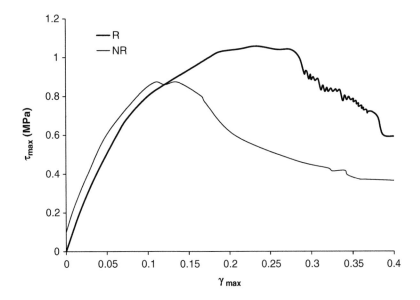

Figure 9.13. Torsion of natural cork stoppers cut with the cylinder axis along the radial (R) and non-radial (NR) directions, represented as the maximum shear stress in function of the maximum distortion (redrawn from Fortes et al., 2004).

9.6. Variation of mechanical properties in cork

One factor of variation of the mechanical behaviour of cork regards the thickness of the original cork plank or its mean annual growth ring width, since this is related to density and, at least to a certain extent, to some features of the corrugation of the cells. Cork from large calliper planks, with a mean annual ring of 6 mm per year, shows lower strength in compression for all strains and lower Young's moduli than thin and mean calliper planks (2–4 mm per year) as shown in Figure 9.14 (Pereira et al., 1992).

The variation of compressive properties between corks from different trees and locations was measured in 200 samples from 10 different provenances (20 trees per site) in Portugal. The average Young's moduli was 10.4 and 9.3 MPa in the radial and non-radial directions, respectively. Most of the variation occurred between trees and the provenance is not a statistical significant factor of variation.

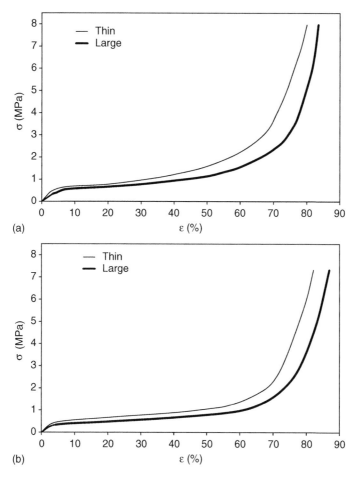

Figure 9.14. Comparison of compression curves of cork from planks with thin and large calibres along the radial (a) and the tangential (b) directions (Pereira et al., 1992).

As regards the commercial quality of cork planks, a comparison between a good (class 1) and average (class 4) quality shows no significant differences in compression except for dimensional recovery, which is higher for the good quality cork samples. In tension, the good quality cork samples show larger strength: in the axial direction, Young's moduli of 31.7 and 22.9 MPa, stress at fracture 1.0 and 0.8 MPa and maximum elongation 6.3 and 5.8%, respectively, for class 1 and class 4 corks; and in the tangential direction Young's moduli of 23.9 and 19.8 MPa, stress at fracture 1.4 and 0.6 MPa and maximum elongation 4.4 and 4.7%, respectively (Anjos, 2000). In flexure the good quality corks also show higher strength.

9.7. Conclusions

Cork is an approximately elastic cellular material that combines an interesting set of mechanical properties. In the mechanical performance of cork, two features have an important role: the ability of cell walls to buckle and strongly collapse without cell wall fracture and the presence of lenticular channels.

In compression, cork has an elastic region followed by a broad plateau where the collapse of cells allows strong dimensional reductions for small stress increases. Although anisotropic, the difference in the compressive behaviour of cork in the different directions is not very large. Cork does not fracture under compression either across cells or across cell walls, and for the highest stresses it undergoes a process of densification. The dimensional recovery after unloading is substantial and the residual strain is small even from initial strains within the collapse region and densification regions. When compressed, cork does not vary (or only little) dimensionally in the perpendicular directions. The boiling of cork in water at 100°C makes cork more uniform and with a better-defined yield-point and plateau region.

Cork only fractures if stressed in tension (also on the tension side in bending) and under shear forces. Pores have a major effect in the tensile strength of cork and fracture initiates always at the vicinity of a pore.

References

Anjos, O.M.S., 2000. Caracterização mecânica da cortiça e sua relação com a qualidade. Ph.D. Thesis. Instituto Superior Técnico, Lisboa.
Anjos, O., Pereira, H., Rosa, M.E., 2006. Effect of quality, porosity and density on the compression properties of cork. Holz als Roh- und Werkstoff, submitted.
Fortes, M.A., Nogueira, M.T., 1989. The Poisson effect in cork. Materials Science and Engineering A122, 227–232.
Fortes, M.A., Rosa, M.E., 1987. Efeito do gás contido nas células na compressão da cortiça. Cortiça 590, 331–334.
Fortes, M.A., Rosa, M.E., Pereira, H., 2004. A cortiça. IST Press, Lisboa.
Gibson, L.J., Easterling, K.E., Ashby, M.F., 1981. The structure and mechanics of cork. Proceedings of the Royal Society of London A377, 99–117.
Mano, J.F., 2002. The viscoelastic properties of cork. Journal of Materials Science 37, 257–263.
Pereira, H., Graça, J., Baptista, C., 1992. The effect of growth rate on the structure and compressive properties of cork Quercus suber L. IAWA Bulletin 13, 389–396.

Rosa, M.E., Fortes, M.A., 1988a. Stress relaxation and creep of cork. Journal of Materials Science 23, 35–42.

Rosa, M.E., Fortes, M.A., 1988b. Rate effects on the compression and recovery of dimensions of cork. Journal of Materials Science 23, 879–885.

Rosa, M.E., Fortes, M.A., 1991. Deformation and fracture of cork in tension. Journal of Materials Science 26, 341–348.

Rosa, M.E., Pereira, H., Fortes, M.A., 1990. Effects of water treatment on the structure and properties of cork. Wood and Fiber Science 22, 149–164.

Chapter 10

Surface, thermal and other properties

The properties of cork that have called attention to this material are not only those related to density, water absorption and mechanical behaviour that are described in the preceding chapters. Other characteristics also contribute to define the complex and unusual nature and performance of cork, and they will be presented here. On the whole, however, they have not been much researched.

The thermal insulation properties of cork are well known in the practice, and they constitute one of the physiological adaptation assets of the cork oak tree since they allow protection against the high temperatures of summer and the occurrence of frequent fires. Copying nature, man has insulated homes and artefacts with cork products since antiquity.

Surface properties are involved in some of the applications of cork, e.g. friction is one important aspect in the performance of wine stoppers, as well as in surfacing. In the latter case wear properties are also of interest. The wettability and the energy characteristics of cork surfaces are important to understand the behaviour with liquids and chemicals, and they constitute background information to develop cork modifications or cork composite products. The results of the few studies available for cork will also be presented here.

Finally cork is a viscoelastic material, and the viscose component is responsible for the damping and energy absorption properties of cork that make it one of the most useful vibrational insulators and shock absorbers.

10.1. Surface properties

10.1.1. Wettability

When a liquid is put into contact with a solid surface, it can spread more or less quickly over the surface or remain confined in drops, depending on the liquid–solid relative chemical affinity and on the surface tension at the liquid–solid–air interfaces. This is usually referred to as the wettability of the solid in relation to a particular liquid.

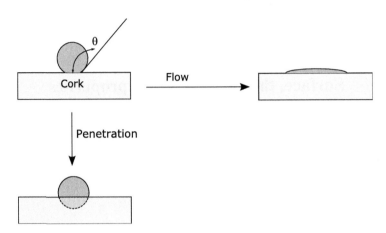

Figure 10.1. Schematic representation of a drop of liquid on a solid surface, and of the corresponding contact angle, with the time evolution due to flow and penetration.

One way to analyse the wettability is by measuring the contact angle (θ) of a drop of the liquid on the solid surface. If the liquid spreads immediately on the surface as a film, $\theta = 0$; when $\theta < 90°$, the solid is wetted by the liquid, and more so as the lower is the contact angle, while for $\theta > 90°$ the solid is not wetted by that liquid. For a given solid surface, the contact angle depends on the chemical nature of the liquid and on its physical characteristics, e.g. viscosity. It is also important to know the evolution with time of a liquid on a solid surface, which combines the liquid flow on the surface and the penetration into the solid (Fig. 10.1).

The contact angle of water on cork is approximately 84°, a relatively high value that is in accordance with the low water affinity of cork (Gomes et al., 1993). The contact angle decreases with time nearly linearly because of water penetration into the cork and the modification of the surface characteristics, and after 1 h it is approximately 40°. The initial contact angle is smaller for less polar liquids, e.g. 61° for di-iodomethane, and nearly 0° for non-polar n-alkanes such as n-heptane, n-decane and n-hexadecane.

The contact angle of resins and adhesive precursors on the cork surface was also studied (Adão et al., 1993; Teixeira et al., 1996). The initial contact angle depends on their chemical nature while the time evolution is related to the liquid viscosity. For instance, with a polyether adhesive the initial contact angle is approximately 27° and decreases rapidly to 7°.

One study has also measured contact angle of various liquids on depolymerised suberin obtained by alkaline methanolysis of cork: it is approximately 75° for water, 45° for formamide and 20° for di-iodomethane (Cordeiro et al., 1997).

10.1.2. Surface energy

The surface properties of cork in terms of surface energy and acid/base characteristics are important for its interaction with chemical systems, namely with adhesives, and they can be used for an optimisation of cork composites. The information available for cork is

Table 10.1. Dispersive surface energy of cork and of depolymerised suberin determined by inverse gas chromatography (IGC) or contact angle (CA) at different temperatures (Cordeiro et al., 1995, 1997; Godinho et al., 2001).

Material	Method	Temperature (°C)	γ_s^D (mJm^{-2})
Cork	IGC	25	41
	IGC	40	38
	IGC	50	35
	IGC	60	34
	IGC	70	31
	CA	24	24
Depolymerised suberin	IGC	25	48
	IGC	50	44
	IGC	60	42
	IGC	70	41
	IGC	80	40
	IGC	90	37

Table 10.2. Surface acid–base characteristics of cork and depolymerised suberin (Cordeiro et al., 1995; Godinho et al., 2001).

Material	K_A	K_B	K_A/K_B
Cork	0.32	0.29	1.1
Depolymerised suberin	0.35	0.15	2.3

shown in Table 10.1, which also includes data for depolymerised suberin, in the form of a melted film.

The total surface energy and the dispersive surface energy were determined on cork by two methods. From contact angle measurements, the total surface energy was calculated as 32 mJ m^{-2} of which 24 mJ m^{-2} corresponded to its dispersive component (Gomes et al., 1993). Using inverse gas chromatography (IGC), a method that avoids the limitations of contact angle measurements on rough and porous surfaces, the determined dispersive component of surface energy was higher, 41 mJ m^{-2} (Godinho et al., 2001). The surface energy decreases with temperature from 38 to 31 mJ m^{-2}, respectively for 40 and 70°C (Cordeiro et al., 1995).

The depolymerised suberin has a higher dispersive surface energy: 48 mJ m^{-2} at 25°C, between 44 and 37 mJ m^{-2} at respectively 50 and 90°C (Cordeiro et al., 1997; Godinho et al., 2001).

The acid/base properties were evaluated using tetrahydrofuran, chloroform and ethyl acetate as polar probes. The results are shown in Table 10.2 together with values for suberin extracted from cork by alkaline alcoholysis. The K_A/K_B value of 1.1 shows that cork has an amphoteric nature with an equal contribution of acidic and basic character which should make it compatible with both acidic and basic polymeric matrices.

The values of K_A and K_B for suberin were 0.35 and 0.15, respectively, with a resulting K_A/K_B of 2.3. The acidic character of this suberin results from the formation of carboxylic end groups by the depolymerisation process and do not represent the acid/base characteristics of *in situ* suberin (see Chapter 3).

10.2. Friction

The resistance to movement offered by cork against a surface is a property of interest in especially two cases: the extraction of cork stoppers from bottles where the cork moves against a glass surface, and the use of cork products as surfacing materials, sometimes with anti-slippery objectives. The properties related to the friction of cork have been little studied and only one publication deals with it in some depth (Vaz and Fortes, 1998).

The friction has two components: adhesion and inelastic loss (Gibson et al., 1981). Adhesion refers to the forces established at atomic/molecular level between two surfaces in contact and energy has to be spent to break such adhesive bonds and separate the two surfaces. This is a surface process and treatments that change the surface, i.e. a film coating, will change or eliminate it. This is what happens in the sliding movement of a cork stopper against the bottleneck, and therefore the extraction force may be largely reduced by use of a finishing coating with paraffin or silicon (see Chapter 12).

Anelastic loss derives from the fact that cork is not perfectly elastic, meaning that the deformation energy caused by a compressive load is not totally recovered when the load is removed. This is what happens when a body with a rough surface slides on cork or a ball rolls on cork (Fig. 10.2). The work done to deform the cork at the front is not recovered as the sliding body passes on; the difference is called the loss coefficient, which is high for cork, and the energy lost is schematically represented in Figure 10.2 as the shaded area. This is not a surface phenomenon and therefore its value is not altered by surface treatments. This anelastic loss is the main process involved, for instance, in the movement of a cork shoe against the floor, or of an object sliding on a cork floor. In this case the friction properties of cork should not be altered by a surface treatment, i.e. a water film or a wax treatment.

The friction coefficient (μ) of one body on a surface is defined as the ratio of the compression force applied against the surface and the friction force tangential to the surface necessary to maintain a certain velocity of the body. Due to the anisotropy of cork the

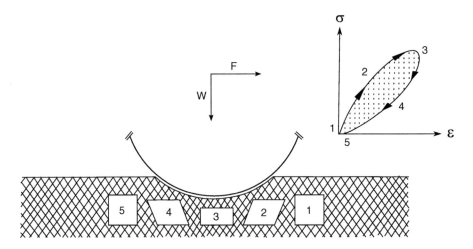

Figure 10.2. Representation of a portion of a rough body sliding on cork: deformation occurring in cork in different positions and stress–strain curve corresponding to the process (adapted from Gibson et al., 1981).

Table 10.3. Variation of friction coefficients of cork in various conditions (Vaz and Fortes, 1998).

	μ_R	μ_{NR}
Compression[a] (MPa)		
0.12	0.74	0.50
0.46	0.55	0.46
Moisture content[b] (%)		
6	0.74	0.57
15	0.56	0.52
Surface of sliding[c]		
Cork (tangential surface)	0.97	0.77
Smooth glass	0.74	0.50
Rough glass		0.67
Smooth steel		0.64
Rough steel		0.71

[a]$v = 0.66$ ms^{-1}, cork moisture: 6% , smooth glass.
[b]$\sigma = 0.12$ MPa, $v = 0.66$ ms^{-1}, smooth glass.
[c]$\sigma = 0.12$ MPa, $v = 0.66$ ms^{-1}, cork moisture: 6%.

orientation of both the sliding surface (tangential and non-tangential, corresponding to a compression force in the radial and non-radial directions), and the direction of sliding in the case of the non-tangential surface (radial and tangential) has to be considered.

The values of the friction coefficients of cork for compression in the radial and non-radial directions, respectively μ_R and μ_{NR}, are summarised in Table 10.3 for various conditions. There is anisotropy regarding the compression surface of cork and the friction coefficient is higher when the sliding plane is tangential. On the contrary, there is no in-plane anisotropy of sliding. The friction coefficient decreases with increasing compressive load and moisture content. The moisture content also reduces the difference between the friction properties of tangential and non-tangential surfaces, i.e. the anisotropy is reduced in humid cork.

The friction increases with the roughness of the counter surfaces. The friction behaviour of cork on other surfaces depends on the adhesion (related to the contact area and friction coefficient of the broken cell walls) and on the deformation by bending of the cell walls (related to the roughness of the counter surface and to the bending stiffness of the walls). Overall the friction coefficients of cork have a high value giving cork its anti-slippery character and an easy-grip handling.

10.3. Wear

Cork is a material resistant to surface wear that withstands without significant fracture or abrasion the repeated rubbing of objects against it. This is the reason why cork-based products, mostly cork agglomerates, are used for flooring in areas subject to intense use, such as hospitals, schools or airport halls. However, quantification of cork wear under controlled conditions is scarce. An early report determines the wear of cork agglomerates in comparison with oakwood and Carrara marble and refers that cork is much more resistant to thickness and mass loss (Natividade, 1938). When cork surfaces are rubbed with a smooth surface, i.e. leather-covered wheels, the cells are squeezed and appear as closed, while with abrasive grits the cell walls are torn and the material seems to be pulled out (Baptista and Vaz, 1993).

The sanding of cork surfaces with abrasive discs or cylinders is one finishing operation for dimensional rectification of stoppers or for surface smoothing of boards of agglomerated cork. The resulting cork particles are very fine, with granulometries well below 0.25 mm and include lenticular material in the form of minute aggregates containing a few cells and cork material with larger particles containing some hundreds or thousands of cells.

10.4. Thermal properties

The heat transfer properties of cork, as of most closed cell foams, are very poor due to the following reasons: the solid fraction is low; the gas enclosed in the cells has a low conductivity; the cells are small which eliminates the gas convection; radiation is reduced through the repeated absorption and reflection at the numerous cell walls.

The thermal conductivity has four contributions: conduction through the solid, conduction through the gas, convection within the cells and radiation through the cells and across the cell voids (Gibson and Ashby, 1997). Heat flow by conduction depends only on the amount of solid and it does not depend directly on cell size. Flow by convection depends on cell size because heat can be transported from one side of the cell to the other by the air convection currents inside the cells. However for small cells below 10 mm, the convection does not contribute significantly to the total heat flow. Flow by radiation depends on the number of times that heat has to be absorbed and re-radiated by cell walls and therefore the smaller the cells the lower is the rate of flow.

If cork is compared with other foams used for insulation, it is seen that they have comparable heat transfer properties but cork has a significantly higher density (Table 10.4). It is the fact that cork has much smaller cells that allows attaining the low thermal conductivity values in spite of the higher density. On the other side, this higher solid fraction imparts to cork other properties, namely regarding its mechanical behaviour, that make it an overall very interesting material (Fernandez, 1987; Pinto and Melo, 1988).

The properties of the cork cell wall material, as of polymers in general, depend on temperature, i.e. they undergo significant changes at a critical temperature, the glass transition temperature, T_g. For a composite of polymers, such as is the case of the cork cell wall material, the glass transition occurs at various interval of temperatures. It is estimated that the glass transition of cork occurs approximately at ambient temperatures, of about 20°C (Mano, 2002).

Table 10.4. Heat conductivity (λ) and thermal diffusivity (α) of cork, of expanded cork agglomerates and of two other insulation foams.

	Density (kg/m³)	λ (W/mK)	α (m²/s)
Cork	140–170	0.040–0.045	10^{-7}–1.5×10^{-7}
Expanded cork agglomerate	100–120	0.035–0.070	1.5×10^{-7}–3.5×10^{-7}
Polyurethane foam	20	0.025	9.0×10^{-7}
Polystyrene foam	25	0.040	1.1×10^{-6}

The thermal expansion of cellular solids corresponds almost exclusively to the expansion of the solid material, and for most polymers the thermal expansion coefficient is approximately 10^{-4} K^{-1}.

In the case of cork, the dimensional variation with changing temperature has to consider another component in addition to the thermal expansion of the cell wall material: it is related to the structural characteristics of the cells and the fact that the cell walls are undulated. Therefore together with the solid thermal expansion, there is a much larger volume increase due to the attenuation of the cell wall corrugations. Heat treatments, especially if in water, result into straightening of the cell wall corrugations that can considerably expand the cells, i.e. the volume increase can attain 15% for a 100°C treatment, as discussed in Chapter 8.

In expanded cork agglomerates the coefficient of thermal expansion at 20°C is 0.3–0.5 $\times$ 10^{-4} (Medeiros, undated). In this case the thermal expansion only refers to the cell wall solid because the cells have not only lost their undulations but chemical changes have also occurred in the cell wall composition and in its structural polymers (Pereira and Ferreira, 1989). The production process and characteristics of the expanded cork agglomerate are detailed further on in Chapter 13.

The specific heat (C_p) is the energy required to raise a unit mass of a material by a unit temperature. The specific heat of cork is practically equal to the specific heat of the solid material since the weight fraction of air is negligible. It should be approximately 2000 J (kg K)$^{-1}$, which is the value for most polymers. In expanded cork agglomerates, the determined value of specific heat is 1700 J (kg K)$^{-1}$ (Andrade, 1962).

The heat conductivity (λ) and thermal diffusivity (α) of cork, of expanded cork agglomerates and of two other insulation foams are summarised in Table 10.4. The conductivity increases with temperature and density. It also increases with moisture content as was shown for insulation expanded cork agglomerates (Matias et al., 1997).

10.5. Heat treatments

When cork is heated there are chemical and physical changes that affect the cell wall mass, their composition and the cellular dimensions. These changes influence the properties, namely the mechanical properties, in an extent that depends on temperature and time of treatment.

The thermogravimetric analysis of cork and of depolymerised suberin is shown in Figure 10.3 (Rosa and Fortes, 1988a; Cordeiro et al., 1998). The mass loss of cork is relatively small (about 6% of the initial mass) until 200°C, but increases afterwards until complete carbonisation at about 450°C. At the temperature when the expanded cork agglomerates are made (over 300°C, see Chapter 13), there is already a significant mass loss of 20–30% (Pereira and Ferreira, 1989). In relation to suberin, the thermal decomposition starts at about 300°C and proceeds rapidly until approximately 470°C, corresponding to 80% volatilisation and the formation of a carbonaceous residue.

The thermal decomposition of the cell walls increases with temperature and time. The cell wall components have different thermal stability, and differ both in initiation temperature and kinetics of degradation (Bento et al., 1992; Pereira, 1992; Rosa and Pereira, 1994). Upon heating at 180°C, there is already a considerable release of acetic

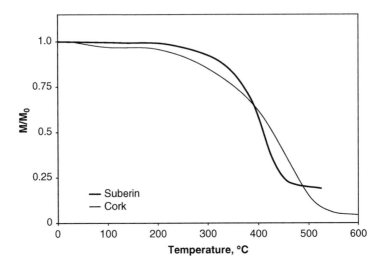

Figure 10.3. Thermogravimetric analysis of cork and of depolymerised suberin (adapted from Rosa and Fortes, 1988 and Cordeiro et al., 1998).

Table 10.5. Dimensional variation of water-boiled cork with heating in air (Rosa and Fortes, 1988; Rosa and Pereira, 1994).

Temperature and Time	L/L_0		
	Radial	Axial	Tangential
100°C, 1 h	0.999	0.994	0.997
100°C, 1008 h	0.985	0.976[a]	
150°C, 24 h	0.988	0.976[a]	
150°C, 1008 h	0.972	0.960[a]	
200°C, 1 h	1.023	1.026	1.023
250°C, 1 h	1.054	1.072	1.084
300°C, 1 h	1.105	1.154	1.175

[a]non-radial.

acid and furfural from degradation of polysaccharides (Salthammer and Fuhrmann, 2000). The chemical changes brought about by the heating may be summarised as follows:

• loss of water occurs in the beginning of the heating;
• extractives are volatilised in the first phase;
• hemicelluloses are the first structural components to be thermally degraded; and
• suberin and lignin are the most stable components.

The cellular structure and dimensions are changed by heating if a sufficiently high temperature is attained (Table 10.5). In previously water-boiled cork, the effect is negligible for heating until 150°C and a small shrinkage in the range of 1–4% is observed in the radial and non-radial directions of the samples (Rosa and Pereira, 1994). For higher temperatures between 200 and 300°C, there is a linear expansion of the cork

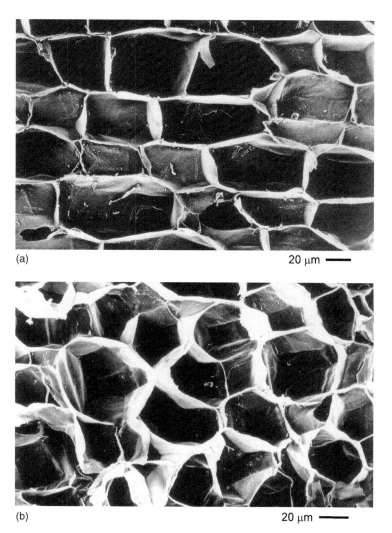

(a)   20 μm ——

(b)   20 μm ——

Figure 10.4. Scanning electron micrographs of the (a) radial and (b) tangential sections of cork after heating in air at 300°C.

samples that shows anisotropy: it is slightly lower for the radial direction (Rosa and Fortes, 1988b). With heating at 200–300°C, the cells swell due to the higher pressure of the gas inside the cells and the cell walls straighten completely and stretch showing a balloon-like aspect as exemplified in Figure 10.4. The fact that the lateral walls of the prismatic cells are larger than the bases explains why the expansion is higher for the non-radial directions.

The compression curves show that in a first step corresponding to the loss of water there is a small increase in strength. Subsequently with the duration of the heating treatment or with higher temperatures the strength decreases, although the stress–strain curves maintain the form shown in Figure 9.3.

10.6. Energy absorption

One of the properties of cork that are put to use in many applications is its capacity of energy absorption. The total energy required for a given deformation (of a unit volume) corresponds to the area under the stress–strain curve. For instance, and in the case of compression, Figure 10.5 shows the total deformation energy of cork as a shaded area, with successively darker shadings for the elastic region, the plateau collapse region and the densification until the maximum strain.

The stress–strain curve shows that the collapse stress of the cells is low, so that the peak stress during impact is limited while large compressive strains are possible, absorbing a great deal of energy as the cells progressively collapse. The total energy of compression to a strain of 83% is approximately 3.4×10^6 J m^{-3} and 2.2×10^6 J m^{-3}, respectively for the radial and non-radial directions. The part corresponding to the deformation until the end of the collapse plateau represents about 1.5×10^6 J m^{-3}. The fact that this high-energy absorption corresponds to relatively low stress values is the reason why cork is used as a shock absorber. Cork is attractive, i.e. for the soles of shoes and flooring, because it is resilient under the foot and absorbs the shocks of walking and has good frictional properties as well, as seen previously. Other applications that use the high damping capacity and friction of cork are in packaging and as handles of tools.

The energy required to deform cork is in part stored within the material and in part dissipated, usually as heat. The dissipated energy results from the fact that cork is not perfectly elastic and it is related with the viscose component of the deformation. The capacity of energy dissipation may be measured by submitting cork to compression and unloading in successive cycles with a given frequency. The loading and the unloading curves of the cycle do not coincide due to the energy dissipation, that means, the energy used to compress cork to the final strain is higher than the energy recovered in unloading.

The energy dissipated in one cycle is given by the area between the two curves (loading and unloading). It is quantified as a loss coefficient, $\eta = D/2\pi U$, where D is the

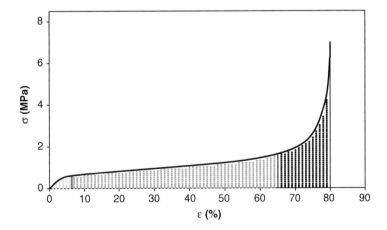

Figure 10.5. Total energy used in the deformation of cork under compression, marked as shaded area: the different shadings correspond to the elastic energy, to the plateau collapse region and to the densification, respectively from lighter to darker shading.

energy dissipated in a complete tension–compression cycle and U is the maximum energy stored during the cycle. The loss factor of cork has been reported at about 0.1 with a weak dependence on frequency in the range 10^{-2}–10 kHz but increasing to nearly 0.3 for high strain amplitudes (Fernandez, 1987; Pritz, 1996). This is a high loss (rubber has a loss coefficient of 0.08) and gives cork good damping and sound-absorbing properties, and a high friction coefficient, as already discussed.

10.7. Electric properties

Another important application of cork is as an electric insulator, since it is a dielectric material. The study of the dielectric properties of cork under the presence of electric static fields was done for the first time by Mano et al. (1995). Cork exhibits a complex pattern of polarisation mechanisms where three different relaxation processes can be distinguished, although they are merged or partially superposed. They are as follows:

- a relaxation mechanism showing a compensation behaviour indicating a high degree of cooperativity attributed to a glass-transition-like relaxation process which took place at about 18°C and with an activation energy of 140 kJmol^{-1}, probably related to suberin (Mano, 2002);
- a low temperature (-100 to -50°C) relaxation probably resulting from slightly hindered internal rotations of polar groups around covalent bonds of the cell wall polymers;
- a higher temperature (30–80°C) mechanism of large intensity without compensation suggesting a low degree of cooperativity and a non-local nature, associated with the melting of the waxes and the desorption of water molecules (Dionísio et al., 1995).

The dipolar relaxation of water molecules absorbed in the cork structure is an important feature of cork that can be changed by heating the cork above 60°C.

The electrical properties of cork are significantly influenced by its water content, even in small amounts of 3.5%. In these conditions the conductivity of cork at 25°C is $\sigma = 1.2 \times 10^{-10}$ S m^{-1} while in dry cork it is $\sigma = 2.9 \times 10^{-14}$ S m^{-1} (Marat-Medes and Neagu, 2003, 2004). The conductivity increases considerably with temperature, by about three orders of magnitude from 25 to 100°C (Lança et al., 2006). Conductivity also increases by nearly four orders of magnitude when cork is mixed with resin in commercial-flooring agglomerates.

10.8. Conclusions

The following properties of cork are of interest for many of its applications:

- surface properties: low wettability in relation to polar liquids, i.e. water, and high affinity for non-polar liquids, i.e. to non-polar resins; high dispersive surface energy and its amphoteric nature with an equal contribution of acidic and basic character make it compatible with both acidic and basic polymeric matrices;
- the friction of cork is high and it has a high anelastic loss;
- it is a material resistant to surface wear and withstands repeated rubbing;

- the heat transfer properties are very poor due to the low solid fraction and small cell size that eliminate gas convection and reduces radiation;
- high capacity of energy absorption but corresponding to relatively low stress values;
- significant dissipation of energy; and
- dielectric properties.

The explanation of these properties ultimately lies in the features of the cellular structure of cork (cell arrangement, cell size and corrugations of cell walls) and in the chemical composition of the cell walls (namely of suberin). The water content of cork influences many of its properties and heating treatments modify cell size and chemical composition.

The well-known applications of cork that make use of these properties include thermal insulation, vibrational absorption (sound and vibrations), shock absorption and friction and wear properties.

References

Adão, M.H., Cabrita, G.M., Gomes, C.M., Almeida, B.S., Fernandes, A.C., Bordado, J.C., 1993. Wetting of cork by polymeric adhesives. Journal of Adhesion Science and Technology 7, 375–384.

Andrade, A., 1962. Isolements thermiques et phoniques. Junta Nacional da Cortiça, Lisboa.

Baptista, A.P.M., Vaz, M.C., 1993. Comparative wear testing of flooring materials. Wear 162–164, 990–995.

Bento, M.F., Cunha, M.A., Moutinho, A.M.C., Pereira, H., Fortes, M.A., 1992. A mass spectrometry study of thermal dissociation of cork. International Journal of Mass Spectrometry and Ion Processes 112, 191–204.

Cordeiro, N., Aurenty, P., Belgacem, M.N., Gandini, A., Pascoal Neto, C., 1997. Surface properties of suberin. Journal of Colloid and Interface Science 187, 498–508.

Cordeiro, N., Pascoal Neto, C., Gandini, A., Belgacem, M.N., 1995. Characterization of the cork surface by inverse gas chromatography. Journal of Colloid and Interface Science 174, 246–249.

Cordeiro, N., Belgacem, M.N., Gandini, A., Pascoal Neto, C., 1998. Cork suberin as a new source of chemicals. 2. Crystallinity, thermal and rheological properties. Bioresource Technology 63, 153–158.

Dionísio, M.S.C., Correia, N.T., Mano, J.F., Moura Ramos, J.J., Fernandes, A.C., Saramago, B., 1995. Absorbed water in the cork structure. A study by thermally stimulated discharge currents, dielectric relaxation spectroscopy, isothermal depolarization experiments and differential scanning calorimetry. Journal of Materials Science 30, 2035–2040.

Fernandez, L.V., 1987. Factores que fazem da cortiça o material isolante mais completo. Cortiça 587, 222–229.

Gibson, L.J., Easterling, K.E., Ashby, M.F., 1981. The structure and mechanics of cork. Proceedings of the Royal Society of London A377, 99–117.

Gibson, L.J., Ashby, M.F., 1997. Cellular solids: structure and properties, 2nd Edition, Cambridge University Press, Cambridge.

Godinho, M.H., Martins, A.F., Belgacem, M.N., Gil, L., Cordeiro, N., 2001. Properties and processing of cork powder filled cellulose derivatives composites. Macromolecular Symposium 169, 223–228.

Gomes, C.M.C.P.S., Fernandes, A.C., Almeida, B.J.V.S., 1993. The surface tension of cork from contact angle measurements. Journal of Colloid and Interface Science 156, 195–201.

Lança, M.C., Neagu, E.R., Silva, P., Gil, L., Marat-Mendes, J.N., 2006. Study of electrical properties of natural cork and two derivative products. Materials Science Forum 514–516, 940–944.

Mano, J.F., 2002. The viscoelastic properties of cork. Journal of Materials Science 37, 257–263.

Mano, J.F., Correia, N.T., Moura Ramos, J.J., Saramago, B., 1995. The molecular relaxation mechanisms in cork as studied by thermally stimulated discharge currents. Journal of Materials Science 30, 2035–2040.

Marat-Mendes, J.N., Neagu, E.R., 2003. The study of electrical conductivity of cork. Ferroelectrics 294, 123–131.

Marat-Mendes, J.N., Neagu, E.R., 2004. The influence of water on direct current conductivity of cork. Materials Science Forum 455–456, 446–449.

Matias, L., Santos, C., Reis, M., Gil, L., 1997. Declared value for the thermal conductivity coefficient of insulation corkboard. Wood Science and Technology 31, 355–365.

Medeiros, H., undated. Steambaked Insulation Corkboard. Methods and Plans for Application in Civil Constructions. Junta Nacional da Cortiça, Lisboa.

Natividade, J.V., 1938. O que é a cortiça. Cortiça 1, 13–21.

Pereira, H., 1992. The thermochemical degradation of cork. Wood Science and Technology 26, 259–269.

Pereira, H., Ferreira, E., 1989. Scanning electron microscopy observations of insulation cork agglomerates. Materials Science and Engineering A111, 217–225.

Pinto, R., Melo, B., 1988. Isolamento térmico: aglomerado negro de cortiçaa e outros materiais isolantes. Cortiça 602, 322–333.

Pritz, T., 1996. Dynamic Young's modulus and loss factor of floor covering materials. Applied Acoustics 49, 179–190.

Rosa, M.E., Fortes, M.A., 1988a. Thermogravimetric analysis of cork. Journal of Materials Science Letters 7, 1064–1065.

Rosa, M.E., Fortes, M.A., 1988b. Temperature induced alterations of the structure and mechanical properties of cork. Materials Science and Engineering 100, 69–78.

Rosa, M.E., Pereira, H., 1994. The effect of long term treatment at 100°C–150°C on structure, chemical composition and compression behaviour of cork. Holzforschung 48, 226–232.

Salthammer, T., Fuhrmann, F., 2000. Release of acetic acid and furfural from cork products. Indoor Air 10, 133–134.

Teixeira, M.J., Fernandes, A.C., Saramago, B., Rosa, M.E., Bordado, J.C., 1996. Influence of the wetting properties of polymeric adhesives on the mechanical behaviour of cork agglomerates. Journal of Adhesion Science and Technology 10, 1111–1127.

Vaz, M.F., Fortes, M.A., 1998. Friction properties of cork. Journal of Materials Science 33, 2087–2093.

Part IV
Cork processing and products

Chapter 11

Cork products and uses

The place of cork in the past is well known and its uses stretch from Antiquity to the present. Some of its historical applications are still in current use today: sealants for liquid containing reservoirs, thermal insulation material, shock absorbers and floating materials. Cork has also a very special place in science starting with the microscopic finding of the cellular structure of plants by Hooke in 1664 (see Chapter 2) but spreading to other scientific areas of chemistry and physics. Still today cork is an amiable material used by researchers for several purposes.

But modern times have brought other ideas of usage for cork products from decorative items to ablative surfacing of space vehicles and medical applications. Several research lines are also under way that explore some of the properties of cork and propose its use in novel products or processes.

11.1. Historical uses of cork

There are archaeological findings and written evidence showing that cork was used from the early Antiquity, mainly as a floating device, a sealant and an insulator. Fishing devices with cork were used already around 3000 B.C. in China, as well as in Egypt and in the countries of Asia Minor. Archaeological remains are numerous in the Mediterranean area, e.g. cork lids in urns in the 3000-year-old conical monuments of Sardinia, engraved cork plaques in Carthage cemeteries, soles of sandals, closures of vials and amphorae from excavations and ship wrecks.

The presence of cork in Greek and Roman times is also well documented by the works of poets and writers. In ancient Greece, the philosopher Teophastro (IV–III century B.C.) refers that cork can be stripped off from the tree and that a new layer of cork is formed with a better quality. Dioscorides, a doctor from the second century, advises some medicinal uses for the cork tissue, namely against baldness: "charred cork rubbed on bald parches with laurel sap makes the hair grow again, thicker and darker than before". Marcus Terence Varrus (116–27 B.C.), a scholar, and Lucius Columela (first century), an agronomist, spoke of the insulation properties of cork and recommended it for bee-hives.

Pliny the Elder (23–79) was a roman officer and an encyclopaedist. He wrote a "Natural History" attempting to describe the full complexity of nature in 37 volumes. In Book 16 (Forest trees and botany) and Book 24 (Drugs obtained from forest trees), he refers to the cork oak and lists the uses of cork: floats for anchor ropes or fishing nets, sealants for barrels and vessels, winter footwear for women, roof coverings for houses and medicinal drugs. The Roman soldiers also used cork in their helmets against the sun heat.

The cork properties were well known in the Roman world and its uses were part of the common life, so that reference to cork appears frequently in the writings of poets from that period.

The poet Horace (65–8 B.C.), writes in the Third Book of the "Odes", in Ode VIII. To Maecenas: "*I made a vow of a joyous banquet …. This day… shall remove the cork fastened with pitch from that jar.*" Horace also uses cork for metaphors. In Ode IX. To Lydia: "*Though he is fairer than a star, thou of more levity than a cork, and more passionate than the blustering Adriatic; with thee I should love to live, with thee I would cheerfully die.*" Also in his "Satires", First Book, Satire IV (where he apologises for the liberties taken by satiric poets in general and particularly by himself), he writes: "*The philosopher may tell you the reasons for what is better to be avoided, and what to be pursued. It is sufficient for me, if I can preserve the morality traditional from my forefathers, and keep your life and reputation inviolate, so long as you stand in need of a guardian: so soon as age shall have strengthened your limbs and mind, you will swim without cork.*" (//www.autho-rama.com/works-of-horace-3.html).

Virgil, a Roman poet (70–19 B.C.) in his famous epic poem "Aeneid" includes an episode where cork plays a significant role in the saving of the baby princess Camilla:

And called Camilla. Thro' the woods he flies;
Wrapp'd in his robe the royal infant lies,
His foes in sight, he mends his weary pace;
With shout and clamours they pursue the chase.
The banks of Amasene at length he gains:
The raging flood his farther flight restrains,
Rais'd o'er the borders with unusual rains.
Prepar'd to plunge into the stream, he fears,
Not for himself, but for the charge he bears.
Anxious, he stops a while, and thinks in haste;
Then, des'perate in distress, resolves at last.
A knotty lance of well-boil'd oak he bore;
The middle part with cork he cover'd o'er:
He clos'd the child within the hollow space;
With twigs of bending osier bound the case;
Then pois'd the spear, heay with human weight,
And thus invok'd my favour for the freight:
'Accept, great goddess of the woods,' he said,
'Sent by her sire, this dedicated maid!

Thro' air she flies a suppliant to thy shrine;
And the first weapons that she knows, are thine.'
He said; and with full force the spear he threw:
Above the sounding waves Camilla flew. (http://classics.mit edu/Virgil/aeneid.11.xi.html)

Plutarch (50–120) makes in his essays various references to the use of cork. Cork as a floating device is referred, for instance, in the work "On the fortune of the Romans" where during a siege of Rome by the Gauls, Caius Pontius bound broad strips of cork beneath his chest and floated to the other side of the river, as well as in "The life of Camillus", one story within his known work "Parallel Lives", Pontius Cominius fastened cork pieces to his body and swam. In another story, "Cato the Young", the floating properties of cork are used for another objective: sailing back from Cyprus with a considerable fortune in coffers, Cato fastened a long rope to each one and at the other end a piece of cork, so that if the ship would miscarry the underwater chest might be discovered. In the Book XIII. Concerning women (of the "Deipnosophists of Athenaeus of Naucratis"), he writes that small women might use cork sewn into their shoes to enhance their stature.

The cork planks were used through Middle Age as an insulator of roofs and walls, for household items and furniture, for sealing reservoirs. However, the use of cork stoppers similar to today's practice only spread in the 17th century when, in 1680, the Benedictine monk Don Pierre Pérignon (1639–1715) started to use cork stoppers in the wine bottles in his Abbey of Hautvillers, near Epernay (Champagne), in France. The legend says that he was inspired by the covers that the pilgrims to Compostela brought in their pans. He initiated what is called the champagne method of wine making and indirectly the cork industry. He substituted the conical wooden plugs wrapped in hemp soaked in olive oil, that were found to pop out frequently due to the pressure from the sparkling wine, by cork stoppers with much better results. This type of wine bottling was adopted by wine producers such as Ruinart in 1729 and Moet et Chandon in 1743.

The "Encyclopédie" of Diderot (1713–1784) and d'Alembert (1717–1783) has given attention to this emerging cork industry. This is a major work that includes meticulous descriptions of the arts, craft and techniques used at the time and accompanied by 11 volumes of illustrations. In one of the volumes, published in 1763, two folios represent the craft of the cork stopper making: one cork shop figuring one woman sorting the finished stoppers and two cork-cutters making the cork strips and rounding the stoppers as well as the knives used for cutting and the stone for sharpening (Fig. 11.1). The Encyclopaedia lists the uses of cork: slippers, roller skates, for suspending fishermen nets, but mostly to seal jars and bottles; charring in closed vessels gives fine and very light ashes called black of Spain that is used by many craftsmen. In the Navy, cork is used to plug the canons to avoid water entering. It also refers two types of cork planks: the white cork (or from France) of medium thickness and a yellowish grey colour, and the black cork (or from Spain), thicker and with a darker external surface, which is the most prized.

The production of cork stoppers for the wine bottling started in France with workshops located in the south and southwest. This craft soon spread to the nearby and cork-producing region of Catalonia, around Girona, in the second half of the 18th century and later on to Portugal. The stoppers were cut manually from the cork planks, so that their

Bouchonnier.

Figure 11.1. The cork cutter. Drawings of the craftsmanship of cork stoppers, included in the Encyclopédie of Diderot and d'Alembert (1763).

axis was in the axial position: first square prismatic pieces with the adequate dimensions were prepared which were then cut with a knife into the required rounded cylindrical form.

The production of cork stoppers increased enormously in the 19th century and concentrated near the raw-material production areas in centres that gained considerable dimension and turned into industrial poles. The cork manufactures benefited from the general technological development that was taking place in Europe and North America, and new machinery was invented for cork production allowing higher production rates. The industry of cork increased in dimension and in some regions it was its main social and economical framework, e.g. in Spain in Palafrugell and other small towns in Catalonia, in Portugal in Silves in the Algarve and Seixal near Lisbon. After the 1st World War, the position of Portugal as the world's leading cork producer was already established and is maintained until today. The corporation that is presently the world's largest network of companies producing cork products, the Amorim Group, is family owned and was started in 1879 by António Alves Amorim with two workers, near O'Porto (Santos, 1997; Oliveira and Oliveira, 2000).

The production of cork stoppers originated a large amount of residual cork material that soon started to be put to use. In 1863, Frederik Walton invented in Great Britain the linoleum: a mixture of oxidised linseed oil, finely granulated cork and gum pressed onto flax sheets that could be used for floor coverings. The turn of the 19th century saw many inventions in relation to cork, mostly from the United States (Gil, 2000, 2005). The crown capsule was invented in 1892 by William Painter with a cork disc as lining. This was the main closure for soft drinks and one large market for cork products. In 1890, started the production of cork parquet for flooring, first by pressing cork granules at 230–300°C and then in 1909 by using a resin for cork binding. The use of glued together cork granules for flooring spread as well as for production of agglomerated cork stoppers.

The expanded cork agglomerate was invented accidentally by the American life-jacket maker John Smith, who patented the process in 1892 as "Joining cork with its natural resins". The first expanded agglomerates for thermal insulation using superheated steam were produced by Armstrong in 1923 with a process similar to what is used today (see Chapter 13).

Cork wall paper was produced already around 1880 cutting very thin sheets of cork that were glued on paper. Thin shavings, the cork wool, were used as a filling material for cushioning. But most famous was the production of thin strips of cork that were glued together forming tape rolls used to cover the mouthpiece of cigarettes (Fig. 11.2).

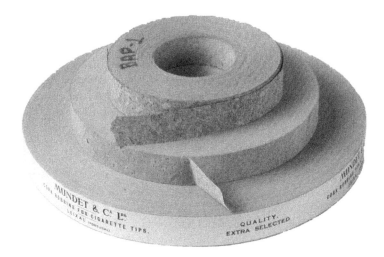

Figure 11.2. Tape of thin cork strips used for cigarette mouthpiece.

The exploitation of cork as a source of chemicals was started in Algeria (1945), Portugal (1949) and in France (1951) for the production of waxes and fatty acids, but this industry was only short-lived.

Cork was also used in the 18th and 19th centuries for architectural models. Cork models as souvenirs and objects of study first appeared in Rome at the beginning of the classicist period and this modelling technique was named "Phelloplastik" by the Würzburg theology professor Franz Oberthür. The Italian Antonio Chichi was the most successful representative of this architectural cork modelling and his models were prized collectors items that adorned art cabinets throughout Europe (Kockel et al., 1998). Because of its porosity, cork proved optically ideal for the portrayal of ancient ruins, and it could also be easily transported. Carl May, a confectioner of the Court who cherished an interest in art, also came up with the idea of replacing the perishable table decorations with cork models, which he placed to inspire erudite conversation about historical monuments and works of art. However, the fascinating art of sculpturing in cork saw only a very short life, following the rise and fall of classicism in architecture. These cork models can still be seen in various collections and museums, e.g. the castle of Johannisburg in Bayern, Germany, advertises the largest collection and includes, for instance, a 3-m diameter model of the Coliseum.

11.2. Cork in science

Cork has gained a leading role in the initial play of plant anatomy when Robert Hooke used it as an example material in his observations with the microscope and recorded that it was formed by minute cells, the later coined cells that are the basis of plant structure (Hooke, 1665).

Hooke, who was employed by Boyle as a research technician, was also appointed in November 1662 as a Curator of Experiments to the recently created Royal Society of London, one of his tasks being the preparation of experiments to be presented in the weekly meetings held by the Fellows (Jardine, 2003). He was ordered to prepare microscopic demonstrations: in 8 April 1663 he showed specimens of wall moss and in 15 April 1663 he demonstrated sections of cork where the cellular structure was visible. The microscope demonstrations of the various materials that followed were a success and in July 1663, Hooke was commanded to present his observations in a book to King Charles II. *Micrographia* was printed by November 1664 with a second edition in 1667, and may be considered the first work in popular science (Ford, 2001). Many of the microscopic drawings made by Hooke are still well known today, namely the two sections of cork already shown in Chapter 2 (Fig. 2.1). Hooke's work inspired the Dutch Thonis Leeuwenhoek (1632–1723) (he later adopted the name as Antoni van Leeuwenhoek) who perfected the microscope and continued the observations, namely of cork, as represented in Figure 2.2.

Cork also attracted the attention of the early chemists. The first recorded study on the chemical composition of cork dates from 1787, made by the Italian chemist Brugnatelli, who treated cork powder with nitric acid and obtained what he called suberic acid (Brugnatelli, 1787). It was Chevreul in two monographies published in 1807 and 1815, who first stated that the acid mixture obtained by chemical degradation of extractive-free cork, that he named suberin, was its main component and the one responsible for the material's properties (Chevreul, 1807, 1815). Many chemists worked since then with cork, isolating different compounds and studying their properties. Their work still goes on, as it has been seen in Chapter 3.

But cork also played a role in other fields of science, even if it is often only the vehicle or the material chosen for the experimental. Galileo (1564–1642) used two balls one of cork and the other of lead in his experiments with the pendulum and the oscillatory movement. In research on electrostatics, Benjamin Franklin (1706–1790) also took cork balls in several experiments as the dielectric material (Millikan, 1943).

Still today cork is often used as an experimental material in the set up of various disciplines, because it is light, impermeable, soft and considerably inert. The following are a few examples:

- Recently, the fall of cork balls in deep mine wells that were deactivated was used as a method for determining the drag coefficient for smooth spheres moving in the laminar regime (Maroto et al., 2005);
- The sound field around a Schroeder diffuser was studied by the distribution of particle velocity using fine cork dust (Fujiwara et al., 2000);
- The boundary element method was used to study the effect of a barrier between two rooms in low-frequency sound and modelled with cork (Tadeu and Santos, 2003);
- Studies on the X-ray dosage in medicine use lung and thorax phantoms mimicking the human body which are made up of cork and water (Blomquist and Karlsson, 1998; Iwasaki, 2002);
- The development of a texture sensor emulating human fingers to be applied in robots used cork surfaces as testing material (Mukaibo et al., 2005).

11.3. Present uses of cork

Most of the uses of cork have been long lived and are today still similar to those made many centuries ago. The properties that are mostly valued in such applications are the low density, the very low permeability to liquids and gases, the chemical and biological inertness, the behaviour in compression, the high friction, the low heat conductivity and the high damping capacity. Often it is a combination of properties that are involved in the performance of cork products.

Along times cork has competed with other materials, namely with synthetic materials during the last century. In a few cases, the technology or the consumer's habits changed and a specific cork appliance was totally abandoned:

- one is the use of thin cork discs as an underlining in crown closures for beverage bottles, first as natural cork, then also as agglomerated cork; once a large market, the crown cork discs were totally substituted by a plastic lining;
- the other is the cigarette mouth-piece, which used a very thin strip of cork around the mouth end; the tobacco industry substituted the cork by paper and production of this product was abandoned in the 1970s.

It is remarkable, however, to see the strong appealing effect that cork had on the consumers. This is why the present-day cigarettes still mimic the appearance of cork in the mouthpiece by selecting a gold brown colour and a spotted pattern resembling the porosity of cork.

The prime objective of the cork industry is nowadays the production of natural cork stoppers. Over 15 thousand million stoppers are produced annually and sold worldwide to the wine industry. Today most stoppers are cylindrical and sharp edged, and only the OPorto wines and liquors such as whiskies and cognacs use edge-bevelled stoppers. However, the stoppers may be dimensioned and shaped following specific customer demands. Examples are the king-sized stoppers used in the champagne bottles opened by the winners of car races.

Cork tiles of different types, dimensions and design are used for floor coverings and also for wall panelling. Finishing may be various: polish, wax, varnish, urethane or vinyl coated. Because of abrasion resistance they are used in all sorts of public buildings such as schools and hospitals, libraries, airports, etc. At home in addition to living and bedroom floors, they can also be used in kitchens and bathrooms due to their resistance to moisture and a non-slippery surface. Most of the cork floors are made up of agglomerated cork but other options are possible: glued together cork pieces that are subsequently laminated into sheets showing either tangential or transverse sections.

Expanded cork agglomerates are used as thermal insulation in roofs, walls, floors and ceilings in private and public buildings. The industrial applications are also numerous: for insulating cold-rooms, storage tanks and for the lagging of pipes. Although in this field the competition with other materials is fierce, the expanded corkboards can boast a very high durability in harsh conditions, resistance to fire and non-toxic volatiles released by combustion. The expanded cork boards are also used for their sound-absorbing properties, in noisy environments (i.e. mess halls, school playrooms) or to insulate from external sounds and reduce reverberation (i.e. conference rooms, movie theatres).

The compression properties and damping capacity of cork are put to use in the expansion or compression joints applied in concrete structures including demanding structures such as dams and tunnels. Anti-vibration layers are also used under machinery and pillars. In several applications, cork is combined with rubber, making the composite known as rubber–cork (see Chapter 13). This material can be used for pavements in industrial, outdoor and sport environments. A "silent slab track" in Dutch railways was developed using the rails embedded in a cork-filled elastomeric material (Lier, 2000).

Sports have a large tradition of using cork in various forms: from the pea in whistles to dartboards and table tennis bats, to balls for hockey, golf, cricket and baseball, badminton volleys as well as golf clubs and baseball bats. The corked baseball bats were found to have a slight increase in the speed of the battered ball, which would translate into an increased distance; they are, however, illegal and several players have been ejected after finding cork in their bats (Hall, 2003).

Gifts and novelties include a large variety of objects such as mats, coasters for plates and glasses, cigarette boxes, desk pads, mouse pads. Combined with an under textile layer it is used in handbags, travelling cases, wallets, purses, document holders and some clothing items such as neck ties, skirts or jackets (Fig. 11.3). It seems that cork objects are becoming fashionable again. Cork is considered to have an organic look, with a dynamic pattern and giving the sense that it comes from something living. Modern furniture design also uses cork, valuing the aesthetics and properties of the material. One recent example is the "Cortiça chaise longue" from the American designer Daniel Michalik (Fig. 11.4). The architectural cork modelling also sees a renaissance with the German artist Dieter Coellen (Fig. 11.5).

Figure 11.3. Shop dedicated to cork objects in Corsica, France.

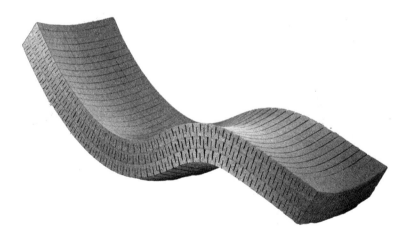

Figure 11.4. Cortiça chaise longue from the American designer Daniel Michalik (2004) (www.danielmichalik.com).

In architecture and building innovative approaches to the use of cork have also appeared, in addition to the traditional applications of cork products for indoor flooring, panelling and insulation: for instance, the prized Portuguese architect Siza Vieira applied expanded corkboards as the external cladding of the Portuguese Pavilion in the Hanover Expo 2000 World Fair, now reproduced as an exhibition hall in the city of Coimbra (Fig. 11.6).

Figure 11.5. A cork model from the German artist Dieter Coellen (www.coellen-cork.com).

Figure 11.6. Portuguese pavilion in the Hanover Expo 2000 designed by the architects Siza Vieira and Souto Moura, now reconstructed in Coimbra, Portugal.

11.4. Cork in space exploration

It is commonly referred when writing on cork uses that it is applied by NASA in space crafts, but little has been compiled on its function. It is, however, interesting to bring some detail into the extremely harsh conditions that are encountered in space exploration and on the role of cork as an ablative insulator in covering space vehicles. These thermal insulation systems are essential for the launch and operation of all spacecraft manned or unmanned.

In the launch phase the vehicle is subject to lift off and ascent acceleration loads, vibration, aerodynamic loads and heating, shock, acoustic loads and loads imposed by flow of liquid fuel and sloshing. Combustion, rocket exhaust plume and aerodynamic heating cause extreme thermal conditions during lift off and ascent. The re-entry further imposes the most severe aerodynamic heating in addition to shock and acoustic loads.

Ablation is a very effective mechanism of minimising the total energy the vehicle absorbs. The heat flux changes the state of the surface substrate either by melting, sublimation or thermal degradation with the surface mass being carried away in the high-speed flow. An ablator like cork forms a char layer, which acts as an insulator while the material underneath continues to decompose and outgas. The gaseous products from decomposition percolate through the char to effectively transpiration cool the surface. In a further phase under high heat flux the char will sublimate.

Some examples of the use of cork containing ablators are given. In 1957, the Jupiter C vehicle demonstrated ablation as an effective thermal protection technique. Scout, the first solid fuel launch vehicle launched the first American satellite Discover I using cork/fibreglass heat shields and cork insulated fins. The Pathfinder used in 1997 for the Mars entry a heat shield with a phenolic honey comb filled with a cork and silica bead filled epoxy. The space shuttle uses cork as a component in several of its parts: for instance, on the solid rocket boosters it uses a sprayable cork/polyurethane foam insulation, and on the external tank the entire outer surface is insulated with a half inch thick cork/epoxy layer covered with 1–2 inches of spray-on foam. The accidents with the space shuttle led to a re-examination of its parts, including its complex thermal insulation system; an ablator made of silicon resins and cork is maintained as one of the components. The description of such an ablator is made in one patented description referring to a mass content of 7.22–7.98% of ground cork (Cosby et al., 2005). The Atlas V rocket launching the probe New Horizons, the NASA mission to Pluto, has it encapsulated inside a fairing containing cork. The European ARD (Atmospheric Reentry Demonstrator) has its conical surface coated with a cork powder and phenolic resin composite.

A recent study was published examining this cork/phenolic resin ablator commercialised as Norcoat*-liège in a temperature range up to 2500°C (Reculusa et al., 2006). The Norcoat*-liège HPK F1 Space Grade contains approximately 50% of cork particles, 30% of a standard phenolic resin and 20% of a fungicide and a mineral fire proof agent. After the thermal treatment the composite is an almost pure carbon material, but the fundamental point is that the original cellular structure is preserved. During heating the cell size increases and the cell wall thickness decreases (400 nm at 1500°C and 100 nm at 2000°C), and the walls only start to break for temperatures above 2250°C (Fig. 11.7). The charred material maintains its closed cell structure, opposite to what happens with classical carbon gels or foams. This presence of closed "macro porosities" of cork seems to be the key to understand its quality as a thermal protection for space vehicles. During ablation, a multi-layered and closed porous system is formed: the external carbonised layer controls the gas circulation and therefore limits the pyrolysis of the inner layers of cork and phenolic resin and the ablation rate.

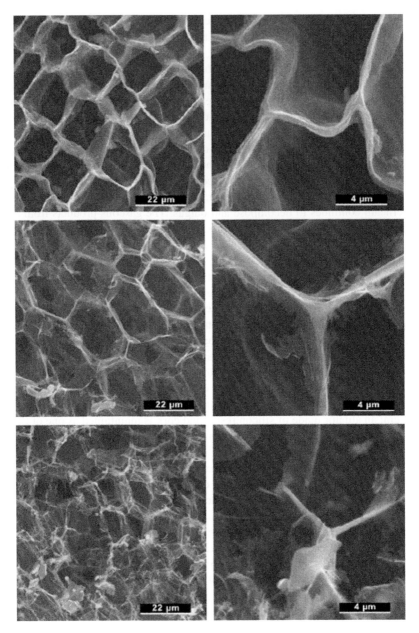

Figure 11.7. Scanning electron microscopy images of Norcoat*-liège samples treated at three different temperatures: 1500°C (top), 2000°C (middle) and 2500°C (bottom) (Reculusa et al., 2006).

11.5. Cork in medicine

Cork is found in medical-related applications mostly because of its mechanical properties or in pharmacology because of its chemical composition.

Therapeutic shoes with medium-density cork inserts may be beneficial in persons with diabetes or in individuals with severe foot deformities (Reiber et al., 2002). Also the inclusion of cork as reinforcement in the medial longitudinal arch of orthotics in combination with a heel pad is a cost-effective and comfortable design that allows treating heel pain and plantar fasciitis (Seligman and Dawson, 2003), while cork soles are used in the footplates for a walking appliance for a paraplegic adult (Henshaw, 1977). Cork discs are also included in mechanical elbows to improve rotational functioning (Ivko, 1999).

A promising use of cork extracts refers to their inclusion in the new generation of anti-aging skin care products, seeking an immediate and visible skin smoothing and lifting effects. The effect of a cork extract applied to the skin was experimented in a double-blind clinical study using a gel formula with 3% cork extract (Coquet et al., 2005). This cork extract refers to a mixture of suberinic fatty acids that were prepared with an ethanolic potassium depolymerisation of previously ethanol-extracted cork, acidification and extraction with an organic solvent. The cork extract provides a remarkable and highly significant tensor and smoothing effect on the skin in a very short time after application. A cork paste applied to wounds prevents dehydration and protects against exogenous agents (Dattatreya et al., 1991).

Cork suberin extracts possess anti-mutagenic properties and can be included in the group of natural anti-mutagenic agents as shown by the results of a study where the response of chloroplast DNA damage with different mutagens (acridine orange, ofloxacin and UV-radiation) in relation to suberin was tested (Krizková et al., 1999). The suberin extract was prepared from cork through transmethylation. The presence of suberin caused the decrease of mutagenic effect of ofloxacin in all concentrations used while the effect was concentration-dependent in relation to the decrease of the mutagenicity of acridine orange and UV-radiation.

The extractives of cork include several compounds that have bioactivity. Various triterpenoids were synthesised from friedelin and 3-hydroxyfriedel-3-en-2-one obtained from cork processing by-products, which show increased insecticidal activity and some possess significant growth inhibitory effects for three human tumour cell lines (breast adenocarcinoma, non-small cell lung cancer and CNS cancer) and of human lymphocyte proliferation (Moiteiro et al., 2001, 2004, 2006). Betulinic acid, also a component of cork extractives, has been ascribed different biological activities: anti-inflammatory, *in vitro* anti-malarial effects, specific cytotoxicity against a variety of tumour cell lines and anti-HIV-1 activity (Cichewicz and Kouzi, 2004).

The protective effect of dietary fibres on carcinogenesis is a controversial matter and the large chemical diversity of these fibres does not help to clarify the issue (Harris and Ferguson, 1999). However, suberised plant cell walls may be particularly significant in protecting against colorectal cancer, and commercial cork and potato skins protect against the development of aberrant crypts, an early marker of colonic carcinogenesis (Ferguson and Harris, 1996, 1998). Cork cells show the best capacity to adsorb hydrophobic carcinogenic heterocyclic aromatic amines and to remove them from the digestive tract (Harris et al., 1996). Natural antioxidant activity has been found in the phenolic extracts of several plants, including cork and potato peel (Kahkonen et al., 1999).

Continued exposure to cork dust and to moulds in cork-processing environments has been reported to cause a respiratory disorder named suberosis, but the role of the cork dust itself and of several frequent micro-organisms is not yet clarified (Deschamps et al., 2003; Morell et al., 2003; Winck et al., 2004).

11.6. Potential applications

Some emerging applications for cork are under development or at a research stage. They explore preferentially one of the properties of cork or, more often, a combined set of properties.

One possible use of cork is in thermoacoustic heat pumps. These are devices that use acoustic power to transfer heat from a low temperature to a high temperature source and consist mainly of an acoustic resonator filled with a gas. The performance of the system is negatively influenced by viscous and thermal relaxation of the solid–gas boundary and minimisation of the thermal-relaxation dissipation is an objective. The combination of cork–helium leads to the lowest damping (lowest thermal losses) and it is suggested to use a thin layer of cork inside a metallic resonator with a very smooth inside surface to minimise viscous losses (Tijani et al., 2004).

Bioabsorption of pollutants using plant and biomass residues in substitution of more expensive materials such as activated charcoal is an on-going active research field. Cork can also be used namely as a biosorbent of heavy metals in aqueous solutions such as $Cu(II)$, $Ni(II)$ and $Zn(II)$ although with less capacity than other lignocellulosic materials (Villaescusa et al., 2000, 2002; Chubar et al., 2004a,b). The important binding sites are the carboxylic groups and the absorption capacity can be increased by pre-treatments such as boiling in a detergent solution, soaking in a hypochlorite solution, or through thermal activation by steam treatment at 700–800°C (Chubar et al., 2004a,b).

Another contamination of wastewaters is related to the use of pesticides and their occurrence in aquifers has become one of the important diffuse pollution risks to drinking water supplies. The sorption behaviour of cork was studied in relation to bifenthrin, one stable and field persistent pesticide (Domingues et al., 2005). The granules of cork effectively remove bifenthrin from wastewaters and show a higher sorption capacity at equilibrium than that obtained with activated carbon. The chemical nature of the cork surface namely regarding the presence of non-polar extractives and of suberin should explain the binding of the non-polar bifenthrin molecule.

Cork pieces are also good absorbers of prenylated flavanones and can be used in cell suspension cultures that produce these compounds without inhibiting cell growth (Yamamoto et al., 1996). They also show a reasonable behaviour as a biofilm support media in packed bed and bubble columns (Wright and Raper, 1998). However, cork substrates may have phytoxicity in relation to horticultural species that are sensitive to phenolic extractives and its use as a potting media component has to be evaluated in relation to the specific species (Ortega et al., 1996). Cork and composted cork are proposed as a container media to improve the water supply to the plant (Carmona et al., 2003).

The preparation of activated carbons from low-cost cork wastes is another line of research with promising results (Carrott et al., 1999, 2003; Carvalho et al., 2003, 2004). Activation by physical processes in carbon dioxide or chemical activation with KOH and K_2CO_3 is possible. The cellular arrangement of cork is maintained after burn-off and activation leading carbon in a form that can be considered as an interconnected membrane of about 1 μm width. The mean width of micropores is in the range of 0.7–2.0 nm and also below 0.7 nm, with a micropore volume that can attain 0.33 cm^3/g. These cork-based activated carbons can be used as molecular sieves and the selectivity shown towards CH_4,

C_2H_6, N_2 and CO_2 suggests its potential use for methane purification, including the separation of ethane from carbon dioxide.

Chars from cork waste can be used as a cation-exchanger after sulphonation and show an equilibrium constant that varies in the order $Na^+ < Ca^{2+} < Fe^{3+}$ (Gómez Corzo et al., 2002).

The low density of cork also suggests its use to prepare low-density slurries made up of water and cork fine particles with densities less than 1 g/ml (Ferrara and Meloy, 1999). The density of the slurry depends on the proportion of cork and can be tailored to be used for low-density solid separation, namely for the separation of low-density polymeric materials: for example, it is possible to separate polymethylpentene (0.83 g/ml) from high-density polyethylene (0.95 g/ml). This type of separations may be useful for the recycling of light polymeric materials as found, for instance, in municipal solid waste. The low density of cork may also be used to make floating capsules containing the fungus *Lagenidium giganteum* to control larvae of *Anopheles quadrimaculatus* in the water column, which proved more effective than sinking capsules without cork (Rueda et al., 1991).

The insulation properties and the low density of cork granules suggest its inclusion in a number of composite materials for building applications, i.e. cork–gypsum (Hernandéz-Olivares et al., 1999), lightened plaster (Rio Merino et al., 2005), lightweight polymer mortar (Nóvoa et al., 2004) and concrete (Aziz et al., 1979).

Cork is a good option for automobile interiors where thermal insulation should be maximised and the thermal distortion minimised, while the energy content of the materials should be low to comply with environmental requirements (Matos and Simplício, 2006). In relation to other materials, cork has advantages related to the easy recyclability, re-usability and health friendliness.

Cork is also a potential component in composite materials, such as in carbon fibre/epoxy lamelates to increase low-velocity impact (Landolt et al., 2005), in polypropylene/nitrile and polyethylene/nitrile rubber blends (George et al., 1999, 2006) and in cellulose derivatives composites (Godinho et al., 2001). Cork particle boards with thermoplastic granulate as binding agent have also been proposed (Gil, 1993). Cork as a source of chemicals has been an appealing idea and recent proposals were made of considering suberin as a valuable precursor to novel macromolecular materials such as polyesters and polyurethanes. Cork powder was oxypropylated and the obtained polyol mixture was tested for polyurethane production (Evtiouguina et al., 2001, 2002). Also depolymerised suberin in the form of a mixture of aliphatic monomers in their methyl ester form was tested for polyurethane polymerisation (Cordeiro et al., 1997, 1999). Depolymerised suberin was also studied as an additive to offset printing inks in replacement of other waxy materials like PTFE oligomers (Cordeiro et al., 2000).

Cork has also been proposed as a renewable component to produce ecoceramics (Silva et al., 2005). Ecoceramics have been so far fabricated with wood materials that are pyrolysed to give a porous carbonaceous structure that is infiltrated with inorganic oxides and non-oxides to form a strong ceramic or composite.

Densification of cork granulates has also been proposed leading to higher density materials. Consolidated cork powder produced by hot pressing (50–55 MPa, 180–250°C) of cork particles (Flores et al., 1992) retains the high thermal resistance, the chemical stability, the low permeability and fire resistance of cork with a density that attained 1200 kg m^{-3}. Previously expanded cork granules from the production of black agglomerates were also densified at 230°C and up to 3 MPa, to yield a 760 kg m^{-3} material (Gil, 1994, 1996).

11.7. Conclusions

The properties of cork that have been recognised since historical times and put to use in various ways are mainly its low density, no permeability to liquids, and its capacity of insulation, shock absorption and elastic compression as well as chemical and biological stability. The combination of these properties continues to be the specific mark of cork, and they ultimately result from the cellular structure and the chemical composition of the material, as detailed in previous chapters.

The past uses of cork have also continued to the present times, now linked with innovation and industrial development: sealants for bottles and reservoirs, flooring and coverings and heat and vibration insulation materials. Some of the applications have developed into specialised fields such as ablators for space vehicles. An active research proposes novel uses, namely related to biosorption, preparation of carbon materials, as a component in mixtures and composites and as a source of chemicals and biological active compounds. The combination of aesthetics, natural origin and the material's properties is also receiving attention in fields such as furniture design, architecture and mode.

Ultimately, novel applications will draw from an increased knowledge on the formation, biochemical synthesis and physical properties of this complex material.

References

Aziz, M.A., Murphy, C.R., Ramaswamy, S.D., 1979. Lightweight concrete using cork granules. International Journal of Lightweight Concrete 1, 29–33.

Blomquist, M., Karlsson, M., 1998. Measured lung dose correction factors for 50MV photons. Physics in Medicine and Biology 43, 3225–3234.

Brugnatelli, D., 1787. Elementi di chimica, Tome II.

Carmona, E., Ordovas, J., Moreno, M.T., Aviles, M., Aguado, M.T., Ortega, M.C., 2003. Hydrological properties of cork container media. Hortscience 38, 1235–1241.

Carrott, P.J.M., Ribeiro Carrott, M.M.L., Lima, R.P., 1999. Preparation of activated carbon "membranes" by physical and chemical activation of cork. Carbon 3, 515–517.

Carrott, P.J.M., Ribeiro Carrott, M.M.L., Moutão, P.A.M., Lima, R.P., 2003. Preparation of activated carbons from cork by physical activation in carbon dioxide. Adsorption Science and Technology 21, 669–681.

Carvalho, A.P., Cardoso, B., Pires, J., Brotas de Carvalho, M., 2003. Preparation of activated carbons from cork waste by chemical activation with KOH. Carbon 41, 2873–2876.

Carvalho, A.P., Gomes, M., Mestre, A.S., Pires, J., Brotas de Carvalho, M., 2004. Activated carbons from cork waste by chemical activation with K_2CO_3. Application to adsorption of natural gas components. Carbon 42, 667–674.

Chevreul, M., 1807. De l'action de l'acide nitrique sur le liège. Annales de Chimie 92, 323–333.

Chevreul, M., 1815. Mémoire sur le moyen d'analyser plusieures matières végétales et le liège en particulier. Annales de Chimie 96, 141–189.

Chubar, N., Carvalho, J.R., Correia, M.J.N., 2004a. Cork biomass as biosorbent for Cu(II), Zn(II) and Ni(II). Colloids and Surfaces. A: Physicochemical Engineering Aspects 230, 57–65.

Chubar, N., Carvalho, J.R., Correia, M.J.N., 2004b. Heavy metals biosorpion on cork biomas: effect of the pretreatment. Colloids and Surfaces. A: Physicochemical Engineering Aspects 238, 51–58.

Cichewicz, R.H., Kouzi, S.A., 2004. Chemistry, biological activity, and chemotherapeutic potential of betulinic acid for the prevention and treatment of cancer and HIV infection. Medical Research Reviews 24, 90–114.

Coquet, C., Bauza, E., Oberto, G., Berghi, A., Farnet, A., Ferré, E., Peyronel, D., Dal Farra, C., Domloge, N., 2005. Quercus suber cork extract displays a tensor and smoothing effect on human skin: an in vivo study. Drugs under Experimental Clinical Research 31, 89–99.

Cordeiro, N., Belgacem, M.N., Gandini, A., Pascoal Neto, C., 1997. Urethanes and polyurethanes from suberin. 2. Synthesis and characterization. Industrial Crops and Products 10, 1–10.

Cordeiro, N., Belgacem, M.N., Gandini, A., Pascoal Neto, C., 1999. Urethanes and polyurethanes from suberin. 1. Kinetic study. Industrial Crops and Products 6, 163–167.

Cordeiro, N., Blayo, A., Belgacem, M.N., Gandini, A., Pascoal Neto, C., LeNest, J.F., 2000. Cork suberin as an additive in offset lithographic printing inks. Industrial Crops and Products 11, 63–71.

Cosby, S.A., Kelly, M., Waveren, B.V., 2005. Silicone–cork ablative material. US Patent 20050096414.

Dattatreya, R.M., Nuijen, S., Vanswaaij, A.C.P.M., Klopper, P.J., 1991. Evaluation of boiled potato peel as a wound dressing. Burns 17, 323–328.

Deschamps, F., Foudrinier, F., Dherbecourt, V., Mas, P., Prevost, E., Legrele, A.M., Belier, S., Toubas, D., 2003. Respiratory diseases in French cork workers. Inhalation Toxicology 5, 1479–1486.

Domingues, V., Alves, A., Cabral, M., Delerue-Matos, C., 2005. Sorption behaviour of bifenthrin on cork. Journal of Chromatography A 1069, 127–132.

Evtiouguina, M., Barros-Timmons, A., Cruz-Pinto, J.J., Pascoal Neto, C., Belgacem, M.N., Gandini, A., 2002. Oxypropylation of cork and the use of the ensuing polyols in polyurethane formulation. Biomacromolecules 3, 57–62.

Evtiouguina, M., Gandini, A., Pascoal Neto, C., Belgacem, M.N., 2001. Urethanes and polyurethanes based on ooxypropylated cork. 1. Appraisal and reactivity products. Polymer International 50, 1150–1155.

Ferguson, L., Harris, P.J., 1996. Studies on therole of specific dietary fibres in protection against colorectal cancer. Mutation Research 350, 173–184.

Ferguson, L., Harris, P.J., 1998. Suberized plant cell walls suppress formation of heterocyclic amine-induced aberrant crypts in a rat model. Chemico-Biological Interactions 114, 191–209.

Ferrara, G., Meloy, T.P.P., 1999. Low dense media process: a new process for low-density solid separation. Powder Technology 103, 151–155.

Flores, M., Rosa, M.E., Barlow, C.Y., Fortes, M.A., Ashby, M.F., 1992. Properties and uses of consolidated cork dust. Journal of Materials Science 27, 5629–5634.

Ford, B.J., 2001. The Royal Society and the microscope. Notes and Records of the Royal Society London 55, 29–49.

Fujiwara, K., Nakai, K., Torihara, H., 2000. Visualization of the sound field around a Schroeder diffuser. Applied Acoustics 60, 225–235.

George, J., Neelakantan, N.R., Varughese, K.T., Thomas, S., 2006. Failure properties of thermoplastic elastomers from polyethylene/nitrile rubber blends: effect of blend ratio, dynamic vulcanization and filler incorporation. Journal of Applied Polymer Science 100, 2912–2929.

George, J., Varughese, K.T., Thomas, S., 1999. Dielectric properties of isotactic polypropylene/nitrile rubber blends: effect of blend ratio, filler addition and dynamic vulcanization. Journal of Applied Polymer Science 73, 255–270.

Gil, L., 1993. New cork powder particle boards with thermoplastic binding agents. Wood Science and Technology 27, 173–182.

Gil, L., 1994. Effect of hot pressing densification on the cellular structure of black agglomerated cork board. Holz als Roh- und Werkstoff 52, 131–134.

Gil, L.M.C.C., 1996. Densification of black agglomerate cork boards and study of densified agglomerates. Wood Science and Technnology 30, 217–223.

Gil, L., 2000. Cortiça: da produção à aplicação. Seixal: Câmara Municipal do Seixal.

Gil, L., 2005. História da cortiça. Santa Maria de Lamas: Associação Portuguesa de Cortiça.

Godinho, M.H., Martins, A F., Belgacem, M.N., Gil, L., Cordeiro, N., 2001. Properties and processing of cork powder filled cellulose derivatives composites. Macromolecular Symposium 169, 223–228.

Gómez Corzo, M., Macías-García, A., Díaz-Díez, M.A., García, M.J.B., 2002. Preparation of a cation exchanger from cork waste: thermodynamic study of the ion exchange processes. Journal of Materials Science and Technology 18, 57–59.

Hall, C.T., 2003. Doctored bats go 2 percent farther. Chronicle Science Writer, www.sfgate.com/cgi-bin/article.cgi?file=/chronicle/archive/2003/06/05..., assessed 5 June 2003.

Harris, P.J., Ferguson, L., 1999. Dietary fibres may protect or enhance carcinogenesis. Mutation Research 443, 95–110.

Harris, P.J., Triggs, C.M., Roberton, A.M., Watson, M.E., Ferguson, L., 1996. The adsorption of heterocyclic aromatic amines by model dietary fibres with contrasting compositions. Chemico-Biological Interactions 100, 13–25.

Henshaw, J.T., 1977. The biomechanical design of a walking appliance for a paraplegic adult. Journal of Medical Engineering and Technology 1, 141–145.

Hernandéz-Olivares, F., Bollatti, M.R., del Rio, M., Parga-Landa, B., 1999. Development of cork–gypsum composites for building applications. Construction and Building Materials 13, 179–186.

Hooke, R., 1665. Micrographia, or Some Physiological Descriptions of Minute Bodies Made by Magnifying Glasses. With Observations and Inquiries Thereon. London: Martyn and Allestry, for the Royal Society.

Ivko, J.J., 1999. Independence through humeral rotation in the conventional transhumeral prosthetic design. Journal of Prosthetics and Orthotics 11, 20–22.

Iwasaki, A., 2002. 10 MV X-ray central-axis dose calculation in thorax-like phantoms (water/cork) using the differential primary and scatter method. Radiation Physics and Chemistry 65, 11–26.

Jardine, L., 2003. The Curious Life of Robert Hooke. The Man Who Measured London. London: Harper Collins Publishers.

Kahkonen, M.P., Hopla, A.I., Vuorela, H.J., Rauha, J.P., Pihlaja, K., Kujala, T.S., Heinonen, M., 1999. Antioxidant activity of plant extracts containing phenolic compounds. Journal of Agricultural and Food Chemistry 47, 3954–3962.

Kockel, V., 1998. Phelloplastica. Modelli in sughero dell'architectura antica nel XVIII sec. nella collezione di Gustavo III di Svezia. Jonsered, Sweden: Astrom Editions.

Krizková, L., Lopes, M.H., Polónyi, J., Belicová, A., Dobias, J., Ebringer, L., 1999. Antimutagenicity of a suberin extract from Quercus suber cork. Mutation Research 446, 225–230.

Landolt, J.H., Magalhaes, P.H., Nóvoa, P.R., Viana, J., Marques, A.T., 2005. Low velocity impact of thermosetting composite systems modified with rubber and cork powders. Science and Engineering of Composite Materials 12, 103–107.

Lier, S.V., 2000. The vibro-acoustic modelling of slab track with embedded rails. Journal of Sound and Vibration 231, 85–817.

Maroto, J.A., Duenas-Molina, J., de Dios, J., 2005. Experimental evaluation of the drag coefficient for smooth spheres by free fall experiments in old mines. European Journal of Physics 26, 323–330.

Matos, M.J., Simplício, M.H., 2006. Innovation and sustainability in mechanical design through materials selection. Materials and Design 27, 74–78.

Millikan, R.A., 1943. Benjamin Franklin as a scientist. Engineering and Science Monthly December: 7–9, 16–18.

Moiteiro, C., Curto, M.J.M., Mohamed, N., Bailen, M., Martinez-Diaz, R., Gonzalez-Coloma, A., 2006. Biovalorization of friedelane triterpenes derived from cork processing industry byproducts. Journal of Agricultural and Food Chemistry 54, 3566–3571.

Moiteiro, C., Justino, F., Tavares, R., Marcelo-Curto, M.J., Florencio, M.H., Nascimento, M.S.J., Pedro, M., Cerqueira, F., Pinto, M., 2001. Synthetic secofriedelane ad friedelane derivatives as inhibitors of human lymphocyte proliferation and growth of human cancer cell lines in vitro. Journal of Natural Products 64, 1273–1277.

Moiteiro, C., Manta, C., Justino, F., Tavares, R., Curto, M.J.M., Pedro, M., Nascimento, M.S.J., Pinto, M., 2004. Hemisynthetic secofriedelane triterpenes with inhibitory activity against the growth of human tumor cell lines in vitro. Journal of Natural Products 67, 1193–1196.

Morell, F., Roger, A., Cruz, M.J., Munoz, X., Rodrigo, M.J., 2003. Suberosis – Clinical study and new etiologic agents in a series of eight patients. Chest 124, 1145–1152.

Mukaibo, Y., Shirado, H., Konyo, M., Maeno, T., 2005. Development of a texture sensor emulating the tissue structure and perceptual mechanism of human fingers. In: Proceedings of the 2005 IEEE International Conference on Robotics and Automation. Barcelona, Spain, April 2005, pp. 2565–2570.

Nóvoa, P.J.R.O., Ribeiro, M..S., Ferreira, A.J.M., Marques, A.T., 2004. Mechanical characterization of lightweight polymer mortar modified with cork granulates. Composites Science and Technology 64, 2197–2205.

Oliveira, M.A., Oliveira, L., 2000. The Cork. Santa Maria de Lamas: Amorim.

Ortega, M.C., Moreno, M.T., Ordovás, J., Agudo, M.T., 1996. Behaviour of different horticultural species in phytotoxicity bioassays of bark substrates. Scientia Horticulturae 66, 125–132.

Reculusa, S., Trinquecoste, M., Dariol, L., Delhaès, P., 2006. Formation of low-density carbon materials through thermal degradation of a cork-based composite. Carbon 44, 1298–1352.

Reiber, G.E., Smith, D.G., Wallace, C., Sullivan, K., Hayes, S., Vath, C., Maciejewski, M.L., Yu, O.C., Heagerty, P.J., LeMaster, J., 2002. Effect of therapeutic footwear on foot reulceration in patients with diabetes. A randomized controlled trial. Journal of the American Medical Association 287, 2552–2558.

Rio Merino, M., Santa Cruz Astorqui, J., Hernández Olivares, F., 2005. New prefabricated elements of lightened plaster used for partitions and extrados. Construction and Building Materials 19, 487–492.

Rueda, L.M., Patel, K.J., Axtell, R.C., 1991. Comparison of floating and sinking encapsulated formulations of the fungus Lagenidium-giganteum (Oomycetes, Lagenidiales) for control of Anopheles larvae. Journal of he American Mosquito Control Association 7, 250–254.

Santos, C.O., 1997. Amorim. História de uma família (1870–1997). Grupo Amorim, Mozelos, Portugal.

Seligman, D.A., Dawson, D.R., 2003. Customized heel pads and soft orthotics to treat heel pain and plantar fasciitis. Archives of Physical Medical Rehabilitation 84, 1564–1567.

Silva, S.P., Sabino, M.A., Fernandes, E., Correlo, V.M., Boesel, L.F., Reis, R.L., 2005. Cork: properties, capabilities and applications. International Materials Reviews 50, 345–365.

Tadeu, A., Santos, P., 2003. Assessing the effect of a barrier between two rooms subjected to low frequency sound using the boundary element method. Applied Acoustics 64, 287–310.

Tijani, M.E.H., Spoelstra, S., Bach, P.W., 2004. Thermal relaxation dissipation in thermoacoustic systems. Applied Accoustics 65, 1–13.

Villaescusa, I., Fiol, N., Cristiani, F., Floris, C., Lai, S., Nurchi, V.M., 2002. Copper(II) and nickel(II) uptake from aqueous solution by cork wastes: a NMR and potentiometric study. Journal of Chemical Technology and Biotechnology 75, 812–816.

Villaescusa, I., Martínez, M., Miralles, N., 2000. Heavy metal uptake from aqueous solution by cork and yohimbe bark wastes. Journal of Chemical Technology and Biotechnology 75, 812–816.

Winck, J.C., Delgado, L., Murta, R., Lopez, M., Marques, J.A., 2004. Antigen characterization of major cork moulds in suberosis (cork worker's pneumonitis) by immunoblotting. Allergy 59, 739–745.

Wright, P.C., Raper, J.A., 1998. Investigation into the viability of a liquid-film three-phase spouted bed biofilter. Journal of Chemical Technology and Biotechnology 73, 281–291.

Yamamoto, H., Yamaguchi, M., Inoue, K., 1996. Absorption and increase in the production of prenylated flavanones in *Sophora flavescens* cell suspension cultures by cork pieces. Phytochemistry 43, 603–608.

Chapter 12

Production of cork stoppers and discs

Cork is an industrial raw material that supports an integrated chain from production to the consumer with an important impact at economic, social and environmental levels in the corresponding regions of production and transformation. The industrial production is organised following basically two streamlines: (a) the production of stoppers of natural cork and the production of discs of natural cork and (b) the production of agglomerates of cork particles, with or without the addition of binding materials. The first objective of the industry at present is the production of stoppers and discs, and reproduction cork is aimed primarily during the production. The agglomerates use all the other types of cork raw materials including the by-products and wastes of the former transformation.

This chapter presents an overview of the industrial organisation in terms of these two main processing lines by separating raw materials and product outputs. The production of stoppers and cork discs is described from the yard storage to the finishing operations. The operation of cork boiling in water, whose effects on the cellular structure have already been presented in Chapter 7, is described in detail giving attention to its recent evolution. The evolution of cork washing and bleaching and cork quality classification are also presented. However, the detailed surface analysis of cork stoppers by image analysis and its relation to commercial quality will be studied later in Chapter 14.

The yields and the quality profiles obtained in the production of cork stoppers and discs are presented following industrial mass balances in relation to the quality of the raw material.

12.1. Industrial production lines

The products of cork are usually classified in industrial environments into products of natural cork and products of agglomerated cork.

The natural cork products are made up by solid cork, mostly as a single piece, and they are produced from the cork planks by cutting operations with a minimum transformation of the cork material, apart from the cork boiling preparation. They include the stoppers for the wine bottles and other sealants, as well as cork discs. The cork discs are in fact an intermediate product that will be assembled with cork agglomerates to make composite stoppers, now called technical stoppers. Also some special products for decoration or design (such as shoe soles) are cut from corkboards.

The agglomerated corks are made up of cork granules that are bound together, and they include three types:

(a) the adhesive bound agglomerates, called composite agglomerates or just cork agglomerates, are made up with cork granules that are glued together with an adhesive and hot pressed; they are used for the production of agglomerated cork stoppers, or for the body of the technical stoppers, production of boards and sheets mostly for surfacing;

(b) the pure expanded agglomerates, also called cork black agglomerates, are made up of cork granules that are bound together without adhesives at a high temperature of about 300°C; these agglomerates are used mostly as insulation materials;

(c) the cork and rubber composites, called rubber cork, are made up of cork and rubber granules after surface activation; they are used as vibration and shock absorbers in surfacing and industrial equipment.

The processes involved in the production of cork agglomerates are described in Chapter 13.

The industrial organisation may be simplified into two main streamlines according to the raw material used, and the types of products, as shown in Figure 12.1. The reproduction cork planks are directed in a first option to the production of natural cork products (cork stoppers and cork discs). Excluded are the refused cork planks (because of excessive defects) or planks without minimum dimensional requirements (the so-called cork pieces, with an area under 400 cm²). This production line generates a large amount

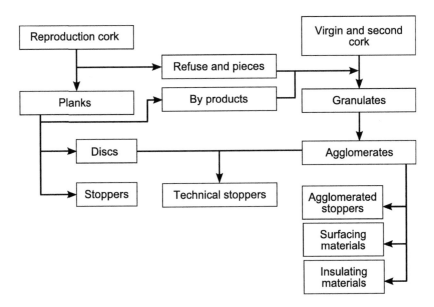

Figure 12.1. Schematic organisation of the material flux in the industry of cork.

of by-products, as detailed later, that constitute a raw material for the other industrial streamline leading to the production of cork agglomerates. In this case the raw material also includes the other types of cork: virgin cork and second cork, as well as corks that are obtained outside the normal cork oak exploitation (i.e. corks from burnt trees after forest fires, under-aged corks obtained from sanitary fellings). All these raw materials are triturated and the granules are used for agglomeration.

In the past there was a segmentation of the industry by sectors: the preparation of cork, the production of stoppers, the preparation of cork granules and the agglomeration. Nowadays there is a trend for a vertical integration namely regarding the preparation and the production of stoppers and discs.

12.2. Post-harvest processing

After reception at the mill, the raw cork planks are stored under ambient conditions and afterwards undergo a processing called preparation which consists basically in the boiling in water, the trimming and the classification of the cork planks (Fig. 12.2).

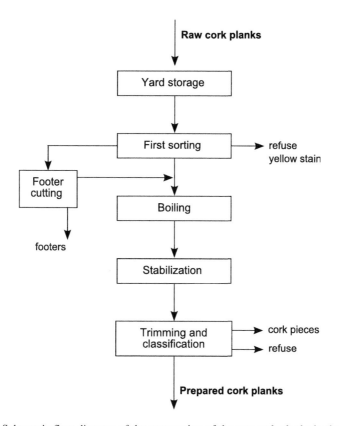

Figure 12.2. Schematic flow diagram of the preparation of the raw cork planks in the mill.

12.2.1. Yard storage

The cork planks after extraction from the tree are usually stacked up in the field before transport to the mill yard, as seen in Chapter 5. In a few cases, and depending on the commercial arrangements, the cork is transported immediately after extraction to the mill avoiding the field storage.

At the mill yard, the reproduction cork planks are either stacked manually with an orderly arrangement with the cork back oriented upwards, or discharged automatically from trucks and piled with a more loosened manner. This is the usual practice in the modern units. Nowadays care is taken to register the origin of the cork planks (at the stand or forest estate level) and the piled raw material is kept separated by provenance. If also stored in that unit, virgin and second corks, cork pieces are kept apart.

The description of the air storage of cork at the mill yard and its preparation for the first industrial operation, i.e. the boiling, has already been made in Chapter 5. From the yard, the cork planks go for processing, baled in prismatic bales or stacked in pallets. Depending on the specific mill, the sorting of the cork planks at the yard for boiling may be either minimal or more demanding regarding their separation by calliper classes (<27 mm and >27 mm classes, for instance) and removal of refuse planks.

12.2.2. Boiling in water

All reproduction cork planks that are used for the production of stoppers and discs undergo a treatment by immersion in hot water at around boiling temperature for 1 h, called cork water boiling.

The boiling of cork is a part of the industrial processing that has undergone significant modernisation in recent times (Fig. 12.3). In the traditional process the bales of cork planks were stacked in large tanks where water was kept under boiling, as shown in Figure 12.3a using a direct fuel furnace (most often wood fuelled); the same water was used for repeated boiling batches, with addition of clean water to compensate for losses due to evaporation and absorption into cork, and the tank was emptied, cleaned and refilled with water in every 4–5 days. Under such conditions the water rapidly turned very dark brown with a vigorous foaming, while the concentration of suspended and dissolved solids in the effluent waste water was high. For instance after 1 day of operation the dissolved solids correspond to 0.24% and the solubilised phenolics to 0.04% in relation to the cork plank mass (Pereira et al., 1979).

The modern cork boiling process and equipment used have seen quite a few changes. The cork planks are now stacked on stainless steel pallets and the water treatment occurs in a closed stainless steel autoclave, as shown in Figure 12.3b. Figure 12.4 shows one simplified diagram of the boiling operation as it occurs in one of the most recent industrial units. The water is kept at 95°C using heat exchangers and it is circulated within the autoclave with filters incorporated into the water lines to remove the suspended solids. The cork pallets are stacked in the empty autoclave, and water is filled in after closure and re-circulated during the 1-h operation. At the end the autoclave is emptied, opened and the pallets taken out. The truncated pyramidal bottom of the autoclave prevents the deposition of solids. The functioning of the autoclave, including its loading and unloading, is mostly automated and computer controlled.

(a)

(b)

Figure 12.3. Equipment used for the water boiling operation of cork planks: (a) traditional tank with boiling water; (b) closed autoclaves with circulation of near boiling water.

The autoclave may also be cylindrical. In some cases the treatment may use steam at 110–120°C, as it is the case shown in Figure 12.5, used for the expansion of thin planks for production of discs for champagne stoppers.

To prevent eventual introduction of chlorinated compounds into cork that may lead to contaminations and taints of wines, the water used is not chlorine-treated. In some cases, such as in the system shown in Figure 12.4, a stripping of volatiles from the hot water is

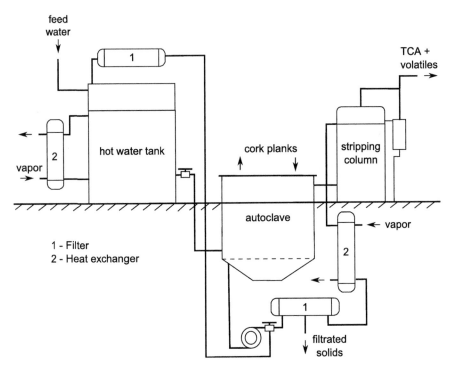

Figure 12.4. Simplified diagram of a modern autoclave for the boiling of cork with water circulation and stripping of volatiles.

Figure 12.5. Autoclave for steam treatment of thin cork planks before production of discs for champagne stoppers; in the background there is a tank with a covered top where the water boiling of cork takes place.

made in a stripping column, since it is known that the cork boiling water also extracts and concentrates 2,4,6, trichloroanisol (TCA) and chlorophenols that may lead to contamination of wine, as further discussed in Chapter 14.

The effluent waters constitute a waste problem with an average chemical oxygen demand (COD) and biochemical oxygen demand (BOD) concentrations of 7.4 and 1.3 g/L respectively, a pH value of 4.8–5.1 and a high toxicity containing phenolic and polyphenolic compounds at respectively 9 g/L (phenol) and 0.5 g/L (as caffeic acid) (Anselmo et al., 2001; Mendonça et al., 2004; Dominguez et al., 2005). Several studies have been devoted to their treatment, and research in this field has been very active in the last years. Processes were proposed such as ultrafiltration (Minhalma and de Pinho, 2001; Acero et al., 2005a), combined flocculation, flotation and ultrafiltration (Minhalma and de Pinho, 2000), fungal degradation and detoxification (Mendonça et al., 2004), flocculation with ferric chloride (Dominguez et al., 2005), chemical oxidation with Fenton's reagent (Guedes et al., 2003; de Heredia et al., 2004b), an integrated coagulation–flocculation and Fenton's reagent process (de Heredia et al., 2004a; Peres et al., 2004), oxidation with ozone (Benitez et al., 2003; Acero et al., 2004), an integrated ozonisation/ultrafiltration (Minhalma et al., 2006) and photo-oxidation (Silva et al., 2004). Several of the major phenolic pollutants present in the cork wastewaters have been studied as model compounds for treatment, such as esculetin (6,7-dihydroxycoumarin; Acero et al., 2005b), gallic acid (Benitez et al., 2005a), ellagic acid (Benitez et al., 2005b) and acetovanillone (Benitez et al., 2005c).

12.2.3. Cork changes with boiling

The main objective of the water boiling is to change the mechanical properties of the cork planks in order to flatten the planks and to facilitate the subsequent cutting processes. The effect of water boiling in the structure and compression properties of cork has been studied in Chapters 8 and 9. The combined heating and the presence of water cause a markedly softening of the material immediately after boiling, as it has been shown in Figure 9.7. This allows the relief of the growth stresses in the cork cells, with the decrease of corrugations of the cell walls and the increase in the uniformity of the cellular structure, as it has been exemplified in Figure 8.9.

As a result of the cell wall straightening there is a dimensional increase of cork: in thickness, the cork planks increase by approximately 15%, corresponding to the radial expansion of cells, whereas in the tangential and axial directions the expansion is about 6% (Rosa et al., 1990). It is expectable to find very different values for the expansion of different corks with boiling since it will depend on the extent of cellular corrugations present in the tissue, as a result of the internal stresses, which are highly variable features. Although average values for the radial expansion of cork planks are usually in the range of 11–15%, individual values from almost 0% to near 40% are also obtained. The porosity of cork as measured by the coefficient of porosity decreases with water boiling (Cumbre et al., 2001). The water boiling of cork is therefore an operation that in practical terms increases the technological quality of the cork planks because it increases the calliper of the cork planks and decreases the dimensions of the lenticular channels. The curvature of the raw cork planks is also practically eliminated and the boiled cork planks are flat boards (Fig. 12.6).

(a)

(b)

Figure 12.6. Pallets of cork planks (a) before and (b) after water boiling.

The dimensional variation of cork with boiling is irreversible at temperature conditions below 100°C. This means that subsequent wetting and drying of cork as it occurs in the washing/bleaching treatment of stoppers do not alter its dimensions.

The boiling of cork only extracts a very small amount of the water-soluble compounds that are present in cork (less than 2% of the total water extractives of cork) because the time and the cork-water surface of extraction are too small. Therefore extraction of chemical compounds from the cork is not the objective of the boiling operation under the conditions used in the industry, in spite of frequent statements in this direction. The same applies to a so-called microbial sterilisation of the cork plank which is also not the case.

The frequency of water renewal and the stripping of volatiles are not intended as a means to increase extraction of material from cork, but more to avoid contamination of cork planks by the boiling water that gradually is concentrating potentially hazardous compounds.

12.2.4. Post-boiling processing of cork planks

After water boiling, the cork planks are left to air dry for some (2–3) days, in what is called a stabilisation step. When using pallets, these are stacked on top of each other, which help to flatten the planks. Traditionally, large piles of boiled cork planks were constructed and the drying period lasted for some weeks. However, this has now changed, and shorter stabilisation periods are practiced which contribute to avoiding microbial growth that could endanger the innocuous character of the cork stopper in relation to wine.

During stabilisation, the moisture content of the cork planks decreases from about 40% to 70% immediately after boiling following drying curves such as those shown in Figure 8.11a. After 2 days, the moisture content should decrease to values about 14–18% that is considered in the industry an adequate working moisture content.

After the drying, the boiled cork planks are individually observed by an expert worker who trims and cuts the plank into sub-planks that are more homogeneous for further processing. The first criterion is to take out parts (or the total plank) that have major defects for processing into stoppers or discs:

• planks with wetcork are separated for kiln drying or long-duration air-drying;
• planks with yellow stain or otherwise mouldy patches are withdrawn from production and are directed for trituration;
• parts with insect galleries are refuse and taken out for trituration;
• planks, or part of planks, with a very large extent of failures, cracks, lung, exfoliation, etc. are refuse and taken out for trituration;
• planks, or part of planks, that are too thin, i.e. a thickness less than 13 mm, are refuse and taken out for trituration.

The trimmer/classifier uses a knife to hand-cut the planks (Fig. 12.7). The objective is to sort out the refuse planks and to separate from the raw planks the parts that are unsuitable for further processing as solid cork. Also if the plank is heterogeneous in terms of thickness and quality, it will be divided to obtain smaller and homogeneous planks, allowing either the punching of stoppers or the production of discs. Large planks are also divided into sub-planks that can be handled manually more easily by the workers. Figure 12.8 shows examples of raw cork planks marked with the cutting lines chosen by the trimmer to cut the final planks. Measurements made in raw cork planks and in the subsequent prepared cork planks show that in the first case the dimensions average 47 and 119 cm in the horizontal and vertical directions respectively (average plank area 3745 cm^2), whereas in the prepared cork planks the average dimensions are 34 and 70 cm respectively (average plank area 1674 cm^2) (Costa and Pereira, 2004). In this operation there is a calculated loss of 5% of the initial raw material as small cork pieces.

An experimental measurement of loss of cork raw material in the process of cutting the prepared planks for further processing was made in relation to thin planks to be used for

(a)

(b)

Figure 12.7. Preparation of the final cork planks: (a) trimming of one edge and (b) the cutting of one raw cork plank into two final cork planks.

the production of cork discs (Fernandes, 2005). The waste material corresponds on average to 19.5% of the initial cork mass, but the between plank yield varies from 0 to 50% losses, depending on the specific form and quality features of the individual planks.

The trimming of the planks is made to straighten the edges and to give better-looking cross sections, and the sorting separates the prepared cork planks by calliper and quality classes. When the production of stoppers and discs is integrated within the mill, trimming and sorting are kept to a minimum in order to avoid raw material losses. Separation into thickness classes and quality classes follow the specific mill requirements: a common practice nowadays is to separate only three calliper classes, 14–22, 22–27 and >27 mm,

Figure 12.8. Examples of raw cork planks that are marked with cutting lines to prepare the final cork planks for subsequent processing.

and two quality classes, the 1st–5th class as the main raw material assortment for the production of stoppers and discs, and a 6th quality of lower value. When the final mill product is the cork board, more care is taken in the regularisation of edges and thickness, and quality classification is described in more detail, e.g. as 1st–3rd, 4th–5th and 6th or following client specifications.

12.2.5. Wet cork processing

The planks with wetcork regions (see Chapter 7) are separated and dried to normal cork moisture contents of about 14% or less. This may occur by natural air-drying during 1–2 years or by forced hot air-drying in kiln chambers. These have been introduced in the more modern units, as an adaptation from wood drying kilns, using drying cycles with increasing air temperature. Conditions may vary from mill to mill: an example is a total duration of 96 h at a final temperature of 90°C; another is a 2-week drying at a final temperature of 80°C.

After drying the cork planks are introduced in the processing line and undergo the normal operational circuit for production of stoppers or discs.

12.2.6. Second boiling

When the time elapsed between the boiling of the cork planks and their processing into stoppers or discs is long enough to dry the cork to equilibrium moisture contents (5–8%), the beneficial effects of a higher moisture content into the material's strength and machining aptitude are lost. This is usually the case if the boiled cork planks are stored for more than 2–3 weeks or if the processing of the planks is not made at the preparation mill.

In this case the planks are put through a second boiling operation with the same procedure as described, but for a shorter time of only 30 min. Since the internal stresses of cork were already relieved in the first boiling, as described above, there occurs no further expansion of cork.

12.3. The production of cork stoppers

12.3.1. Process operations

The prepared cork planks that have a thickness of more than 27 mm produce cork stoppers following a general industrial flowsheet as shown in Figure 12.9. The unit operations may be grouped into production of stoppers and finishing treatments.

The first operation is the cutting of the cork planks into parallel strips with a width slightly more than (by 1–2 mm) that corresponding to the length of the stopper that is going to be produced (for instance 38 mm, which is one of the most common length dimension for wine stoppers). The cutting lines are made horizontally in the cork plank,

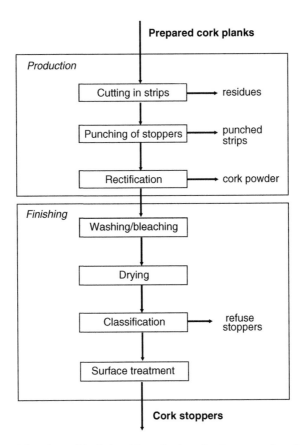

Figure 12.9. General flowsheet of the industrial production of cork stoppers from boiled cork planks.

meaning the direction perpendicular to the axial direction of the plank. This cutting is made with vertical rotating blade discs either with a manual feed or with an automatic functioning (Fig. 12.10). In this operation, scraps from eventual incomplete bottom or strips constitute material for trituration.

The stoppers are bored out from these cork strips by punching with a sharp cutting cylinder with an inner diameter equivalent to the diameter of the stopper to be produced. The punching is done with the cylinder axis in the axial direction of the cork strip (Fig. 12.11). Therefore the two circular tops of the stopper correspond to transverse sections of cork and this is the reason why the thickness of the cork plank has to be more than 27 mm in order to have a cork width more than 24 mm which is one of the most common diameters for wine stoppers.

The automation of the punching operation may differ from a fully automated punching to a semi-automated system where the piston-like movement of the punching cylinder is automatic, but the positioning of the cork strip is done manually by the operator. This

(a)

(b)

Figure 12.10. Cutting of cork strips from the prepared cork planks with (a) a manual feeding and (b) an automated operation.

(a)

(b)

Figure 12.11. Punching out of cork stoppers from a cork strip: (a) a semi-automated punching equipment where the piston-like movement of the cutting cylinder is automatic, but the positioning of the cork strip is done by the operator and (b) an automatic equipment where the punching and the forward movement of the strip is automatic; in this case the feeding of the cork strips is manual.

allows to avoid punching stoppers from positions which are defective, for instance by the presence of a fracture or by insufficient local thickness of the cork strip. The automated cork punching allows overall higher gross yields of stoppers, whereas the semi-automated punching may decrease the number of refused or lower quality stoppers. The two strips shown in Figure 12.12 have been punched out automatically and semi-automatically respectively. Earlier the control of the punching movement was done also by the operator using a foot clutch (manual punching) but this is no longer used, or only rarely, due to the much lower output yields.

The punched cork strips are by-products that are forwarded for trituration into granulates and production of cork agglomerates.

The dimensions of the stoppers are rectified by abrasion of the tops (for length conformity) and the lateral cylindrical surface (for diameter conformity). An intermediate operation of sorting our defective cork stoppers may be incorporated before the finishing operations.

Figure 12.12. Microwave drying of cork stoppers in a continuous fed tunnel (Delfin process).

12.3.2. Finishing and surface treatments

The finishing operations of the cork stoppers include a washing and bleaching stage with its subsequent drying, the classification into commercial quality classes and the final surface treatment with paraffin or silicone (Fig. 12.9).

The water washing of cork stoppers is aimed at cleaning surface dust or loosened lenticular material. The washing aqueous solutions may have the addition of oxidant chemicals to add a disinfection effect to the surface lenticular channels and in some of the cases also a bleaching of the cork surface.

The bleaching of stoppers is mostly a cosmetic operation aimed at giving a lighter colour shade to the surface of the stopper, and it is not a necessity in what regards the material's performance. Nowadays the standard cork bleaching is done by using a hydrogen peroxide aqueous solution (about 10% of 130 vol. H_2O_2 solution) with 1% sodium hydroxide, with a subsequent neutralization with a 1% citric acid solution. Washing with other chemicals such as sodium metabisulphite (1–2% concentration), sulphamic acid or peracetic acid is made more rarely and on client's demand. The traditional bleaching with chlorine-based chemicals such as calcium hypochlorite is at present abandoned due to the risks of chlorinating the phenolic compounds of cork that could be precursors to formation of taint compounds.

After the bleaching, the stoppers are dried at 40–60°C under air pressure or under reduced pressure in drying chambers until a final moisture content of 5–8%. Drying is also done using microwaves in a tunnel with continuous feeding (Delfin process) as shown in Figure 12.13.

In the classification the stoppers are separated by quality class according to the macroscopic appearance of their surface (see Chapter 7). Earlier, this was a manual operation where each stopper was visually appraised and sorted; 40 years ago this was still

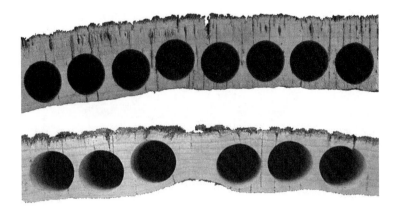

Figure 12.13. Cork strips after the punching out of the stoppers using an automated equipment (top) and a semi-automated procedure (bottom).

one of the striking images of the large industrial mills: a hall full of women sorting out stoppers, each one surrounded by nine different baskets into which they threw with an incredible speed and efficiency the classified stopper! Nowadays this operation is carried out automatically with sorting machines based on machine vision (Fig. 12.14a): CCD cameras observe the circular tops and the lateral surface of each stopper, and image analysis techniques allow to characterise their main porosity features and to refer them to pre-defined quality classes. The quality discrimination of the wine stoppers goes from the premium qualities (such as Flower, Extra and Superior) to successively lower qualities (from 1st–6th) and to a refuse lot. However, the porosity variables defining each quality grade are not standardised nor quantified in the client's purchase order which in most cases is based on a reference sample. The surface characterisation of cork stoppers of different quality grades is described further in detailed in Chapter 14. Often, and especially in the case of the best-quality classes, a final check of the quality classification is made manually by observation of the stoppers as they pass on to a moving belt (Fig. 12.14b). The manual sorting is still found in some cases, namely in the smaller industrial units (Fig. 12.14c).

Final printing of stoppers with the client brand may personalise the stopper and increase the traceability between supplier and client. This is made before the final surface treatment.

The surface treatment has the objective to coat the stopper with a lubricant film to reduce friction, thereby allowing an easier introduction and extraction into and out the neck of the bottle. Paraffin and silicone of food grade quality are usually used. The finished stoppers are packed in polyethylene bags which are evacuated and injected with SO_2. Gamma-irradiation of the finished stoppers may also be used for sterilisation (Botelho et al., 1988).

The stoppers with poor aspect due to a conspicuous porosity (5th and 6th, sometimes also 4th quality grades) are often taken out for a colmation of the superficial lenticular channels. These are filled with a mixture of cork powder (obtained from the dimensional rectification of the stoppers) and a resin (usually polyurethane or a rubber binder). The result is a more homogeneous cork surface.

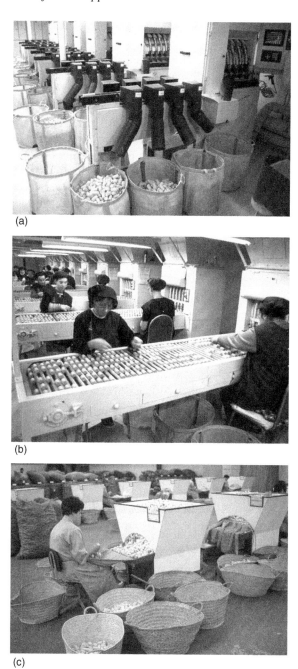

(a)

(b)

(c)

Figure 12.14. Classification of cork stoppers: (a) automatic sorting machine based on machine vision techniques; (b) manual verification of classification of one quality grade; (c) manual classification of stoppers.

12.3.3. Production yields

The yield of stoppers is directly related to the area of the cork planks, and it is clear that large and straight-sided planks will maximise, from a geometrical point of view, the potential yield of stoppers that can be punched out (Fig. 12.15). The estimated area that will be left over from the sides of cork planks and strips for geometrical reasons is on average 17.4% but very variable between planks (Costa and Pereira, 2004).

Seen the production process of cork stoppers, it is clear that the amount of residual material obtained from the initial prepared cork planks is high. The mass balance for production of 38 mm × 24 mm (length × diameter) cork stoppers from standard cork planks (34–40 mm) shows that the total commercial stoppers represent on average 24.2%, the rest being 5.0% of refused stoppers (incomplete and defective stoppers), 66.6% of punched out cork strips and 4.2% of cork powder (Pereira et al., 1994).

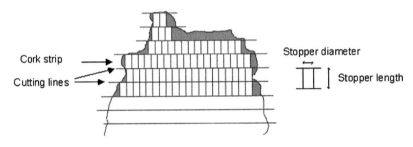

Figure 12.15. Theoretical utilisation of a cork plank for the production of stoppers. The shaded areas correspond to left over area.

Table 12.1. Production yield of cork stoppers from cork planks of three different thickness classes (22–27, 27–32, 32–40 and >40 mm) and three quality grades (1st–2nd, 3rd–4th and 5th–6th).

Thickness class (mm)	Quality grade	Cork stoppers (%)	Refuse stoppers (%)	Punched strips (%)	Trimmings and powder (%)
22–27	1st–2nd	22.5	0.9	68.0	8.8
	3rd–4th	20.3	1.2	69.2	9.4
	5th–6th	15.6	2.0	70.7	11.8
27 – 32	1st–2nd	23.7	2.4	67.5	6.4
	3rd–4th	24.5	2.2	65.7	7.7
	5th–6th	21.7	2.5	74.1	1.7
32–40	1st–2nd	24.0	2.3	72.4	1.3
	3rd–4th	20.5	3.9	70.4	5.2
	5th–6th	19.1	4.5	72.7	1.7
>40	1st–2nd	26.4	1.3	68.6	3.7
	3rd–4th	23.2	1.6	66.4	8.8
	5th–6th	15.8	4.8	71.3	8.2

A more detailed study on the mass balances of the production of stoppers from planks of different thickness and quality classes has the results shown in Table 12.1. On average, the obtained commercial stoppers amount to 22.1% of the initial boiled cork planks, with 2.8% of refused stoppers, 70.1% of punched cork strips and 5.0% of scraps and powder. There is a large variation between individual planks but, on average, yields for different cork plank qualities and thickness classes are not very different,

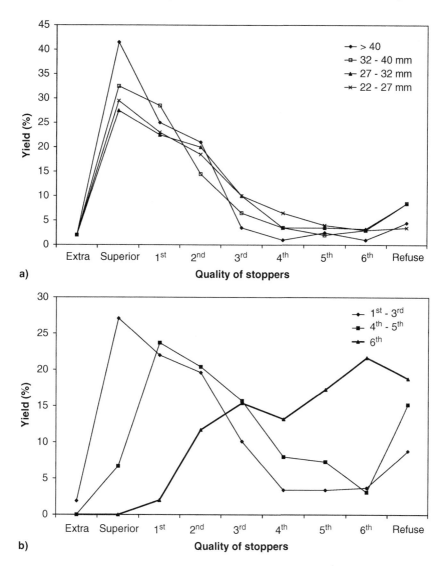

Figure 12.16. Quality profile of the stoppers classified in eight quality classes (extra, superior, 1st–6th) and refuse obtained from the punching of (a) cork planks of the same quality class (1st–2nd) with four callipers (22–27 mm, 27–32 mm, 32–40 mm and > 40 mm) and (b) cork planks with the same calliper (27–32 mm) with three quality grades (1st–2nd, 3rd–4th and 5th–6th).

although with a tendency for the 5th–6th quality planks to yield a lower amount of commercial corks.

Maybe more important than total mass yield of cork stoppers is their quality profile, i.e their distribution into quality classes. This is a factor of determining importance for the economical performance of the industrial production since the price value for the different cork stoppers grades is very different: for instance an extra stopper is over three times more expensive than a 3rd quality stopper. Figure 12.16 shows the quality profiles of cork stoppers (classified as extra, superior, 1st–6th and refuse) obtained from different cork planks. There is no difference in the quality profiles obtained from cork planks of similar quality but different thickness, as exemplified for the 1st–2nd quality planks of the four calliper classes, However, the difference of quality profiles is striking when comparing cork planks of different quality grades: for instance from the stoppers obtained from 27–32 mm planks, there are 29% of extra-superior stoppers when using good quality planks (1st–2nd), only 7% of medium quality planks (3rd–4th) and 0% of low quality planks (5th–6th).

12.4. The production of discs for technical cork stoppers

12.4.1. Process operations

The prepared cork planks with a thickness that do not allow the boring out of stoppers (less than 27 mm) are used do produce cork discs following a general industrial flowsheet as shown in Figure 12.17. Cork discs are also among the most valuable products on a unit mass base and they are used for production of technical stoppers, namely stoppers for champagne type and sparkling wines.

Figure 12.18 shows a sequence of the main operations used industrially in the production of cork discs. The production is automated to a significant extent. The first operation is the cutting of the cork planks in the axial direction into parallel rectangular strips. The strips are then laminated using rotating steel blades, first by taking out a fine slice corresponding to the belly layer and then slices with the thickness of the desired cork discs (i.e. 6–8 mm) until only the cork back remains. Usually two cork slices are obtained, but three slices may be cut if the thickness of the cork strip allows it. The belly slice constitutes a good quality by-product for trituration since it contains only cork tissue, whereas the cork back remains have lower quality since most of their mass corresponds to the ligneous outer layer of the plank. The discs are automatically punched out from each cork slice using a battery of cutting cylinders. In this case the direction of cutting is the radial direction of the cork and the circular faces of the discs corresponding to tangential sections of cork. The diameter of the discs depends on the objective of production: discs for champagne stoppers are larger than discs for technical stoppers, i.e. about 34.5 and 26.5 mm respectively. The by-products of this operation are the punched out cork slices, named in the industrial jargon as cork "laces", which constitute a high-quality material for preparation of cork granulates.

The discs are washed and dimensionally rectified by sanding of the surface. One new technological introduction into disc production is an autoclave washing under cycles of air pressure to cause a hydrodynamic extraction of TCA and other volatiles (the INOS II process). This process also allows a pressure-variation induced cleaning of the lenticular channels which are made more accessible due to their short length from face to face of the

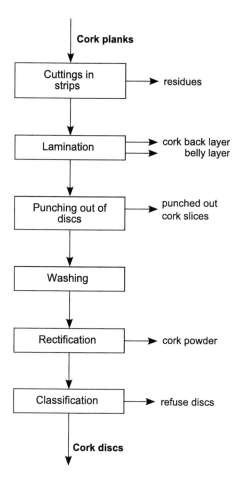

Figure 12.17. General flowsheet of the industrial production of cork discs from boiled cork planks.

discs. Also an increased extraction of water-soluble materials is made in the autoclave, and the content in tannins and other phenolics decrease in the cork discs.

The discs are classified into quality classes: for instance into Extra, Superior, A, B and C classes and refuse. Machine vision systems are used that observe both faces of the discs and characterise their porosity features, sorting them according to the best face. When the mill integrates the production of the final stopper, a marking of the worse face may be incorporated so that the further processing knows which face should be glued to the body of the stopper.

A washing and bleaching with hydrogen peroxide may be carried out as described for cork stoppers.

12.4.2. Production yields

The production of cork discs has the following by-products that constitute raw material for production of cork granulates (a) from the strip cutting operation, the remains of cork

(a)

(b)

(c)

(d)

(e)

Figure 12.18. Photographs taken in industrial mills show different operations of production of cork discs from the boiled cork boards: (a) strips that are cut longitudinally from the cork planks; (b) the cork strips are sliced with rotary disc machines that are manually fed; (c) the cork slices are fed to a disc punching machine; (d) the discs are image analysed for classification and (e) final manual classification; the discs are marked on the worst face with a dark strip.

planks; (b) from the lamination into cork slices, the belly layer and the cork back remaining layer; (c) from the punching out of the cork slices, their remaining material; (d) from classification, the refused discs.

The mass balance for production of cork discs for champagne stoppers from prepared boiled cork planks with a calliper below 27 mm, as measured in one production mill (Fernandes, 2005), shows that the total cork discs represent on average 17.9%, the rest being 9.4% of the belly layer, 19.9% of the punched out cork slices, 35.2% of the cork bark layer and 18.6% of plank stripping remains and losses. Further the produced discs showed upon quality classification represent yields of 15.0% champagne discs, 2.6% discs to be rectified for technical stoppers and 2.4% refuse discs.

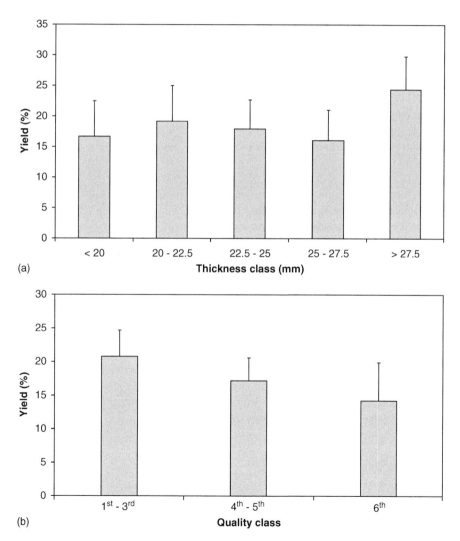

Figure 12.19. Production yield of cork stoppers from cork planks of (a) different thickness classes and (b) three quality grades.

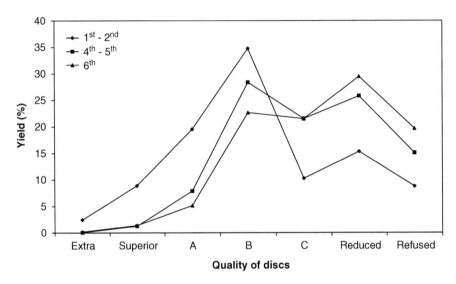

Figure 12.20. Quality profile of the cork discs classified in five quality classes of champagne discs (extra, superior, A, B and C), discs to be reduced into a smaller diameter ("reduced") and refuse obtained from the punching of thin cork planks with three quality grades (1st–3rd, 4th–5th and 6th).

A more detailed study on the mass balances of the production of discs from planks of different thickness and quality classes has the results shown in Figure 12.19. The thickness of the cork plank has an effect on increasing yield when they allow the slicing of one more cork slice: the yield of cork discs was 25% for the planks over 27.5 mm. As regards the influence of the quality on the cork plank, there is a decrease of yields with 20.8% for the 1st–3rd class, 17.2% for the 4th–5th class and 14.2% for the 6th class.

The quality profile of the cork discs is also an important factor. Figure 12.20 shows the quality profiles of cork discs including the champagne discs classified as extra, superior, A, B and C, the defective discs that can be further punched for smaller diameters to be used in technical stoppers (here named "reduced") and the refuse discs obtained from different cork planks. There is a difference of the quality profiles when comparing cork planks of different quality grades, with more good discs and less defective discs obtained from good planks: for instance, the extra-superior discs represent 11.4% of the total discs when using good quality planks (1st–3rd), and only 1.3% for medium quality planks (4th–5th) while the refuse discs are respectively 8.8 and 15.1%.

12.4.3. Assembling stoppers

The cork discs are used as components of the so-called technical stoppers to be glued onto a body of agglomerated cork. The stoppers for champagne and sparkling wines have in general two discs glued together to the bottom of the cylindrical body which will be in contact with the wine. The stoppers for still wines usually have one disc glued to each of

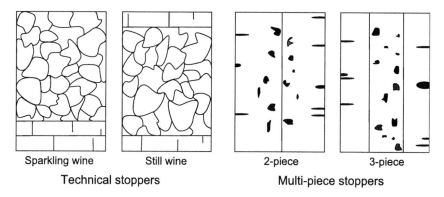

Figure 12.21. Schematically representation of several possibilities of assembling of technical stoppers using cork discs and an agglomerated body and of multi-piece cork stoppers.

the circular bases of the agglomerate body (named $1+1$ stoppers, or Twintop^R). However, in some cases only one disc is glued to the bottom of the stopper.

There is also the possibility, which is in commercial practice, to construct a natural cork stopper by gluing together several cork pieces: for instance two "half-stoppers" or a three-piece stopper. These are called multi-piece cork stoppers.

Figure 12.21 shows schematically several possibilities of assembling technical stoppers and multi-piece stoppers. Several other options are possible and are sometimes client- or product-specific.

12.5. Conclusions

The cork industry is directed to the production of cork stoppers and discs using cork planks of adequate thickness and quality, and to the production of cork agglomerates based on the granulation of by-products and refuse cork planks and other cork raw materials.

In the production of cork stoppers and discs, the structure and properties of cork are kept unaltered apart from the changes that occur in the boiling of cork with hot water. One striking feature of this industrial sector is that the mass yields of natural cork products are low: cork stoppers represent 22–24% of the prepared cork planks and discs represent 18% of the prepared cork planks. When referred to the cork planks before preparation, these yields are smaller and even more if put in relation to the crude total raw material as extracted from the forest. These figures show the importance of maximising yields already from the field extraction onwards along the processing line. It is also of high economic relevance the fact that the quality of the cork planks influences the quality distribution of the products and therefore their value.

The industry has seen a significant technological evolution in recent times, regarding the general production organisation, the boiling operation, the washing and bleaching process, the machine vision for classification and the general increase in automation. Special measures and some technological innovation have been introduced to prevent concentration on TCA and other taint compounds, as well as to avoid microbial growth.

References

Acero, J.L., Benitez, F.J., de Heredia, J.B., Leal, A.I., 2004. Chemical treatment of cork-processing wastewaters for potential reuse. Journal of Chemical Technology and Biotechnology 79, 1065–1072.

Acero, J.L., Benitez, F.J., Leal, A.I., Real, F.J., 2005a. Removal of phenolic compounds in water by ultrafiltration membrane treatments. Journal of Environmental Science and Health- Part A – Toxic/Hazardous Substances & Environmental Engineering 40, 1585–1603.

Acero, J.L., Benitez, F.J., Real, F.J., Leal, A.I., Sordo, A., 2005b. Oxidation of esculetin, a model pollutant present in cork processing wastewaters by chemical methods. Ozone – Science and Engineering 27, 317–326.

Anselmo, A.M., Gil, L., Mendonça, E., 2001. Águas residuais da cozedura da cortiça – caraterização e perspective ambiental. Água e Ambiente 35 Separata, 1–4.

Benitez, F.J., Acero, J.L., Garcia, J., Leal, A.I., 2003. Purification of cork processing wastewaters by ozone, by activated sludge, and by their two sequential applications. Water Research 37, 4081–4090.

Benitez, F.J., Real, F.J., Acero, J.L., Leal, A.I., Garcia, C., 2005a. Gallic acid degradation in aqueous solutions by UV/H_2O_2 treatment, Fenton's reagent and the photo-Fenton system. Journal of Hazardous Materials B126, 31–39.

Benitez, F.J., Acero, J.L., Leal, A.I., Real, F.J., 2005b. Purification of ellagic acid by UF membranes. Chemical Engineering and Technology 8, 1035–1040.

Benitez, F.J., Real, F.J., Acero, J.L., Leal, A.I., Cotilla, S., 2005c. Oxidation of acetovanillone by photochemical processes and hydroxyl radicals. Journal of Environmental Science and Health – Part A – Toxic/Hazardous Substances & Environmental Engineering 40, 2153–2169.

Botelho, M.L., Almeida-Vara, E., Tenreiro, R., Andrade, M.E., 1988. Searching for a strategy to gamma-sterilize Portuguese cork stoppers. Preliminary studies on bioburden, radioresistance and sterility assurance level. Radiation Physics and Chemistry 31, 775–781.

Costa, A., Pereira, H., 2004. Caracterização e análise de rendimeto da operação de traçameno na preparação de pranchas de cortiça. Silva Lusitana 12, 51–66.

Cumbre, F., Lopes, F., Pereira, H., 2001. The effect of water boiling on annual ring width and porosity of cork. Wood and Fiber Science 32, 125–133.

de Heredia, J.B., Dominguez, J.R., Lopez, R., 2004a. Treatment of cork process wastewater by a successive chemical-physical method. Journal of Agricultural and Food Chemistry 52, 4501–4507.

de Heredia, J.B., Dominguez, J.R., López, R., 2004b. Advanced oxidation of cork-processing wastewater using Fenton's reagent: kinetics and stoichiometry. Journal of Chemical Technology and Biotechnology 79, 407–412.

Dominguez, J.R., de Heredia, J.B., González, T., Sanchez-Lavado, F., 2005. Evaluation of ferric chloride s a coagulant for cork processing wastewaters. Influence of the operating conditions on the removal of organic matter and settleability parameters. Industrial Engineering and Chemical Research 44, 6539–6548.

Fernandes, R.M.O.S., 2005. Estudo da influência do calibre e da qualidade das pranchas de cortiça delgada no rendimento do processo fabril de produção de discos de cortiça natural. Graduation thesis. Instituto Superior de Agronomia, Lisboa.

Guedes, A.M.F.M., Madeira, L.M.P., Boaventura, R.A.R., Costa, C.A.V., 2003. Fenton oxidation of cork cooking wastewater – overall kinetic analysis. Water Research 37, 3061–3069.

Mendonça, E., Pereira, P., Martins, A., Anselmo, A.M., 2004. Fungal biodegradation and detoxification of cork boiling wastewaters. Engineering and Life Sciences 4, 144–149.

Minhalma, M., de Pinho, M.N., 2000. Flocculation/flotation/ultrafiltration integrated process for the treatment of cork processing wastewaters. Environmental Science and Technology 35, 4916–4921.

Minhalma, M., de Pinho, M.N., 2001. Tannin-membrane interactions on ultrafiltration of cork processing wastewaters. Separation and Purification Technology 22–23, 479–488.

Minhalma, M., Domínguez, J.R., de Pinho, M.N., 2006. Cork processing wastewaters treatment by an ozonization/ultrafiltration integrated process. Desalination 191, 148–152.

Pereira, H., Melo, B., Pinto, R., 1994. Yield and quality in the production of cork stoppers. Holz als Roh – und Werkstoff 5, 211–214.

Peres, J.A., de Heredia, J.B., Dominguez, J.R., 2004. Integrated Fenton's reagent-coagulation/flocculation process for the treatment of cork processing wastewaters. Journal of Hazardous Materials B107, 115–121.

Rosa, M.E., Pereira, H., Fortes, M.A., 1990. Effects of hot water treatment on the structure and properties of cork. Wood and Fiber Science 22, 149–164.

Silva, C.A., Madeira, L.M., Boaventura, R.A., Costa, C.A., 2004. Photo-oxidation of cork manufacturing wastewater. Chemosphere 55, 19–26.

Chapter 13

Cork agglomerates and composites

The industrial sector dealing with the production of cork agglomerates is an important component within the total cork chain since it provides an outlet for a large amount of residual by-products that are produced during the manufacture of stoppers and discs (see Chapter 12) as well as for the substantial amount of refuse raw cork planks that are produced in the forest and other types of cork raw materials such as virgin cork. The cork materials that are used have different composition in terms of purity and characteristics of the cork tissue according to their origin and, therefore, are directed to the different productions lines according to the technical demands of the products.

The starting point in the flow sheet for the production of agglomerates is the trituration of granules to adequate size and density. These are directed to three processing lines: the production of agglomerates with a resin binding the granules, the production under high temperature steaming and without external binding agent of expanded agglomerates and the copolymerisation with rubber to make the rubber–cork composite. These processing lines are not integrated and they are pursued in individual plants.

Most of the uses of cork agglomerates are related to cork's thermal insulation and energy absorbing properties coupled with considerable physical, chemical and biological stability. They are, therefore, extensively used in domestic and industrial environments for insulation and surfacing purposes. The changes that occur in the structure and properties of the cork agglomerates, namely in the case of expanded cork agglomerates where there is an important thermochemical effect on cork, will also be presented in this chapter.

13.1. Cork raw materials and types of agglomerates

The cork raw materials that are used in the production of agglomerates and composite materials include:

(a) by-products from the processing line of production of stoppers and of discs, as detailed in Chapter 12; these corks have been boiled;

(b) raw corks that are obtained directly from the forest or are separated in the mill yard and do not enter the processing line for the production of stoppers and discs; therefore they are raw corks that have not been boiled.

From the products that will be produced the most demanding in terms of purity and iniquity are the agglomerates that will constitute agglomerated wine stoppers and will be part of the assembly of technical stoppers. For these reasons, only boiled cork materials are used. The details and uses of the different raw materials for production of agglomerates are detailed below, by types of material.

13.1.1. Types of cork agglomerates

The products that are made after milling of the various raw materials may be classified into three major groups:

- cork agglomerates made up with cork granules bound by a resin that are directed to production of stoppers (agglomerated stoppers and technical stoppers), as well as of flooring and surfacing material among other various uses;
- cork–rubber composites made by the binding of rubber and cork granules (rubber–cork or cork–rubber) and
- pure cork expanded agglomerates where the cork granules are bound together without extraneous adhesives.

Other composites or mixtures are possible, namely in light concrete, as it has been referred to in Chapter 11.

13.1.2. Processing by-products

The by-products obtained from the prepared cork planks (boiled reproduction cork), during processing into stoppers and discs are the following:

(a) from the preparation of cork planks: trimmings and small pieces of cork from plank cutting as well as refuse cork planks;
(b) from the production of stoppers: punched strips and refuse stoppers and
(c) from the production of discs: punched cork slices, belly layer, cork back layer and refuse discs.

These materials that remain as left-over or waste materials from these productions correspond to large amounts representing more than 80% of the prepared cork planks as seen in the previous chapter. All these by-products are milled into granules and they are good quality raw materials since the proportion of cork tissue is high. Therefore, they will be used preferentially for the production of granulates directed for agglomerated stoppers. One exception is the cork back layer from the lamination of the thin planks since it contains only a minor proportion of cork tissue.

From the rectification of dimensions of stoppers and discs, a fine cork powder is also obtained. In part this material is used in a mixture with polyurethane or other elastomers to fill up the lenticular channels at the surface of low-quality stoppers.

13.1.3. Raw cork materials

There are substantial amounts of cork materials of various types that are not boiled and go directly for production of granulates. They can be summarised in the following categories:

- reproduction cork: planks of insufficient quality (refuse planks) or dimension (pieces or extra thin planks); stem footers; cork with yellow stain or otherwise mould stained;
- second cork and
- virgin cork: from the first extraction of young trees, from the increase of stripping height and from the cork removal of pruned branches.

Occasional cork materials including reproduction and virgin cork are also available from accidental tree death and subsequent felling, for instance due to sanitary reasons or fire occurrences.

All the cork materials that are taken out from death trees or cut branches have a low quality as regards the yield in cork granules because they include a large quantity of inner bark and wood attached to the cork layer. They are directed to the less demanding sector, the production of expanded insulation agglomerates. All the other raw materials are used to produce cork agglomerates as boards and sheets, mostly for flooring and surfacing as well as for production of rubber composites.

13.2. The trituration of cork

Size reduction and particle classification are the important unit operations in the milling of cork, which constitutes the first step in the production of cork agglomerates. Usually this is a line that is integrated with the subsequent processing and therefore the product specifications dictate the operational parameters regarding particle distribution and density of the different granulate assortments.

The type and the quality of the raw material also dictate the processing needs in terms of cleaning operations and number or type of milling cycles. Usually the first milling is carried out in hammer or star type mills while the subsequent grindings are done in knife or disc grinders. The fine fractions are separated and the oversized particles are recycled. The separation of granulometric fractions is done by classification, by particle size, using sieving equipment and by density, using densimetric tables (Fig. 13.1). The final granulometric distribution depends on the specific product requirements, but usual particle dimensions of granulates are in the range 1–10, 2–5, 1–2, 0.5–1 and 0.2–0.5 mm.

Figure 13.2 shows an example of an industrial flow sheet for the preparation of cork granules. The granules are made up mostly of cork tissue without significant amount of lenticular material. The separation by density classes allows sorting out heavier particles, that is to say, particles which contain ligneous inclusions. Drying of raw materials or of intermediate granulates using hot air may be incorporated along the process since the size reduction and milling equipment do not allow excessive moisture contents. Fines are removed along the process and regrinding of large particles through recycling is carried out as needed. The final product corresponds to granulates calibrated by particle size and density as specified by the user.

Figure 13.1. Partial view of an industrial classification of cork granulates showing a row of densimetric separation tables.

The yields obtained in the production of granulates depend on the raw material that is used. The highest yields are obtained using punched residues without cork back that correspond to 65–85%, while the punched strips with cork back yield about 50% granulates (Graça et al., 1985). The cork fines that are obtained as a waste material are estimated in approximately 22% (Saraiva and Soares, 1980).

13.3. The agglomeration of cork with adhesives

The agglomeration of the cork granules may be made using an adhesive under a moderate pressure and heating as adequate for the specific polymer curing and bonding. The adhesives that are used include thermosetting polymers, such as urea–formaldehyde, melamine or phenolic adhesives, or thermoelastic polymers such as polyurethanes. The former are used for flooring agglomerates, while polyurethanes are used for stoppers and softer surfacing materials. There are some concerns with the release of volatiles from phenolic resin bound composites in indoor environments, including cork composites for which emissions of phenol and furfural have been measured (Horn et al., 1998). The use of polyurethanes has, therefore, increased greatly in the last years.

The agglomeration of cork particles can be made basically in three ways:

(a) by extrusion;
(b) by moulding in large blocks or in cylinders of variable length and
(c) by continuous forming of corkboards and hot pressing.

The production of agglomerated stoppers is made by extrusion or by individual moulding in long cylinders or also in individual mouldings for the case of production of the

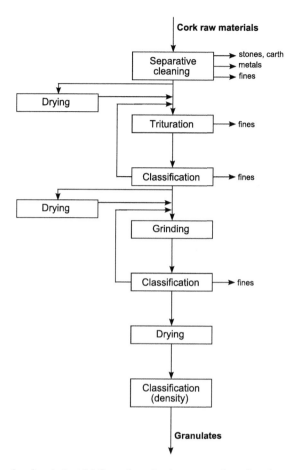

Figure 13.2. Example of an industrial flow sheet for the preparation of cork granulates.

bodies to be used for sparkling wines. The individual moulding is being preferred because it allows a better control of particle and density distribution in the stopper, avoiding stratification signs and giving homogeneity and isotropic behaviour to the stopper. The cork particles are mixed with the adhesive and a lubricant, pressed into the mould, and this is heated to a temperature and during a period adequate for the polymerisation of the resin. A rest period for complete adhesive setting, cooling and dimensional stabilisation is required. The finishing operations include rectification to dimensions, cutting, gluing to cork discs in the case of technical stoppers. Figure 13.3 exemplifies the extrusion of rods and the gluing of the agglomerate body to the cork discs.

The flooring and surfacing agglomerates are produced from rectangular prismatic blocks that are laminated into boards. Cylindrical blocks are produced for rotary lamination to produce a continuous cork sheet. The process used for the production of wood particleboards was also adapted for cork agglomerates: the cork granules are mixed with the adhesive and plasticisers and fed to a continuous mat that is pressed and heated in a hot plate press.

(a)

(b)

Figure 13.3. Production of cork agglomerated stoppers: (a) extrusion of rods and (b) gluing the agglomerate body to discs.

There is a large diversity in the operating conditions and on the formulations of adhesives and of cork particle assortments used depending on product specificities and plant practices. The adhesive content is usually in the range of 3–8% and the polymerisation temperature and time depend on the adhesive chemical composition (from about 100 to 150°C, during less than 1 h to over 20 h). The selected distribution of the granulometric fractions used as cork raw material and the compression used in the moulds originate different densities for the agglomerates. Agglomerated stoppers have densities in the range

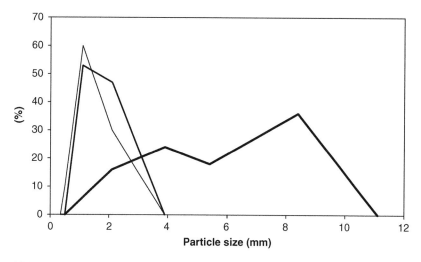

Figure 13.4. Examples of distribution of particle size in the granulated raw materials for three types of agglomerates.

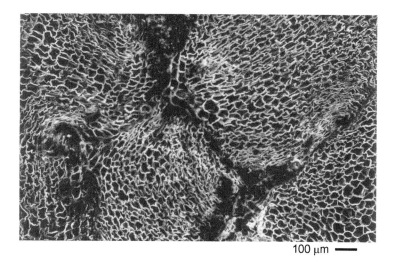

100 μm ▬

Figure 13.5. Cellular structure of cork agglomerates showing the junction of different particles.

of 200–300 kg m^{-3}, as well as agglomerates for surfacing, partitioning and insulation purposes. The flooring cork agglomerates have higher densities that can attain over 500 kg m^{-3}. Figure 13.4 shows an example of granulometric distributions of three types of cork agglomerates, calculated based on data from Lissia et al. (1972).

At the cellular structural level, the compression of the cork particles during agglomeration causes the partial densification of cells at the grain boundary, with cell collapse and corrugation (Fig. 13.5). The distribution of cell collapse is not uniform and increases with the compression extent, which also translates into higher density values of the materials.

13.4. Cork composites with rubber

The cork–rubber or rubber–cork composites are made by mixing and binding cork granules with natural or synthetic rubber. They combine the properties of the two materials, taking advantage of some and compensating for others that are more disadvantageous. Cork compressibility and recovery characteristics, and the very small lateral flow under compression, compensate for the relatively large positive Poisson coefficients of rubber and its small recovery after deformation. Also, the chemical and thermal stability of cork overcome the fact that rubber is easily oxidised, thermally degraded and suffers aging alterations. On the contrary, the comparatively smaller resistance of cork and its dimensional variations are compensated by the rubber component.

The production of cork–rubber composites resembles rubber processing since the binding of cork to rubber is made by chemical cross-linking reactions. The cork and rubber particles have in general fine granulometries and are thoroughly mixed with cross-linking agents and catalysts, as well as other additives such as antioxidant or colouring materials. The mixture is homogenised and repeatedly roll pressed to obtain a homogeneous paste that is injected or compression moulded before the temperature induced cross-linking polymerisation.

The cork–rubber composites are used for technologically demanding applications such as gaskets and sealing systems in the automobile industry, vibration and acoustic insulation for industrial machinery, civil engineering and railways, gaskets for electrical transformers, heaters and gas meters or heavy-duty flooring and footwear. This type of material was developed in the United States in the beginning of the 1960s as a sealing gasket for oils. The product properties depend on the proportion of the cork and the rubber components and therefore tailor-made formulations are possible for specific applications. The advantages that the rubber–cork products can boast of are a wide range of fluid compatibility with chemical resistance over long periods, surface sealing, compression with negligible lateral flow, resistance to compression over long periods, vibration and shock absorption, high friction and resistance to abrasion.

Figure 13.6 shows the structure of rubber–cork composites with two different cork proportions and particle granulometries. The cork particles are imbedded in the rubber matrix, even in the low rubber proportion composites. The cork granules are oriented randomly in the composite and the different aspects of their cellular structure may be observed on a section (Fig. 13.6b). The cells are buckled and distorted to varying degrees, but the effect of compression during the production process is higher in the finer particles, while the interior of larger granules remains substantially unchanged.

13.5. Expanded insulation cork agglomerates

The expanded cork agglomerates are cork-only materials since the cork granules are thermally self-bonded without addition of external adhesives. The adhesion is obtained due to the formation of chemical compounds resulting from the chemical degradation of extractives and of the structural components of cork. These agglomerates are also called black agglomerates due to their temperature-induced dark brown colour.

The expanded agglomerates are less demanding in terms of raw materials than the other agglomerates. Therefore, the raw materials used have a comparatively lower content

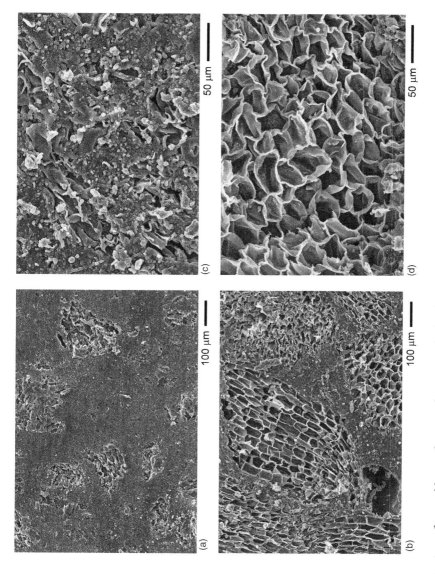

Figure 13.6. Structure of two rubber–cork composites as seen in the electron scanning microscope: (a) a composite with a low proportion of cork and fine cork particles; (b) a composite with a high cork proportion and larger cork particles; (c) detail of the distribution of the rubber component on the cork cells at the surface of rubber–cork sheet and (d) cellular structure of the cork granules.

of cork and a higher proportion of cork back, wood and inner bark remains, such as it is the case with virgin cork from prunings, the cork back layer from the production of discs, or reproduction corks obtained from dead trees (Pereira and Baptista, 1993). The granules that are used have dimensions above those of the other agglomerates, in the range of 4 mm to over 22 mm.

The expanded agglomerates are produced in blocks by heating the cork granules in autoclaves with a superheated steam at temperatures around 300–350°C and 40 kPa for approximately 20 min. The steam is injected through openings in the bottom face and bottom edge of the lateral faces of the autoclave. The autoclaves are rectangular and have the possibility of compressing the granules by hydraulically moving the bottom upwards to reduce the height of the final agglomerated block. The extent of height reduction represents the compression applied before steaming and it has the effect of compacting the granules inside the autoclave. The extent of this pre-compression allows the control of the final product density. The void content in an uncompacted mixture of cork granules is high, in the range of 45–50%. Some compaction is needed in order to guarantee a sufficient bonding of the granules and strength of the agglomerate, but its extent may vary depending on the density of the material that is to be produced. After the autoclaving, the agglomerated blocks are cooled in the first step by the injection of water and subsequently by keeping them in closed containers (to avoid fire ignitions) until temperature stabilisation. Figure 13.7 shows one autoclave and the agglomerated blocks.

Cork is chemically and physically altered by heat treatments, as discussed in Chapter 10. The thermochemical behaviour of cork has shown that degradation increases above 200–250°C (Pereira, 1992). Therefore, at the temperature of steaming in the autoclave, there is a substantial chemical alteration of cork and in the conditions of the production of expanded agglomerates, cork looses approximately 25–30% of its mass due to volatilisation of extractives and structural components. This mass loss is directly in relation with the temperature of steaming, as shown in Figure 13.8 from experimental determinations of agglomeration (Pereira and Baptista, 1993). The chemical analysis of expanded agglomerates shows that suberin and lignin are more thermally stable and most degradation occurs for the extractives and polysaccharides. In fact the composition of the agglomerate is 10.4% of extractives, 44.5% of suberin, 34.4% of lignin and 9.6% of polysaccharides, while virgin cork from prunings and refuse reproduction cork, two of its main raw materials, have respectively 14.8 and 13.4% of extractives, 28.9 and 31.0% of suberin, 31.0 and 31.5% of lignin and 19.3 and 20.9% of polysaccharides (Ferreira and Pereira, 1986).

Upon heating there is also a major effect on the cellular structure of cork: the cells expand, the cell walls stretch and decrease in thickness. The cells loose partially their prismatic form and acquire rounded forms, more of a balloon type; the arrangement in radial rows becomes less clear and the material is more isotropic. The cell expansion measured in virgin cork from prunings and in refuse reproduction cork shows average increases of 34% in the diameter of the tangential section and 34% of cell height, while the increase is smaller for previously boiled cork such as obtained from the cork back layer (Ferreira and Pereira, 1986). The cell volume increase is over 100% and this is the reason why this type of cork agglomerates is called expanded agglomerates. However, this effect is not spread homogeneously through the cork granules and many granules, or parts of granules,

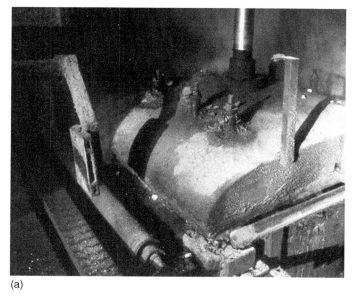

(a)

(b)

Figure 13.7. Autoclave used in the production of (a) expanded cork agglomerates and (b) hot agglomerated blocks after injection of cooling water.

are less altered and the typical cell arrangement is still observed with a clear distinction between tangential sections and transverse/radial sections.

In the junction of granules, the cells are compressed against each other, and a band of collapsed cells is found at the boundary of adjoining granules with a variable width depending on the applied compression, but usually not exceeding 10–20 cell layers per granule (Pereira and Ferreira, 1989). This provides a good example how cork is able to absorb compression energy and collapse without long-range deformation. The bonding between the cork granules results from their pressing against each other and in well-bonded granules it is not possible to distinguish which cells belong to each granule.

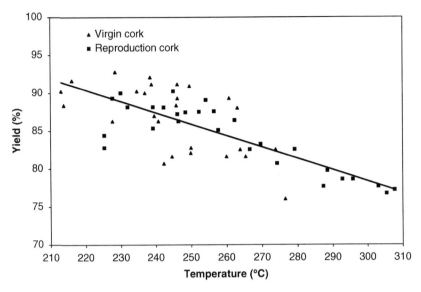

Figure 13.8. Mass yields of cork agglomerates produced by a 30-min steaming at different temperatures using virgin cork and reproduction cork (adapted from Pereira and Baptista, 1993).

Figure 13.9 shows photographs of expanded cork agglomerates where these different structural aspects may be seen: the cell expansion in the interior of the granules due to the heating effect, as observed in different sections, and the cell collapse and deformation at the external part of the granules due to compression under high temperature. Due to the irregular shape of the cork granules, some intergranular spaces remain in the agglomerate in an extent that varies with the pre-compression applied and the resulting density of the agglomerate: in a section of a 130 kg m^{-3} board, the voids corresponded to 16% of the total. This is why the expanded agglomerates have a rather low bending strength and are not resistant to friction and wear.

The adhesion corresponds to bending strengths of approximately 0.2–0.4 MPa and is determined mostly by temperature. Below 250°C, the agglomeration does not occur or only insufficiently, and it is necessary to have an adequate chemical degradation of the cell wall corresponding to mass losses over 15% to allow the full cellular expansion. In practice, this can be followed by the colour of the agglomerate: a light brown shade indicates an insufficiently "cooked" agglomerate, a black colour an "over-cooked" agglomerate with carbonised granules.

The tannins and phenolic compounds of cork are involved in the chemical adhesion between the granules, with a possible reaction with the aldehydes formed by the degradation of hemicelluloses. Waxes should have a limited role in the process since they are degraded already at temperatures below 150°C (Rosa and Pereira, 1992). Expanded agglomerates can be produced from wax-free cork granules (obtained by solvent extraction) but not from tannin-free cork granules after water extraction (Pereira and Baptista, 1993).

The volatiles that are produced from the thermochemical degradation of cork are carried out from the autoclave in the steam flow and are cooled and condensed afterwards. Some of the compounds solidify in the pipes and equipment as sticky dark solids.

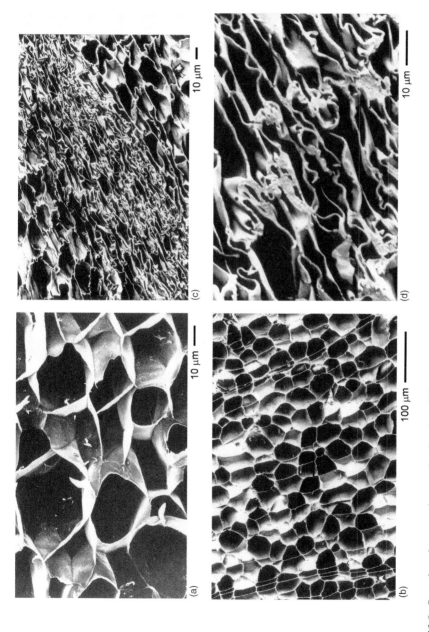

Figure 13.9. Scanning electron micrographs showing different aspects of the cellular structure of expanded cork agglomerates: (a) expansion of cells with a ballooning effect; (b) one complete growth ring of virgin cork showing the partial loss of the radial alignment of cells; (c) junction between two cork granulates and (d) undistinguishable boundary between two granules.

Isolation of chemicals such as triterpene friedelanes has already been carried out and further biovalorised (Kane and Stevenson, 1960; Moiteiro et al., 2006).

The expanded cork agglomerates are used mostly for thermal insulation, acoustic insulation and vibration absorption. The density of the agglomerates depends on their use: 80–100, 100–150 and 175–320 kg m^{-3} for acoustic, thermal and vibration-type agglomerates, respectively. They are produced as boards, tiles, bricks or tube coverings with different dimension and thickness. The expanded granules (non-bonded) are also used for thermal and sound insulation as a filling in partitions, walls and floors, for instance. An example of an unconventional use of expanded cork agglomerates as cladding material for external use has been already referred to in Chapter 11 and shown in Figure 11.6.

The thermal properties of the expanded cork agglomerates are well known and have been referred to in Chapter 10. They are at the basis of the large-scale utilisation of these cork agglomerates for thermal insulation. One important feature of the expanded cork-boards is their durability under use conditions due to their chemical and biological inertia, as well as low water absorption and comparatively high mechanical resistance under a broad range of temperatures. Many insulation corkboards installed in the beginning of last century, i.e. in cold storage, are still in performing conditions.

13.6. Conclusions

The cork production and processing chains generate a very substantial amount of materials that are the raw materials for the production of cork agglomerates. Totalling the refuse reproduction cork planks and the virgin cork with the by products of the production of stoppers and discs, it is estimated that over 85% of all cork production at the end will be granulated and agglomerated.

The binding of the cork granules may occur by the use of: (a) adhesives (i.e. polyurethanes and phenolic resins), leading to the so-called cork agglomerates; (b) by copolymerisation to rubber, with production of the rubber–cork composites and (c) by thermal induced adhesion compounds in cork, leading to the expanded cork agglomerates.

There is a large diversity in product characteristics, regarding physical and mechanical properties as well as aesthetical appearance that is achieved at the production level by the selection of the raw materials, the size and density of the cork granulate used, the type and content of adhesives and additives, and the processing conditions. For instance, the production of cork agglomerates for stoppers, including the agglomerated body of technical stoppers, is one of the demanding processing lines that requires quality in the raw materials and process control as regards prevention of contamination leading to off-flavours and taints in the wine. In addition, the use of agglomerates and rubber–cork for heavy duty and technologically demanding applications has increased.

References

Ferreira, E.P., Pereira, H., 1986. Algumas alterações anatómicas e químicas da cortiça no fabrico de aglomerados negros. Cortiça 576, 274–279.

Graça, J., Barros, L., Pereira, H., 1985. Importância da produção de cortiça de qualidade para a indústria transformadora. Cortiça 566, 697–707.

Horn, W., Ulrich, D., Seifert, B., 1998. VOC emissions from cork products for indoor use. Indoor Air 8, 39–46.

Kane, V.V., Stevenson, R., 1960. Friedelin and related compounds. 3. The isolation of friedelane-2,3-dione from cork smoker wash solids. Journal of Organic Chemistry 25, 1394–1396.

Lissia, F., Pes, A., Puliga, B., 1972. Aglomerados brancos de granulados de cortiça. Cortiça 460, 328–333.

Moiteiro, C., Curto, M.J.M., Mohamed, N., Bailen, M., Martinez-Diaz, R., Gonzalez-Coloma, A., 2006. Biovalorization of friedelane triterpenes derived from cork processing industry byproducts. Journal of Agricultural and Food Chemistry 54, 3566–3571.

Pereira, H., 1992. Thermochemical degradation of cork. Wood Science and Technology 26, 259–267.

Pereira, H., Baptista, C., 1993. Influence of raw material quality and process parameters in the production of insulation cork agglomerates. Holz als Roh- und Werkstoff 51, 301–308.

Pereira, H., Ferreira, E., 1989. Scanning electron microscopy observations of insulation cork agglomerates. Materials Science and Engineering A111, 217–225.

Rosa, M.E., Pereira, H., 1992. The effect of long term treatment at 100ºC–150ºC on structure, chemical composition and compressive behaviour of cork. Holzforschung 48, 226–232.

Saraiva, L.M.C., Soares, J.M.A., 1980. Desperdícios no sector corticeiro. Cortiça 501, 199–202.

Chapter 14

Wine and cork

Cork has been used for sealing wine reservoirs since ancient times from Phoenician and Roman amphora to the present glass bottles. Still wines use cylindrical stoppers that are fully inserted in the bottle, Its use as a closure has been so dominant that it coined the English word for stopping bottles, while sparkling wines have stoppers that are introduced in the bottle only to their half-length. Port wines, whiskies and other liquors have cork stoppers topped by a plastic protuberant body.

The properties required from a wine closure are its sealing capacity, which means that no leakage occurs either at the glass interface or through the stopper, inertness towards the liquid content, namely in relation to flavour, durability along storage time and possibility of removal with adequate effort. Cork fulfils these conditions and has an air permeability that seems to allow an adequate slow oxygen transfer to the aging wine. The visual quality of cork stoppers is an important factor to establish their commercial value and the extent of the lenticular porosity in tops and in the lateral surface of the stoppers determine their grading. The subjectivity of this appreciation may be overcome by using quantified image features.

Certification of the quality for cork stoppers is made in relation to standard properties such as dimensions, mechanical strength, density or residual contaminations, as well as regarding the good practices followed in industrial production. This certification has increased mainly triggered by the necessity met by the cork industry to demonstrate that the taints and off-flavours that occasionally spoil the wine are only marginally of the cork's responsibility. Of special importance is a mouldy taste given by the volatile trichloroanisole (TCA) that is sensorial perceived at extremely low levels even by the common drinker. TCA may be carried by corks through the microbial methylation of trichlorophenol (TCP) or by its absorption from contaminated cellar environments or from the wine. Process changes and innovative prevention or curative methods were introduced by the cork industry in the recent years to circumvent contamination sources during processing.

The increase in wine production and consumption in the new production areas such as North and South America and Australia, which are less committed to tradition, has permitted the consideration of other closures, namely synthetic stoppers mimicking the cork stoppers, or other more radically different closures such as the aluminium screw caps or the bottling in plastic bags or in tetrapak packages. Although the discussion on the best closure type is vivid among wine bottlers and oenologists, the equilibrium reached along time between wine making characteristics and its aging using a cork stopper seems the safest way to have a quality wine.

14.1. History and types of wine stoppers

The use of cork for sealing wine containing reservoirs is recorded from antiquity, and ceramic amphorae closed with cork plugs were among the archaeological findings from Egyptian, Greek and Roman times. The relatively large openings of these ceramic containers required the cork plug to be cut in the cork plank radial direction, therefore being used much in the way it functions on the tree, with the cork back to the outside. The sealing performance was enhanced by using leather or hemp coverings, and pitch or resin coatings were added as additional barriers to leakage.

Cork stoppers were also used to close several types of vials, and in the case of small openings the corks could be manufactured with their axis in the axial direction in the plank. In all cases the cork closures had a conical form to help the introduction into the reservoir's neck or opening.

There are little references on cork wine closures in medieval times, although their use should have continued much in the same manner. There are records of exports of cork from the producing countries to Northern Europe: for instance, the trade between Portugal and Flanders is documented since the early 15th century. It is also curious to see the depiction made by van Leeuwenhook of a wine cork stopper that he used for his microscopic observations and reported in a letter to the Royal Society of London, in 1705 (Fig. 2.2): the cork stopper is cylindrical and oriented as today's stoppers.

The credit for the wide spread use of wine bottling in glass containers closed with a cork stopper is usually attributed to the abbot Pierre Pérignon (1639–1715), from the French Champagne region. He adapted a cork stopper for his sparkling wines, strapped to the bottleneck to withstand gas pressure, with a considerable success that led to the adoption of this method by the neighbouring wine producing houses and later on to the wide spreading of wine glass bottles with cork closures. The champagne cork stoppers were produced initially with solid cork; the present construction of a body of agglomerated cork with discs of natural cork glued to the bottom part started only in the beginning of the 20th century.

Figure 14.1a represents schematically the stoppers used at present for sparkling wines: the cylindrical stopper assembled with agglomerated cork and discs is introduced in the bottle to about half of its length and is secured in place by wire braces. After unbottling, there is a partial recovery of dimensions of the part of the stopper that was inserted in the bottle's neck; the recovery is higher in the natural cork discs than in the agglomerated cork, leading to the typical mushroom shape of the unbottled champagne stoppers.

Another type of stoppers is used for port wine and liquors such as whiskies. In this case it is important for the user to have the possibility of putting again the stopper after pouring a part of the content. Therefore, the stopper is shorter, only moderately compressed

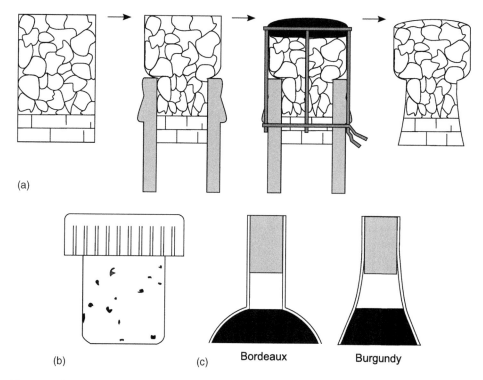

Figure 14.1. Schematic representation of: (a) a stopper for champagne and sparkling wines before insertion, inserted in the bottle and after removal from the bottle; (b) a stopper for port wine; (c) the Bordeaux and Burgundy (Bourgogne) type of wine bottles.

against the bottleneck, with a rounded or chamfered bottom edge and it is glued to a plastic or wooden top that allows an easy removal and re-insertion (Fig. 14.1b).

The wine bottles now use cylindrical stoppers with sharp circular edges. The following types are in use: solid natural cork stoppers (one-piece); assembled solid natural cork stoppers (multi-piece); technical stoppers, with an agglomerated body and discs of natural cork glued to the tops; and agglomerated only stopper (see Fig. 12.21). Although cork stoppers may be produced at client's demand with different dimensions and forms, there is a considerable standardisation of glass bottles and therefore of stoppers' dimensions. Figure 14.1c shows schematically the two most important forms of wine bottles, the Bordeaux and the Bourgogne (Burgundy) type. Table 14.1 shows current dimensions of cork stoppers, with the most usual being the 24 mm × 36 mm and the 24 mm × 45 mm (diameter × length).

14.2. Visual quality of cork stoppers

14.2.1. The surface of natural cork stoppers

The cork stoppers are punched out from strips cut transversely in the cork planks, as detailed in Chapter 12. Therefore the direction of punching, corresponding to the stopper

Table 14.1. Dimensions of diameter and length of cork stoppers used currently for the bottling of wines.

Length (mm)	Diameter (mm)									
	20	21	22	23	24	25	26	30	31	32
27	✓	✓	✓			✓				
33	✓	✓	✓			✓	✓			
38			✓	✓	✓	✓	✓			
45			✓	✓	✓	✓	✓	✓		
49						✓	✓	✓	✓	
54						✓	✓	✓	✓	✓

cylinder axis, is axial. On observing the surface of a cork stopper, the following features can be singled out:

- the circular tops correspond to transverse sections of cork where the lenticular channels cross the surface as thin dashes set perpendicular to the growth rings; usually a stopper contains 3–5 growth rings; and
- the lateral surface of the cylindrical body of the stopper has different aspect; it contains two tangential sections and two radial sections, respectively parallel and perpendicular to the growth rings, and all the in-between sections; therefore the aspect of the lenticular channels vary from an approximate circular shape in the tangential section that become elongated horizontally until a strip form in the radial sections; the macroscopic aspect of the lenticular cannels in the different sections of cork was detailed in Chapter 7.

Figure 14.2 shows the photograph of a cork stopper including the tops and the lateral surface, and arrows mark the orientation of the sections, as discussed. The cylinder axis of the cork stopper is an approximate symmetry axis. However a closer observation shows that the lenticular porosity displayed by the half cylinder closer to the belly of the cork plank is smaller than that of the opposed half cylinder. This is the result of the radial variation of the porosity in the cork planks, smaller near the belly and increasing to the outside, as referred to in Chapter 7. This is the reason why the stoppers are punched out as close to the belly as possible, to get benefit of the lower porosity in this part of the cork plank. In this way not only the enlarged dimensions of the lenticular channels but also the eventual cracks that may develop near the cork back due to the tangential growth stresses of the cork in the tree's stem are avoided. Figure 14.3 clearly shows this difference. It represents a cork strip after the boring out of the stoppers cut through the middle; the porosity of the belly side is lower that that of the cork backside.

14.2.2. The quality classification of cork stoppers

The natural cork stoppers are graded into quality classes in function of the homogeneity of their external surface. The discontinuities and defects that may appear in the cork planks were presented in detail in Chapter 7, and these may also be found on the corks.

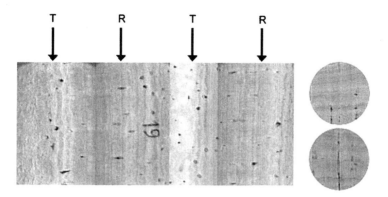

Figure 14.2. Photographs of the surface of a cork stopper, including the two circular tops and the total lateral surface (as a composition of successive pictures taken around the cylinder). Arrows mark out the position of the different cork sections (T – tangential, R – radial).

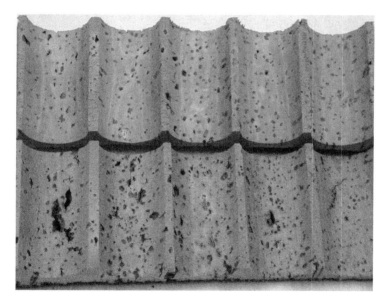

Figure 14.3. A cork strip of medium quality after the punching out of the stoppers, sectioned in the middle, showing the belly side (upper photograph) and the cork back side (lower photograph).

Some of the defects may be enough to fully disqualify a stopper from its wine sealant functions: this is the case of the presence of insect galleries, of a microbial stain (i.e. yellow stain), or of a previously undetected wet cork spot. Other disqualifying defects may result from the cutting process, such as the inclusion of a portion of the cork back or of the belly. But mostly the quality classification of the stoppers is based on the extent of the lenticular porosity shown by their surface, as seen by human or machine vision (see Fig. 12.12).

The traditional quality grading is made into a total of eight classes: extra, superior, 1st, 2nd, 3rd, 4th, 5th and 6th. The best extra and superior classes should include stoppers with only few and very small lenticular channels, while lower quality stoppers of 5th and 6th classes already contain a noticeable porosity with appreciable dimensions. At present the lowest 6th grade is not retained and those stoppers will either be considered refuse material for trituration or be colmated. The colmation of cork stoppers is an operation made for giving a better appearance to high porosity stoppers and involves the stuffing of the pores with a composite of polyurethane and fine cork dust. This is done for the worse quality grades of 5th and 6th, and in some cases also for the 4th grade.

The different quality grades of cork stoppers have quite different market values. Price, although vary as a result of business opportunities and commercial agreements, should range somewhere in the relative rating, from extra to the 6th class, of 100-78-59-45-32-23-17-3, with 100 set as an index for the extra stoppers.

The quantitative description of the porosity features representative of the different quality classes has been made only very recently by analysing tops and the lateral surface of stoppers from extra to the 5th class (Costa and Pereira, 2005, 2006). In fact the classification is made commercially using visually set references, and the image processing of the automatic vision machines is tuned accordingly.

The measurement of the dimensions and distribution characteristics of the pores in cork stoppers of different quality classes has shown differences (both in tops and in the lateral surface) such as the coefficient of porosity (area of pores in % of total area), maximal pore area and number of pores per unit area. This is shown in Figure 14.4: the mean values of such variables increase steadily from extra to the lower quality classes, both in tops and the lateral surface of the stopper.

For instance, the following quality classes will have the average values in their lateral surface:

- extra stoppers, a porosity coefficient of 1.4%, resulting from pores less than 3.1 mm^2, and 3.1 mm in width or length;
- 2nd class stoppers, a porosity coefficient of 4.2%, with pores less than 8.7 mm^2, 5.3 mm in width and 4.6 mm in length; and
- 5th class stoppers, a porosity coefficient of 6.2%, with pores up to 26.5 mm^2, 7.3 mm in width and 11.9 mm in length.

As regards tops, the extra, 2nd and 4th classes have mean porosities of 1.1, 4.0 and 8.2%, respectively.

Other variables are constant throughout the grades: this is the case of all pore characteristics related to shape (i.e. aspect ratio, esfericity) or to the distribution (i.e. nest neighbour distance).

14.2.3. Decision rules for quality grading

It is clear from Figure 14.4 that despite the regular trend shown by the mean values of pores characteristics from best to worst classes, there is a large dispersion of individual values. It is empirically known that the quality classification of stoppers is subjective and therefore prone to vary between individual experts and even for the same person in

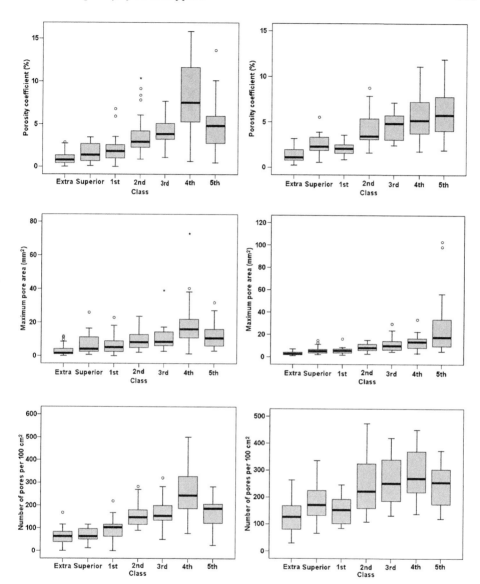

Figure 14.4. Box plots of pores characteristics (porosity coefficient, maximum pore area and number of pores per 100 cm²) measured in tops (left) and in the lateral surface (right) of stoppers of different quality grades (extra, superior, 1st, …, 5th). Mean of 24 stoppers per quality grade, standard deviation and outliers (Costa and Pereira, 2005).

different occasions. Sometimes the grading may be adjusted to the exigency threshold of the client, and therefore made either by the lower or the higher limits. This subjectivity has been quantified using cork stoppers (Melo and Pinto, 1989), cork square prisms with the stopper dimensions (Barros and Pereira, 1987) and with cork planks

(Macedo et al., 1998) and a large percent of mismatch in different classifications was obtained. The mismatch is higher in the mid-quality range since it is usually consensual what is a very good or a bad stopper but it is harder to unambiguously identify a 2nd from a 3rd class stopper or a 3rd from a 4th class stopper.

The analysis of the porosity characteristics of the surface of cork stoppers of different quality grades using principal component and discriminant analysis shows that the main variables explaining the grading are the coefficient of porosity and maximum pore dimensions such as length, width and encircling rectangle in the lateral surface, while in the tops the number of pores is also relevant (Costa and Pereira, 2006).

These results may be used as background information to establish quantified threshold limits for quality classes. It is clear that the coefficient of porosity is probably the most important single variable since it holistically expresses the extent of porosity in the surface and it is correlated to most mean and maximal dimensional features of pores (Costa and Pereira, 2005). However the existence of large pores is a factor of quality devaluation due to their high visual impact, and it should be taken into account in the grading. The question also arises on the relative importance of tops and of the lateral surface of the stoppers. The area of the lateral surface is larger by a factor of eight than the area of one top. On bottling, the stopper is compressed and the top surface substantially reduced. Table 14.2 exemplifies for a stopper of 24 mm diameter and 38 mm length the surface values and their reduction upon insertion into a 18.5 mm diameter bottle neck: the area of the top in a bottle is reduced to 24% of the initial area and the lateral surface by 23%. Therefore the impact of pores in the tops is less, as shown in Figure 14.5, where a stopper is schematically represented before and after insertion in the bottle.

A proposal of decision rules for defining the quality grades of cork stoppers may be made using the coefficient of porosity and the maximal pore length as decision variables, pore area for the lateral surface and the porosity coefficient and the number of pores for the tops, as shown in Table 14.3 (Costa and Pereira, 2006). The threshold limits for the porosity coefficient were, respectively from extra to 5th grade, 2.1, 2.4, 3.0, 3.8, 4.8, 6.4 and 13.0%. The classification of the stopper uses a weighed valorisation of tops and lateral body as

$$Q_s = f(Q_{top}, Q_{lat})$$

with $Q_{s\,i} = \{Q_{top\,i} \cdots Q_{top\,i+2}, Q_{lat\,i}\}$ V $Q_{s\,i} = \{Q_{lat\,i} \cdots Q_{lat\,i-1}, Q_{top\,i+2}\}$

where Q_s is the quality grade of the stopper, Q_{top} the quality grade of the tops, Q_{lat} the quality grade of the lateral surface and i is the grade ($i=1$ extra, etc). For instance, a

Table 14.2. Surface dimensions of a top and the lateral surface of a 24 mm × 38 mm (diameter × length) stopper, before and after insertion into a bottleneck with 18.5 mm inside diameter.

	Initial stopper	Bottle inserted stopper
Top (mm²)	351	269
Lateral surface (mm²)	2864	2207

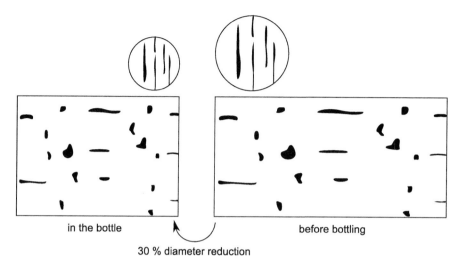

Figure 14.5. Schematic representation of tops and lateral surface (pores in black) of a 24 mm × 38 mm stopper in original dimensions and after insertion in a bottleneck with a diameter reduction of 30%.

Table 14.3. Decision rules for the classification of the lateral surface and the tops of cork stoppers based on threshold limits for selected variables.

	Stopper variable	Extra	Superior	1st	2nd	3rd	4th	5th
Lateral surface	Porosity coefficient (%)	2.1	2.4	3.0	3.8	4.8	6.4	13.0
	Maximum length (mm)	4.0	5.0	6.0	8.0	10.0	12.0	18.0
	Maximum area (mm²)	4.0	6.0	8.0	12.0	16.0	22.0	50.0
Tops	Porosity coefficient (%)	2.1	2.4	3.0	3.8	4.8	6.4	13.0
	Number of pores	5	7	9	12	15	20	38

Source: Costa and Pereira (2006).

stopper will be considered as 1st quality if (1) the lateral surface is classified as 1st and tops are classified from extra to 3rd, or (2) the tops are classified as 3rd quality and the lateral surface is classified as extra or superior.

Figure 14.6 shows a set of stoppers classified using these decision rules from extra (left) to 5th (right) quality grades. However the discrimination into seven quality grades is probably too high and lacking in practical relevance, both visually as referred to above, and in sealing performance as discussed later on. A grading into three quality classes, i.e. premium, good and standard should be better adapted to reality and performance requirements (Costa and Pereira, 2005).

The classification based on image analysis and computerised vision systems with selection of quantified features for grading ensures an increased class uniformity and transparency in trade. The adoption of porosity quantifiers in the reference terms of contract agreements could clarify the market (as already done for dimensional and resistance features, as detailed later on) and therefore avoid subjective appreciations and discussions.

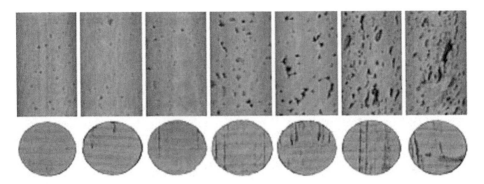

Figure 14.6. Photograph of a set of cork stoppers with different quality grades from extra, at the left, to 5th at the right.

14.2.4. The quality grading of cork discs

The discs of natural cork used in the technical stoppers are seen as tangential sections of cork i.e. the pores appear with their characteristic near circular form, as a result of the corresponding production process as detailed in Chapter 12. In the industrial processing the cork discs are observed using image vision equipment and sorted automatically into quality classes (extra, superior, A, B and C). A manual validation of the classification is also frequently made, as it has been shown in Figure 12.18. It is the quality of the cork discs that determine the quality of the technical stopper, or of the bottom disc in the case of champagne stoppers.

The same type of reasoning that was made for the cork stoppers can be applied to the cork discs and quantified thresholds may be used to discriminate the different grades (Lopes and Pereira, 2000).

14.3. Cork properties as closures

It is asked from a bottle closure that: (a) it does not allow any leakage from the liquid content either through the closure's material or at the interface between closure and bottle; (b) it is innocuous to the liquid and does not negatively alter its chemical and sensorial characteristics; (c) it stays in the bottle during storage; (d) it is durable and maintains the physical and chemical characteristics during storage; and (e) it can be removed for consumption in a practically easy way. To fulfil these requirements it is necessary to have a sufficient compression against the bottleneck and contact between closure and the bottle surface to avoid liquid percolation as well as material's impermeability to avoid diffusion through the closure. The performance of cork stoppers in this regard is detailed below.

14.3.1. Compression of stoppers

The cork stoppers have a diameter well above the inside diameter of the bottleneck. Therefore they will be compressed when inserted in the bottle exerting a pressure against the bottle wall and under a strain directly related to the difference between both diameters:

$$\varepsilon = 1 - \frac{D_c}{D_b}$$

where D_c is the cork stopper diameter and D_b the bottle inside diameter. Given the geometry of the stopper and its structural features, as described previously, the compression stress is applied around the interface in different directions, from radial to tangential (Fig. 14.7).

The strains in the cork approximate 30%; for instance, a 24 mm cork stopper inserted into a 18.5 mm bottleneck will have a strain of 23%, while the strain will be 29% for a 26 mm cork stopper. These strain values are located in the stress-strain curves of cork (see Fig. 9.3) in the plateau region corresponding to the physical phenomenon of the buckling of cells and to stress values in the range of 1 MPa (Fig. 9.7). There is anisotropy in the stress distribution around the stopper with the radial direction corresponding to the maximal stress and the tangential direction to the minimum value. However after insertion in the bottle, there will be a quick stress relaxation. Measurements of stress in a cork stopper inserted into a cylinder (corresponding to a strain of 25%) showed an initial stress of 1.2 and 0.9 MPa for the radial and tangential sections, respectively, that decreased to 0.8 MPa in less than 1 h in both directions (Fortes et al., 2004). Therefore, the compression properties of the stopper in use are uniform in spite of the anisotropy of cork.

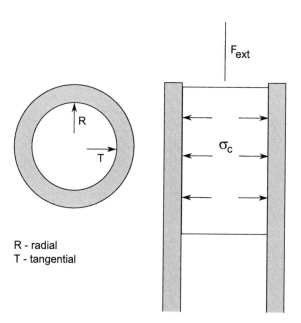

Figure 14.7. Schematic representation of the forces applied on a cork inserted in a bottleneck.

14.3.2. Extraction force

The stopper is usually pulled out from the bottle by applying an extraction force in the longitudinal direction of the bottleneck after fixing the cork to a pulling device (usually a cork screw), as in Figure 14.7. The extraction force depends on the compressive stress against the bottle and on the sliding friction between cork and glass as given by the corresponding friction coefficient (see Chapter 10). It may be calculated as $F_{ext}=\mu_0 S \sigma_c$, where F_{ext} is the extraction force, S the contact surface (inside perimeter $\times$ height), and σ_c the compressive stress of cork.

The extraction force will therefore increase with the increase of dimensions of the stopper: a longer stopper will increase the contact surface and a larger diameter stopper will increase the compressive stress. The adhesion of the two surfaces may be substantially reduced by surface treatments, such as the presently used silicon and paraffin coatings of the cork stoppers, therefore reducing μ_0 and the extraction force.

One variable that strongly influences the extraction force of stoppers is their moisture content, which should be 6–10% on bottling. Figure 14.8 shows an example for the extraction force of 24 mm $\times$ 45 mm stoppers with and without surface treatment and with different moisture contents. The moister stoppers can be extracted with appreciable less effort: 207 and 459 N, respectively for 10 and 6% moisture content of surface treated stoppers. This behaviour is the reason why it is much easier to uncork bottles with stoppers that are somewhat impregnated with wine. On the contrary, the commercial quality does not influence directly the extraction force of cork stoppers.

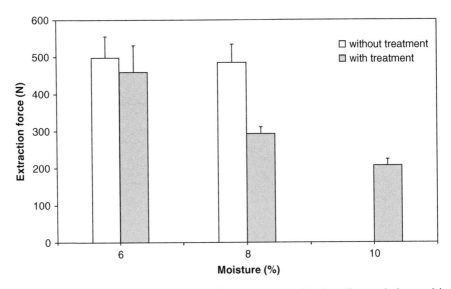

Figure 14.8. Extraction force of 24 mm $\times$ 45 mm stoppers of 3rd quality grade inserted in a 18.5 mm diameter Burgundy type bottle, with and without surface treatment and with different moisture contents (drawn with data from Fortes et al., 2004).

It is usually considered that the extraction force of natural cork stoppers should be above 200 N and around 350 N (it is the practice in the industry to report these values in daN, as 20 and 35 daN).

14.3.3. Sealing quality

The sealing performance of a bottle stopper involves two types of possible liquid passages: one is by leakage between the stopper and the bottleneck, involving the penetration and advance of the liquid in the interface; the other is by diffusion through the cork material.

The sealing capacity of a stopper against the glass surface is evaluated by penetration of the liquid in the interface and it is usually measured by applying liquid pressures over the atmospheric pressure and observing the depth of penetration of the liquid (Fortes et al., 2004). The depth of penetration x varies with time as $x = C t^{\alpha}$, with α between 0.2 and 0.4 and C depending on the applied pressure gradient and the orientation of the stopper surface. The depth of penetration is not uniform around the stopper: it is highest in the regions corresponding to the radial section and lowest in the tangential sections. The commercial quality of the stopper and the surface treatments do not influence the velocity of penetration.

In order to have a sealed joint between the cork surface and the glass it is necessary to have a sufficient contact of the cell walls of cork to avoid a direct path for the passage of liquid from the interior to the outside surface. Using the theory of percolation and a honeycomb model it can be estimated that the fraction of the interface surface that has to have contact to avoid leakage should be around 65%. The contact between the cell walls at the surface of the stopper and the surface of the glass is not complete because both show rugosity at macroscopic and microscopic levels. For cork, the microscopic rugosity is given by the sectioned cell walls and has therefore a scale of about 10–30 μm; the macroscopic rugosity is in the order of hundreds or thousands micrometers and results from the cutting process. In order to overcome this surface rugosity and to establish a degree of cork–glass contact enough to avoid percolation, it is necessary to compress the cells against the bottleneck. This is what is done in the present practice of bottling.

The penetration through the cork is governed by the diffusion of the liquid in contact with the surface of the stopper. The penetration depth is given by $x = (D t)^{0.5}$, with D as the diffusion coefficient (see Chapter 8). At room temperature and with a D of about 10^{-11} m^2 s^{-1}, Fortes et al. (2004) estimated that it would take less than one year for water to diffuse from the bottom to the top of a 45 mm long stopper. Even if the diffusion coefficient at the storage wine temperature is much lower (as reported by Marat-Mendes and Neagu, 2004), and therefore the penetration much slower, it is clear that the cork stopper absorbs water/wine. Every wine drinker has noticed that the bottom part of the stopper in contact with the wine is moist in comparison with the upper part. The moisture content of stoppers from different types of wine bottles bought at the supermarket was determined along its length from the interior to the exterior in slices corresponding each to a thickness of 20% of the stopper's length. In all cases there was a gradation from the bottom part to the upper part of the stopper: the moisture content varied in individual bottles between 28 to 63% in the bottom, and subsequently between 14 and 24%, 11 and 19%, 9 and 17%, and between 7 and 11% at the upper part.

14.4. Wine off-flavours and TCA taint

Wine is a chemically complex solution that is based on a natural biological raw-material (the grapes), on a microorganism-mediated biochemical transformation (the fermentation), and on an evolution with time also through chemical and biochemical reactions that contribute to the formation of the wine taste and aroma. As a consequence, wine cannot be completely characterised by analytical data only, and sensory evaluation is mandatory, while variability is also an intrinsic characteristic of wines.

14.4.1. Wine TCA taint

The ultimate evaluation of a wine is made by the consumer though taste and aroma at the time of opening the bottle and drinking. Occasionally there are disagreeable off-flavours and taints and in the worst cases wine may be undrinkable. These taints are usually classified into different categories by sensorial analogies (Ebeler, 2001; Pisarnitskii, 2001): oxidised, acetic, mouldy, earthy, smoky, musty, corky, plastic, styrene-like, etc. One of the important taints in wine is a mouldy taste and aroma due to the presence of haloanisoles (halogenated methylbenzenes), namely of the 2,4,6-trichloroanisole (TCA) with the structure shown in Figure 14.9. Traditionally this taint has been termed as corkiness or cork taint.

The extent of TCA incidence in the bottled wines is a matter where reliable inventory data are missing and values from under 1% to nearly 10% have been referred to. The estimates of how much TCA taint is there vary widely with the wine expert and the background and reference framework for the estimates. Statements on the incidence of

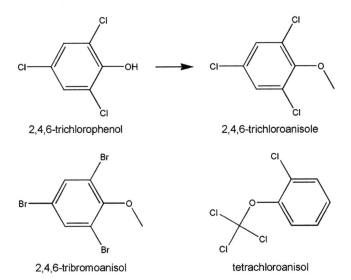

Figure 14.9. Chemical structures of TCA (trichloroanisole), TCP (trichlorophenol), TBA (tribromoanisole) and TeCA (tetrachloroanisole).

wine taint such as considerably less than 5% of the bottles, or not even seeing 0.5%, or close to 3–5% are often made at seminars, while ranges of 2–7% (Butzke et al., 1999) or 2–5% (Simpson, 1990) were also reported. A recent study of a large sample of commercial wines of different country origins and with different bottle closures showed an overall incidence of wine taints of 6% (Soleas et al., 2002).

However, the perceived taint may originate from compounds other than TCA, namely other haloanisoles such as tetrachloroanisole (TeCA) at concentrations above 10 ng/L, or tribromoanisole (TBA) that has recently been identified as a potent odour agent (Fig. 14.9). Other haloanisoles, as well as guaiacol, geosmin, 1-octen-3-one, 1-octen-3-ol, 2-methylisoborneol may be involved (Amon et al., 1989; Pena-Neira et al., 2000) and TCA is not the only cause, or the major cause, for the cork taint (Silva Pereira et al., 2000; Maga and Puech, 2005). In some cases of tainted wines, the quantity of TCA is not enough to explain it: for instance, in a sample of wine bottles classified by a panel as tainted, only 27% contained TCA in concentrations above 2 ng/L (Soleas et al., 2002).

14.4.2. Detection of TCA

TCA is sensed in very low concentrations with detection thresholds in the range of 1–4 ng/L and recognition thresholds somewhat higher in the range of 4–10 ng/L. There is quite some variation between the expert evaluation panels, depending on their experience and the specific wine, e.g. 1.4 ng/L (Duerr, 1985), 2 ng/L (Soleas et al., 2002), 3 ng/L (Pollnitz et al., 1996), 4 ng/L (Amon et al., 1989), 2–5 ng/L (Liacopoulos et al., 1999). The consumer has higher thresholds and there is quite a range of individual sensitivity depending on such things as age, smoking habits or wine drinking experience: an example is the required 210 ng/L by inexperienced tasters and the 17.4 ng/L by more experienced tasters with a Sauvignon blanc wine (Suprenant and Butzke, 1997), or the consumer rejection threshold of 3.1 ng/L reported for Chardonnay white wine (Prescott et al., 2005). A general detection by the consumer requires a TCA concentration value of 10 ng/L although it can be detected already at concentrations below that value.

In general it is easier to recognise TCA in white wines than in red wines, where the bottle aging plays an important role and adds to the complexity of the wine sensory composition. This is the reason why the studies on TCA taint and its causes as well as the public discussion of wine TCA taint have focused mainly on white wines.

The first studies of TCA in wine encountered major methodological difficulties related to the quantification thresholds and to the accuracy and reproducibility of results, because of the very low concentrations of TCA that need to be determined. Since the early determinations of TCA (e.g. Tanner and Zanier, 1981), a large effort was put into developing suitable procedures for extraction and preparation of extracts for analysis, and for their separation and quantification (e.g. Fischer and Fischer, 1997; Evans et al., 1997; Taylor et al., 2000; Gómez-Ariza et al., 2005; Insa et al., 2006), and detection and quantitation thresholds of 0.1 and 2 ng/L respectively were achieved (e.g. Soleas et al., 2002). The question now lies more on the statistical significance and the sampling intensity of the experimental designs, since most previous studies and results were based on a small number of samples and of experimental treatments.

14.4.3. Origin of TCA and of TCA-taint

It is known that TCA is formed by the biomethylation of 2,4,6-trichlorophenol (TCP), which can be carried out by several micoorganisms (Jager et al., 1996; Alvarez-Rodriguez et al., 2002; Coque et al., 2003). The presence of TCP may occur by direct absorption since this compound was used formerly in agriculture as a pesticide, or by chlorination of the phenolic compounds (lignin and phenolic extractives) that are chemical components of wood, bark and other plant parts. Other halogenated compounds are used as herbicides and biocides; for instance, tribromophenol is largely used in biocides. Due to long distance transport in the atmosphere, there is a generalised presence of chlorinated compounds in nature both in terrestrial and aquatic environments. Paperpulp bleaching also is a source to the presence of chlorinated phenols in the environment and in packaging materials. Similarly wood preservative treatments are the cause for the presence of chlorinated and brominated phenols in wooden pallets and other utensils.

The presence of TCA and other chloroanisoles is not restricted to wine and it has been detected in many natural products: for instance, in several foods, such as dried raisins (Aung et al., 1996, 2004), green coffee (Spadone et al., 1990; Cantergiani et al., 2001), cocoa liquors (Denis et al., 2001), eggs (Engel et al., 1966) and chicken (Curtis et al., 1974), in essential oils (Stoffelsma and Roos, 1973), in sake (Miki et al., 2005), cognac (Cantagrel and Vidal, 1990) and water (Nystrom et al., 1992; Karlsson et al., 1995), and in the soil (Schmitzer et al., 1989).

14.5. Cork and wine taints

The accumulation of chloroanisoles and chlorophenols in cork may have the following causes: the use in forest of polychlorinated phenolic biocides, the chlorine bleaching of cork, the hypochlorite washing of wine barrels, the use of chlorinated phenolic biocides in wooden pallets, cartons and packaging material and containers and the environmental contamination of wine cellars. The research has shown that contamination may occur at the cellar, during the handling and storage of the materials as well as during wine bottling. Therefore the contamination sources (e.g. halogenated biocides) have to be eliminated, and microbial sources, much as the cork industry has pursued in the last years. Adequate handling of cork stoppers is also required due to their capacity to absorb halogenated anisoles from the environment.

The concerns on taints and the following attention given to closures and their role in the bottle until consumption, had the effect to foster research on several aspects of wine aging, namely on the extent of oxygenation and its chemical and sensorial role. Numerous differences exist among wines with different closures in all the attributes, including the level of SO_2 and of oxidation and sensory quality (Godden et al., 2001, Francis et al., 2003).

14.5.1. TCA in cork stoppers

TCA is a non-polar compound that has a large affinity for lipids such as those found in cork (monomeric extractives and polymeric suberin) or in plant cuticles (cutin). The cork therefore has the ability to absorb TCA as well as other chloroanisoles, namely from the wine (Capone et al., 1999) or from the environment (Barker et al., 2001).

The location of TCA in the cork stoppers was investigated as being primarily found on their external surface (Howland et al., 1997; Barker et al., 2001), thereby implying that most of the contamination occurred after the punching of the stoppers from the cork planks. Highly TCA-contaminated cork stoppers, either because of yellow stain in the raw cork planks or after spiking, allow TCA migration into the wine although in variable extents, e.g. 0.6–25% in bottling experiments (Juanola et al., 2005). However TCA does not permeate through the cork and studies with deuterium labelled TCA which was added to the surface of wine corks in bottles did not contaminate wine after more than 3 years of storage (Capone et al., 2002), thereby showing that cork closures are excellent barriers to the transmission of TCA from external sources. Therefore the cork stoppers are a good sink for TCA and other haloanisoles, and they act as an effective barrier against environmental contamination. Instead of being the causes for TCA taint in wine, the cork stoppers may contribute to protect the wine against it, namely when using stoppers from certified industries that have very low contents of naturally absorbed TCA.

In the numerous recent studies involving cork stoppers, it is striking how little effort has been paid to characterise the corks used experimentally, namely in relation to the presence and extent of the lenticular channels. Given the variability encountered in this so-called porosity of cork, it is clear that it may contribute to the variation of the results found for the behaviour of cork stoppers as usually referred to by most researchers. Since the lenticular filling material and the cork have quite different structure and chemical composition, it is probable that they will behave differently in relation to absorption and permeation of TCA and other compounds.

14.5.2. The processing of cork and TCA

The cork industry has adopted measures to avoid TCA and TCA-leading compounds that are three-fold directed: (a) to prevent contamination of cork with chlorinated compounds; (b) to avoid microbial growth on cork; and (c) to introduce corrective stripping from natural TCA contamination.

As regards the cork raw material, the planks are screened before processing for elimination of those that contain signs of microbial activity (such as the yellow stain) and the bottom strip of the cork plank in contact with the forest soil (the footer) is cut off, since it was found that the eventual contamination with TCP or TCA was more probable in this part. The screening for yellow stain or other microbial attacks continues also in the process line, especially at the preparation stage of plank trimming and sorting (see Fig. 12.7), and such planks are eliminated from the production of stoppers, discs or from trituration for production of agglomerated stoppers. Also the pre-processing storage of the raw cork planks has eliminated their contact with the soil and the industrial yards are nowadays cemented or using a plastic barrier layer between the cork piles and the ground (see Fig. 5.8).

Microbial contamination is also reduced by decreasing the time and by controlling the conditions of storage between the boiling of cork planks and their processing into stoppers, thus eliminating the most obvious source of microbial activity in the industrial processing. The long drying period between boiling and plank processing in moist and closed environments, with strong microbial development of various fungal species on the cork back of the planks, as it was the rule in previous times, no longer applies. The motto in the cork industry is now "avoid microbial development".

The other rule in the present day cork mills is to eliminate all halogenated compounds (namely the chlorinated materials) from the premises and the processing. The most obvious change was to substitute calcium hypochlorite as the bleaching agent of cork stoppers for hydrogen peroxide, thereby eliminating the eventual chlorination of phenolic substances in cork (lignin and phenolic extractives). This means that the main reason given for the presence of TCA in wine in almost all the publications as the washing and bleaching of cork with chlorine-containing solutions no longer applies with the present industrial conditions. The use of non-chlorinated water for the boiling of cork further eliminates another possible contamination source, coupled with changes in equipment and processing conditions. In addition all products that may contain halogenated compounds were substituted, such as wooden pallets, cardboard and paper, and cleaning materials.

The appearance of the modern cork industries is very different from past plants. Clean and controlled environments are in place, handling and storage of cork raw materials and products avoid contaminations, process automation increased significantly, and quality control and certification are now the rule. Figure 14.10 shows an example of industrial premises where automation was implemented to a substantial degree.

The reduction of the natural contamination of the raw cork planks by TCA that occurred in the tree is also reduced by additional treatment such as vapour stripping, microwave evaporation, pressure gas extraction or beta-irradiation in some of the large industrial units. The use of enzymes, e.g. Corkzyme® for peroxide degradation and Suberase® for phenol polymerisation, is also made in some cases.

Several of the major industrial groups have developed and protected such curative treatment systems. The Amorim group, the largest cork industrial group, has patented the

Figure 14.10. General view of one hall of an industrial unit producing cork stoppers where the degree of automation is very high (see Colour Plate Section).

ROSA process for the treatment of cork granules used for technical stoppers and is developing its application to stoppers. It is a controlled steam distillation process where pressurised steam extracts the volatile compounds in cork, namely TCA, as reported by 70–80%. For cork discs, the process is the INOS II, where the discs are immersed in water and subjected to varying pressures to achieve a hydrodynamic extraction of cork-soluble compounds and a cleaning of the lenticular pores from loosened particles. There is a reduction of TCA and other halogenated compounds, as well as of tannins. Sabaté, the world's second largest cork producer, has developed the DIAMOND process, a supercritical carbon dioxide extraction for extracting TCA at a temperature over 31°C and a pressure over 74 bar and claims to remove over 95%. Juvenal & Oller, with the DELFIN process (see Fig. 12.12) use microwaves to dry the cork and remove volatiles killing TCA-producing microbes in stoppers and granulates. Process TF99.9 from GANAU uses a physical–mechanical process to clean cork and claim to be 99.9% TCA free.

14.6. Certification and quality control

Cork stoppers are controlled in relation to physical and chemical properties that potentially impact their performance as wine closures, and quality control is in place in the medium and large industrials plants. Procedures are given by standard protocols, i.e. ISO standards or country specific standards as well as any determination asked by the client.

Natural cork stoppers are controlled in relation to dimensions (length, diameter and cross-diameter ratio), moisture, density, extraction force, resistance to leakage, amount of soluble releasable compounds and of residual oxidants and presence of microorganisms. Agglomerated and technical stoppers are additionally tested for torsion strength variables and resistance to disaggregation.

The process of certification is now well established, both in relation to general quality and industrial environment as well as to the specific certification of industrial good practices for cork stoppers production. This was developed by the European Cork Federation (C.E. LIÈGE) who started its activity in 1987 and put in place a certification system named SYSTECODE of conformity to its International Code of Cork Stopper Manufacturing Practice (Celiège, 2002). This system assures, since 1999/2000, a voluntary process of application in the industries that manufacture cork stoppers, now with a total of 419 industries certified (status in 2005). The code, now in its 4th edition, has been revised to accommodate recent knowledge and technological innovations, as well as European policies regarding materials in contact with foodstuffs. Under the mandatory regulations are, for instance, the ruling out of chlorinated bleaching and all the microbial and TCA prevention measures referred previously.

14.7. The use of non-cork sealants

The occurrence of the so-called mouldy taints is of considerable concern to the wine industry since its incidence represents a substantial economic loss, and a first reaction was to blame the cork closure. The term cork taint favoured it. In some cases there was a search for alternative closures.

The alternatives for wine packaging are: (a) synthetic stoppers based on PE (polyethylene), SBS (styrene-butadiene-styrene) and EVA (ethylvinyl alcohol); (b) screw caps either traditional or of the more recent Stelvin type; (c) a plastic bag inserted within a cardboard box, the Bag-in-Box system; and (d) tetrapak type packages (Ross, 2002). Different trade names of synthetic stoppers were introduced, i.e. Supremecork®, Normacorc®, Betacorque®. One company introduced the Altec® stopper using a composite of very fine cork particles with synthetic granules. Other approaches were also to cover the cork stopper with a synthetic coating totally or just in the bottom part to avoid direct contact between wine and cork (the Procork® stopper is one example).

Most of the commercial experimenting and switch to wine packaging other than the cork stopper in a glass bottle has been with white and rosé wines, and in wine companies from the United States, Australia and New Zealand. For red wines, the in-bottle aging requires a long-term interaction with the stopper that has been optimised along time for the use of cork stoppers. Therefore other closures for red wines may necessitate adjustments in wine making.

Wine closures are a passionate and vivid discussion theme in wine magazines and among wine experts and critics. Any search on the matter rapidly shows the usually strong opinions and beliefs of both sides. As for the public, it is seen in several surveys that it prefers cork.

14.8. Conclusions

Cork stoppers are extremely successful bottle closures that have been in use from ancient times. The cork stoppers fulfil the requirements for a sealant of wine bottles:

- they have mechanical properties that allow insertion into the bottle and compression against its neck;
- they seal adequately the bottle to the passage of wine;
- they can be removed with a moderate extraction force;
- they are physically and chemically stable along time;
- they are innocuous and do not alter the wine; and
- they maintain their properties under diverse conditions of temperatures and environment.

The cork stoppers are also evaluated in regard to their macroscopic appearance in relation to the porosity shown by their surface. Quality grading is an important industrial operation and the commercial value of different grades is substantial. Although the industrial classification of stoppers and discs is presently made automatically by vision machine systems, the underlying variables and class limits are not stated. This quantification can be made by using measures of the coefficient of porosity and of maximal pore dimensions.

Mouldy taints in bottled wine are an important problem for the wine industry. TCA causes or contributes to the formation of this taint but it is not the responsible compound in numerous cases, and other halogenated anisoles and other compounds are involved. The presence of TCA indicates a previous contamination with chlorophenols, and elimination of chlorinated treated materials is therefore required. However the presence of chlorinated compounds is widespread on earth and on natural products including food and water. Therefore the cellar and wine bottling facilities also have to be strictly controlled for halogenated phenols and anisoles and for microbial contamination.

The cork industry has eliminated what was considered one of the major sources of chlorophenols, the hypochlorite bleaching. The storage and processing of cork for the production of stoppers in today's facilities also avoid microbial development and contaminations. This is a recent development carried out approximately during the last five years and therefore most studies on TCA are outdated in what refers to the natural cork stoppers.

Although synthetic closures and screw caps now have a solid share of the wine closures market, the cork stoppers remain as those preferred by the consumer. It should not be forgotten that the evolution of wine making has been made along times considering the bottle aging of wine with a cork closure and other solutions of wine packaging would require adaptation and research.

References

Alvarez-Rodriguez, M.L., Lopez-Ocana, L., López-Coronado, J.L., Rodriguez, E., Martinez, M.J., Larriba, G., Coque, J.J.R., 2002. Cork taint of wines: role of filamentous fungi isolated from cork in the formation of 2,4,6-trichloroanisole by O-methylation of 2,4,6-trichlorophenol. Applied Environmental Microbiology 68, 5860–5869.

Amon, J.M., Vandepeer, J.M., Simpson, R.F., 1989. Compounds responsible for cork taint in wine. Australian and New Zealand Wine Industry Journal 4, 62–69.

Aung, L., Jenner, J., Fouse, D., 2004. Detection of 2,4,6-trichloroanisole in microorganism-free irradiated non-processed dry-on-the-vine raisins by solid phase microextraction and GC-MS. Journal of Stored Products Research 40, 451–459.

Aung, L., Smilanik, J.L., Vail, P.V., Hartsell, P.L., Gomez, E., 1996. Investigation into the origin of chloroanisoles causing musty off-flavor of raisins. Journal of Agriculture and Food Chemistry 44, 3294–3296.

Barker, D.A., Capone, D.L., Pollnitz, A.P., McLean, H.J., Francis, I.L., Oakey, H., Sefton, M.A., 2001. Absorption of 2,4,6-trichloroanisole by wine corks via the vapour phase in an enclosed environment. Australian Journal of Grape and Wine Research 7, 40–46.

Barros, L., Pereira, H., 1987. Influência do operador na classificcação manual da cortiça por classes de qualidade. Cortiça 582, 103–105.

Butzke, C.E., Evans, T.J., Ebeler S.E., 1999. Detection of cork taint in wine using automated solid-phase microextraction in combination with GC-MS-SIM. In: Waterhouse, A.L., Ebeler, S.E. (Eds), Chemistry of wine flavour. ACS Symposium Series, American Chemical Society, Washington DC, pp. 208–216.

Cantagrel, R., Vidal, J., 1990. Research on compounds responsible for cork taint in cognacs. In: Charalambous, G. (Ed.), Proceedings of the 6th International Flavor Conference. Elsevier, Amsterdam, pp. 139–157.

Cantergiani, E., Brevard, H., Krebs, Y., Feria-Morales, A., Amadó, R., Yeretzian, C., 2001. Characterisation of the aroma of green Mexican coffee and identification of mouldy/earthy defect. European Food Research and Technology 212, 648–657.

Capone, D.L., Skouroumounis, G.K., Barker D.A., McLean, H.J., Pollnitz, A.P., Sefton, M.A., 1999. Absorption of chloroanisoles from wine by corks and other materials. Australian Journal of Grape and Wine Research 5, 91–98.

Capone, D.L., Skouroumounis, G.K., Sefton, M.A., 2002. Permeation of 2,4,6-trichloroanisole through cork closures in wine bottles. Australian Journal of Grape and Wine Research 8, 196–199.

Celiège (2002) International cod of cork stopper manufacturing practice, 4th Edition, www.celiege.com.

Coque, J.J.R., Alvarez-Rodriguez, M.L., Larriba, G., 2003. Characterization of an inducible chlorophenol O-methyltransferase from *Trichoderma longibrachiatum* involved in the formation of chloroanisoles and determination of its role in cork taint of wines. Applied Environmental Microbiology 69, 5089–5095.

Costa, A., Pereira, H., 2005. Quality characterization of wine cork stoppers using computer vision. International Journal of Vine and Wine Sciences 39, 209–218.

Costa, A., Pereira, H., 2006. Decision rules for computer-vision quality classification of wine natural cork stoppers. American Journal of Enology and Viticulture 57, 210–219.

Curtis, R.F., Land, D.G., Griffiths, N.M., Gee, M.G., Robinson, Peel, J.L., Dennis, C., Gee, J.M., 1974. 2,3,4,6-tetrachloroanisole association with musty taint in chickens and microbiological formation. Nature 23, 223–224.

Denis, C.N.S., Visani, P., Trystram, G., Hossenlopp, J., Houdard, R., 2001. Feasability of off-flavour detection in cocoa liquors using gas sensors. Science des Aliments 21, 537–554.

Duerr, P., 1985. Wine quality evaluation. In: Proceedings of the International Symposium on Cool Climate Viticulture and Enology, Eugene, USA, 25–28 June.

Ebeler, S., 2001. Analytical chemistry: unlocking the secrets of wine flavor. Food Reviews International 17, 45–64.

Engel, C., de Groot, A.P., Weurman, C., 1966. Tetrachloroanisole: a source of musty taste in eggs and broilers. Science 154, 270–271.

Evans, T.J., Butzke, C.E., Ebeler, S.E., 1997. Analysis of 2,4,6-trichloroanisole in wines using solid-phase microextraction coupled to gas chromatography-mass spectrometry. Journal of Chromatography 786, 293–298.

Fischer, C., Fischer, U., 1997. Analysis of cork taint in wine and cork material at olfactory subthreshold levels by solid phase microextraction. Journal of Agriculture and Food Chemistry 45, 1995–1997.

Fortes, M.A., Rosa, M.E., Pereira, H., 2004. A cortiça. IST Press, Lisboa.

Francis, L., Field, J., Gishen, M., Coulter, A., Valente, P., Lattey, K., Hoj, P., Robinson, E., Godden, P., 2003. The AWRI closure trial: sensory evaluation data 36 months after bottling. The Australian and New Zealand Grapegrower and Winemaker 475, 59–64.

Godden, P., Francis, L., Field, J., Gishen, M., Coulter, A., Valente, P., Hoj, P., Robinson, E., 2001. Wine bottle closures: physical characteristics and effect on composition and sensory properties of a Semillon wine. 1. Performance up to 20 months post bottling. Australian Journal of Grape and Wine Research 7, 64–105.

Gómez-Ariza, J.L., García-Barrera, T., Lorenzo, F., 2005. Optimisation of a two-dimensional on-line coupling for the determination of anisoles in wine using ECD and ICP-MS after SPME-GC separation. Journal of Analytical Atomic Spectrometry 20, 883–888.

Howland, P.R., Pollnitz, A.P., Liacopoulos, P., McLean, H.J., Sefton, M.A., 1997. The location of 2,4,6-trichloroanisole in a batch of contaminated wine corks. Australian Journal of Grape and Wine Research 3, 141–145.

Insa, S., Besalu, E., Iglesias, C., Salvado, V., Antico, E., 2006. Ethanol/water extraction combined with solid-phase extraction and solid phase microextraction concentration for the determination of chlorophenols in cork stoppers. Journal of Agriculture and Food Chemistry 54, 627–632.

Jager, J., Diekmann, J., Lorenz, D., Jakob, L., 1996. Cork-borne bacteria and yeasts as potential producers of off-flavours in wine. Australian Journal of Grape and Wine Research 2, 35–41.

Juanola, R., Subirà, D., Salvadó, V., Garcia-Regueiro, J.A., Anticó, E., 2005. Migration of 2,4,6-trichloroanisole from cork stoppers to wine. European Food Research and Technology 220, 347–352.

Karlsson, S., Kaugare, S., Grimvall, A., Boren, H., Savenhed, R., 1995. Formation of 2,4,6-trichlorophenol and 2,4,6-trichloroanisole during treatment and distribution of drinking water. Water Science and Technology 31, 99–103.

Liacopoulos, D., Barker, D., Howland, P.R., Alcorso, D.C., Pollnitz, A.P., Skouroumounis, G.K., Pardon, K.H., McLean, H.J., Gawel R., Sefton M.A., 1999. Chloroanisole taint in wines. In: Blair, R.J., Sas, A.N., Hayes, P.F., Hoj, P.B. (Eds), Proceedings of the 10th Australian Wine Industry Technical Conference. 2–5 August, Sydney, pp. 224–226.

Lopes, F., Pereira, H., 2000. Definition of quality classes for champagne corkstoppers in the high quality range. Wood Science and Technology 34, 3–10.

Macedo, J., Lopes, F., Pereira, H., 1998. Influência do método de amostragem na avaliação da qualidade da cortiça no mato. In: Pereira, H. (Ed), Cork Oak and Cork. Proceedings of the European Conference on Cork Oak and Cork. Centro de Estudos Florestais, Lisboa, pp. 93–98.

Maga, J.A., Puech, J.L., 2005. Cork and alcoholic beverages. Food Reviews International 21, 53–68.

Marat-Mendes, J.N., Neagu, E.R., 2004. The influence of water on direct current conductivity of cork. Materials Science Forum 455/456, 446–449.

Melo, B., Pinto, R., 1989. Análise de diferenças nos critérios de classificação qualitativa das rolhas. Cortiça 601, 293–302.

Miki, A., Isogai, A., Utsunomiya, H., Iwata, H., 2005. Identification of 2,4,6,trichloroanisole (TCA) causing a musty/muddy off-flavor in sake and its production in rice koji and moromi mash. Journal of Bioscience and Bioengineering 100, 178–183.

Nystrom, A., Grimvall, A., Krantzrulcker, C., Savenhed, R., Akerstrand, K., 1992. Drinking-water off-flavor caused by 2,4,6-trichloroanisole. Water Science and Technology 25, 241–249.

Pena-Neira, A., Fernandez de Simón, B., García-Vallejo, M.C., Hernandéz, T., Cadahía, E., Suarez, J.A., 2000. Presence of cork-taint responsible compounds in wines and their cork stoppers. European Food Research and Technology 211, 257–261.

Pisarnitskii, A.F., 2001. Formation of wine aroma: tones and imperfections caused by minor components (Review). Applied Biochemistry and Microbiology 37, 552–560.

Pollnitz, A.P., Pardon, K.H., Liacopoulos, D., Skouroumounis, G.K., Sefton, M.A., 1996. The analysis of 2,4,6-trichloroanisole and other chloroanisoles in tainted wines and corks. Australian Journal of Grape and Wine Research 2, 184–190.

Prescott, J., Norries, L., Kunst, M., Kim, S., 2005. Estimating a "consumer rejection threshold" for cork taint in white wine. Food Quality and Preference 16, 345–349.

Ross, J.P., 2002. Natural cork, corkiness, synthetics and screw caps. Enology International, pp. 8, www.enologyinternational.com/cork, accessed 13.04.2006.

Sanvicens, N., Varela, B., Marco, M.P., 2003. Immunochemical determination of 2,4,6-trichloroanisole as the responsible agent for the musty odor in foods. 2. Immunoassay evaluation. Journal of Agriculture and Food Chemistry 51, 3932–3939.

Schmitzer, J., Bin, C., Scheunert, I., Korte, F., 1989. Residues and metabolism of 2,4,6-trichlorophenol carbon 14 in soil. Chemosphere 18, 2383–2388.

Schwarz, L., Bowyer, M.C., Holdsworth, C.I., McCluskey, A., 2006. Synthesis and evaluation of a molecularly imprinted polymer selective to 2,4,6-trichloroanisole. Australian Journal of Chemistry 59, 129–134.

Silva Pereira, C., Figueiredo Marques, J.J., San Romão M.V., 2000. Cork taint in wine. Scientific knowledge and public perception: a critical review. Critical Review in Microbiology 26, 147–162.

Simpson, R.F., 1990. Cork taint in wine: a review of the causes. Australian and New Zealand Industrial Journal 5, 286–297.

Soleas, G.J., Yan, J., Seaver, T., Goldberg, D., 2002. Method for the gas chromatographic assay with mass selective detection of trichlorocompounds in corks and wines applied to elucidate the potential cause of cork taint. Journal of Agriculture and Food Chemistry 50, 1032–1039.

Spadone, J.C., Takeoka, G., Liardon, R., 1990. Analytical investigation of Rio off-flavour in green coffee. Journal of Agriculture and Food Chemistry 38, 226–233.

Stoffelsma, J., de Roos, K.B., 1973. Identification of 2,4,6-trichloroanisole in several essential oils. Journal of Agriculture and Food Chemistry 21, 738–739.

Suprenant, A., Butzke, C.E., 1997. Implications of odor threshold variations on sensory quality control of cork stoppers. *American* Journal of Enology and Viticulture 48, 269.

Tanner, H., Zanier, C., 1981. Zur analytischer differenzierung von Muffton und korkgeschmack. Schweizerrisch Zeitschrift Obst Weinbau 117, 752–757.

Taylor, M.K., Young, T.M., Butzke, C.E., Ebeler, S.E., 2000. Supercritical fluid extraction of 2,4,6-trichloroanisole from cork stoppers. Journal of Agriculture and Food Chemistry 48, 2208–2211.

Index

Printed and bound by CPI Group (UK) Ltd, Croydon, CR0 4YY

08/05/2025

01864806-0003